RELIABILITY
MATHEMATICS
Fundamentals; Practices; Procedures

BERTRAM L. AMSTADTER

Manager of Reliability, Power Systems Operations
Aerojet Nuclear Systems Company

McGRAW-HILL BOOK COMPANY

New York   St. Louis   San Francisco   Düsseldorf   London
Mexico   Panama   Sydney   Toronto

Sponsoring Editor Tyler G. Hicks
Field Editor Dale L. Dutton

RELIABILITY MATHEMATICS

PREFACE

This book is an outgrowth of the author's experience in reliability and is intended to serve the needs of the undergraduate or practitioner who has had only an introductory course in statistics or in a related subject such as statistical quality control. Knowledge of calculus is desirable, but an understanding of basic reliability methods can be gained if the reader will accept the formulas without proof. Advanced courses in theoretical statistics and higher mathematics are not necessary to obtain full benefit from the text material.

The text emphasizes the application of statistical methods to reliability functions and directly related activities. It provides to professional design, test, and reliability personnel a summary of the methods and procedures of reliability statistics. In this sense it is an applied rather than a theoretical statistics text, and it does not include the complete derivations of all formulas and tables which are presented. Rather, it furnishes useful tools to technical personnel who can benefit from them and gives guidelines for the selection, application, and utilization of these tools. At the same time, it makes available to supervisory and management personnel, and to college and university students, a single ready reference to the various methods used for treating particular reliability statistics problems. It also contains new, more efficient techniques not previously presented in a text.

An important aspect of the approach of this book is its orientation toward practicality rather than rigorous theoretical exposition. This should not be interpreted as the sacrifice of accuracy or as the compromise of sound theoretical principles. However, these principles are utilized selectively and the major emphasis is on the practical approach of overall program efficiency rather than on mathematical efficiency. Extreme precision, for example, may be reduced in favor of ease of test performance and analysis. The advantages and limitations of such trade-offs are discussed as they relate to particular statistical applications.

The chapters are grouped according to subject matter rather than statistics. The first chapters present some basic concepts and formulas which will be used repeatedly throughout the remaining chapters or which are necessary for understanding underlying principles of the techniques presented later. The concept of *confidence*, as well as that of *probability*, is first introduced here. A number of statistical distributions most frequently used in reliability are described next. They are generally presented without complex discussion of their mathematical derivations. Instead, emphasis is placed on the meanings of the various parameters defining them, on when they are used, and on simple methods of selection, utilization, and interpretation of results. A following chapter explains their use in some detail and provides the reader with the opportunity to relate the statistical tables included in Appendix B to the text material and to gain familiarity with these tables. More detailed discussions are deferred until later chapters when various statistical methods and procedures are explained to the extent that the explanations contribute to the understanding of the reliability principles. A number of auxiliary statistical methods which are frequently employed in reliability analyses are also covered.

Another chapter discusses data considerations and related questions which frequently arise and which are best resolved prior to the application of statistical methods. Such considerations as the need for defining a failure are covered in this chapter. Although it may appear that a text dealing with the statistical treatment of data need not be concerned with non-mathematical questions, the validity of the data and of its analysis directly depend on the satisfactory resolution of such questions prior to the data generation. It is the study of actual data—its generation, collection, analysis, and interpretation—in which we are interested.

The remaining chapters are organized according to the applications of the methods which are discussed. Thus, predictions of reliability, apportionment of reliability goals or objectives, and reliability growth are treated together in successive chapters, while assessment, evaluation, and demonstration of reliability are discussed in other chapters. Even when the methods may be similar and may even utilize the same sets of tables, the application remains the primary consideration and the text is arranged accordingly. For example, the assumption of a normal distribution might be made when predicting, from historical data, the probability of a component's performance characteristics meeting design specifications. A normal distribution might also be used when assessing the results of a stress-to-failure test and evaluating the probability of a failure, using actual test data. Although both analyses utilize the same statistical distribution, they are discussed in separate chapters which are grouped according to the applications rather than by the statistical method.

These chapters on applications cover, in turn, the following subjects: reliability prediction, apportionment, and growth; reliability assessment and demonstration; and system reliability considerations. Some of the techniques are the author's own developments and have not as yet had wide circulation. However, they provide simple methods of performing certain reliability calculations which heretofore have required much more time, and they have proved quite helpful in those programs where they have been used.

The Appendixes include, in addition to some derivations too lengthy to be included in the text, a series of tables. Some of these cover purely statistical distributions and others have been specifically prepared for reliability applications. References are provided throughout the book for those wishing to investigate the mathematical bases of those equations and formulas which are stated without full theoretical proofs. Other references are given for those who wish to study further either in the field of theoretical statistics or, perhaps more important, in the other two major areas of reliability which are introduced in the first chapter: reliability engineering and operations analysis.

Many of the tables in the text and in the Appendixes are reprinted by permission of the authors and their publishers. I wish to thank Dr. B. Epstein and the American Statistical Association for the formula for confidence limits from life test data which is presented in Chap. 13 and which appeared in Life Testing, *J. Am. Statist. Assoc.,* vol. 48; I also wish to thank L. H. Miller and the Association for permission to use the table Percentage Points of Kolmogorov Statistics, from vol. 51 of the same journal. I am indebted to the literary executor of the late Sir Ronald A. Fisher, F.R.S., and to Oliver and Boyd Ltd., Edinburgh, for values of the t and chi-square distributions which appear in Tables B.3 and B.4 and which were taken from "Statistical Methods for Research Workers," and also to L. B. Owen, the Atomic Energy Commission, the Sandia Corporation, and Addison-Wesley Publishing Company, Inc., for additional values in these two tables and for values of the F distribution in Table B.6 obtained from the "Handbook of Statistical Tables." I am also indebted to Profs. E. S. Pearson, M. Merrington, and C. M. Thomson for additional values in Table B.6 from *Biometrika,* vol. 33.

The curves in Table C.4 are reprinted from Reliability Assessment Guides for Apollo Suppliers by permission of North American–Rockwell Corporation and the National Aeronautics and Space Administration, and were developed from formulas appearing in Use of Variables in Acceptance Sampling by W. A. Wallis, printed in "Techniques of Statistical Analysis," McGraw-Hill Book Company, 1947, and in "Statistical Theory with Engineering Applications" by A. Hald, John Wiley & Sons, Inc., 1952. Figure 14.9, Weibull probability paper, is reprinted by permission of Technical and Engineering Aids for Management, Lowell, Massachusetts.

The approach used in the development of Eq. (8.12) for the mean time to failure is taken from "Reliability Theory and Practice" and is used by permission of I. Bazovsky and Prentice-Hall, Inc. The Method of Bounds, developed by the author and presented in Chap. 10, is presented by permission of Aerojet-General Corporation, North American–Rockwell Corporation, the National Aeronautics and Space Administration, and the Institute of Electrical and Electronics Engineers, Inc. Parts of Sec. 6.3.3 on Personnel Training are based on Sec. 4.2 of "Reliability Engineering" and are used by permission of ARINC Research Corporation and Prentice-Hall, Inc.

I wish to express my sincere thanks to Mr. Ralph A. Tiemann, who painstakingly reviewed the text and checked many of the calculations, and who suggested numerous modifications to improve the technical accuracy and readability. However, any errors which may still exist either in theory or in mathematics should be charged to me. I also wish to thank Mr. Norman K. Rumpf and his staff who prepared the computer programs used for calculating a number of the reliability tables. Finally, my sincere gratitude goes to my secretary, Mrs. Mary-Lu Gouze, whose excellent typing and figure preparation did much toward readying the manuscript for publication.

Bertram L. Amstadter

CONTENTS

1

INTRODUCTION

1.1 THE RELIABILITY CONCEPT

The concept of reliability is not new. People have long been concerned with the reliability of the products they use and of the friends and associates with whom they are in contact. Although the term "reliable" may not have been used specifically, its meaning was intended. The familiar complaint "things don't last as long as they used to" is a comparison, although a subjective one, of past and present reliabilities. When we say that someone is reliable, we mean that he can be depended on to complete a task satisfactorily and on time. These descriptions of reliability are qualitative, and they do not involve numerical measures.

It was not until the Korean War that quantitative reliability became widely used and statistical methods were applied to its measurement. The definition of reliability as *the probability that a device will operate adequately for a given period of time in its intended application* evolved and is now widely accepted. Variations have been defined for single-operation items such as explosive devices and for characteristics which are not time dependent, but essentially this definition applies. The definition includes the term *probability,* which indicates the use of a quantitative measure. Probability is the likelihood of occurrence of a particular form of an event (in this case, adequate operation). It can be determined for any of the innumerable consumer or military equipments which are of interest. Only the methods of measuring the probability differ for the various types of equipment.

In addition to the probabilistic aspect, the reliability definition involves three other considerations: satisfactory operation, length of time, and intended application. There must be a definition of what constitutes satisfactory operation. Certainly, equipment does not necessarily have to be totally inoperative for it to be unsatisfactory. If the compression in two cylinders of an automotive engine is low, the performance of the engine will be less than satisfactory. On the other hand, 100 percent compliance of all desired characteristics may not be a realistic definition of satisfactory performance and something less than 100 percent may be acceptable. The United States has had many manned space flights which were considered very successful even though not every item of equipment performed perfectly. Just what constitutes satisfactory performance must be defined if a measure of reliability is to be meaningful.

The length of time of operation is more definitive. A mission is defined as covering some specific length of time. A warranty is written for a specified number of months or years. Once the criteria of satisfactory performance have been defined, the operation of the equipment can be compared with the criteria for the required time period. Even in this area, however, there may be some flexibility. The criteria of acceptability may change as a function of time so that what is considered satisfactory at the end of the operating period may be something less than what was satisfactory at the beginning. A new automobile should not use any oil between oil changes, while the addition of 1 quart of oil every 1,000 miles may very well be acceptable for a 5-year-old car.

The last consideration—intended application—must also be a part of the reliability definition. Equipment is designed to operate in a given manner under particular sets of conditions. These include environmental conditions (temperature, pressure, humidity, acceleration, vibration, shock, acoustic noise, etc.) and operating conditions (voltage, current, torque, and corrosive atmosphere) which will be encountered in manufacturing, transportation, storage, and use. If the equipment fails or degrades excessively when operated in its intended environment, it is unsatisfactory, whereas if it is subjected to stresses in excess of those for which it was designed, failures or degradation may not be reasonable measures of unreliability.

This text will be concerned principally with probabilistic considerations— their definition, procedures for calculation, means of combining probabilities of components into system estimates, and other phases of the numerical aspects of reliability. The existence of adequate criteria of satisfactory performance, time span, and application will, in general, be assumed. Restatement or reexamination of these criteria will be made, however, when applicable to the discussion of the mathematical aspects.

1.2 THE NEED FOR RELIABILITY

The importance of obtaining highly reliable systems and components has been recognized in recent years. From a purely economic viewpoint, high reliability is desirable to reduce overall costs. The disturbing fact that the yearly cost of maintaining some military systems in an operable state has been as high as ten times the original cost of the equipment emphasizes this need. The failure of a part or component results not only in the loss of the failed item but most often also results in the loss of some larger assembly or system of which it is a part. The old story of the loss of a horseshoe nail is truly applicable. A leaky brake cylinder can result in a costly repair bill if it causes an accident. A space satellite may be rendered completely useless if a switch fails to operate or a telemetry system becomes inoperative.

Safety is an equally important consideration. The leaky brake cylinder could result in serious personal injury as well as creating undue expense. The collapse of a landing gear on an aircraft resulted in the loss of the plane although no passengers were injured. However, the consequences could easily have been much more serious. The untimely deaths of three of our astronauts serve to point out the importance of reliability to personnel safety.

Also caused by reliability (or unreliability) are schedule delays, inconvenience, customer dissatisfaction, loss of prestige (possibly on a national level), and, more serious, loss of national security. These conditions also involve cost and safety factors. Cost, for example, is inherent in every failure, as is inconvenience or delay. Most failures also involve at least one of the other considerations. Even the prosaic example of a defective television component involves cost, inconvenience, loss of prestige (of the manufacturer or previous serviceman), and customer dissatisfaction.

The need for and importance of reliability have been reflected in the constantly increasing emphasis placed on reliability by both the government and commercial industry. Most Department of Defense, NASA, and AEC contracts impose some degree of reliability requirements on the contractor. These range from the definition of system reliability goals to the requirement for actual demonstration of achievement. Compliance with a reliability document such as DOD specification MIL-STD-785, NASA specification NPC 250-1, or USAF specification MIL-R-26484, which define in detail the requirements for an acceptable reliability program, has been contractually required on most government-funded technical programs. Many of these have specific funds allocated to the reliability effort, require the development and maintenance of a reliability program plan, and specify the preparation of periodic reliability reports.

However, commercial industry has been slower in implementing reliability programs. In the past, some producers of commercial goods believed that the expense of a reliability effort outweighed its value and therefore gave it only token recognition. While the public was being told about high quality and high reliability through television and other advertising media, very little actual effort may have been expended on the reliability program. Fortunately, this attitude is gradually changing as more companies are becoming aware of the economic advantages that an efficient reliability program affords.

1.3   AREAS OF RELIABILITY ACTIVITY

The growth, recognition, and definitization of the reliability function were given much impetus during the 1960s. Reliability has become a recognized engineering discipline, with its own methods, procedures, and techniques. In arriving at this status, it encountered growing pains similar to those that quality assurance experienced in the two previous decades. Convincing corporate management that reliability was economically desirable sometimes required an effort comparable to that expended on the performance of the reliability tasks themselves. Hence, the development of reliability in the area of management and control included justification of its existence as well as application of engineering management principles to the organization and direction of the reliability activities.

During the growth process, three main technical areas of reliability evolved: (1) reliability engineering, covering systems reliability analysis, design review, and related tasks; (2) operations analysis, including failure investigation and corrective action; and (3) reliability mathematics. Each of these areas developed its own body of knowledge and, although specific demarcations cannot be drawn between one activity and another, in actual practice the reliability functions are often organized into these divisions. Some activities relating to the first two of these main divisions are not always delegated to the reliability organization; e.g., responsibility for design review is sometimes delegated to the design organization itself. In these instances, reliability is then usually responsible for monitoring the performance of the activities to ascertain that they are carried out according to management or contractual directives. However, the third function—reliability mathematics—is seldom delegated outside the reliability organization. First, the methods, although not unique to the reliability function, are not familiar to most personnel in design, test, and other organizational entities; and second, when the term "reliability" is mentioned, the mathematical aspects are the ones usually thought of. In fact, its primary definition is given in mathematical terms involving probability.

Design aspects of reliability cover such functions as system design analyses, comparison of alternate configurations, drawing and specification reviews, compilation of preferred parts and materials lists, and the preparation and analysis of test programs. Some of the specific activities include failure-modes-and-effects analyses, completion of design review checklists, and special studies and investigations. Environmental studies are frequently included in the reliability design function, as are supplier reliability evaluations. In reliability organizations associated with complex systems, e.g., refineries or spacecraft, the design reliability functions might be separated into systems concepts, mechanical design, and electrical design groups. A fourth group could include such functions as supplier reliability and parts application. Regardless of the particular organizational structure, however, almost all the individual activities make some use of numerical procedures.

The operations reliability functions relate to manufacturing and assembly operations, test performance, failure analysis and corrective action, operating time and cycle data, field operations reports, and other activities associated with the implementation and test of the design. Personnel in this function help to ensure that the design intent is carried out and they report discrepancies in operations and procedures as well as performance anomalies. The operations reliability group provides much of the data on actual equipment reliabilities that are used by the other reliability groups.

The statistical group is usually the smallest but can provide equal benefits to the overall program. In addition to accomplishing the reliability numerical activities of prediction, apportionment, and assessment, this group (or individual) provides support to the reliability design and operations groups and directly to the design and test engineers. Statistical designs of experiments, goodness-of-fit tests, system prediction techniques, and other mathematical methods are developed and applied to engineering problems. The methods and procedures discussed herein are applicable to both classes of activities, and it is hoped that reliability design and operations personnel and members of the engineering organizations as well as reliability statisticians find them informative and useful.

1.4 CAUSES OF UNRELIABILITY

There are a number of reasons why equipment and systems may be unreliable. These can range from errors in the original concept to operational mistakes, and they may include both deficiencies in the equipment itself and discrepancies in its application and use. The broad spectrum from concept through field application is the major portion of the more comprehensive discipline of *systems effectiveness.* We will be primarily concerned with technical reasons for actual effectiveness being less than planned effectiveness, and specifically with those areas of unreliability that can be evaluated and corrected through the use of statistical methods. First, however, we will briefly examine some of the principal causes of unreliability.

Design deficiencies are the causes that are usually thought of first. Actual errors or oversights do not constitute a large percentage of these deficiencies. Rather, unknown environments, lack of capability of state-of-the-art parts and equipment, unsuspected materials incompatibilities, and unavoidable complexities can all contribute significantly to failure of a design to meet reliability requirements. Unforeseen deleterious conditions often arise in programs which are being carried out in new fields, e.g., nuclear power systems.

There are two aspects of unknown environments: (1) new environments may be encountered which have not yet been studied extensively, and (2) actual levels of familiar environments such as earth temperatures that will be encountered may not be known. Among the environments which are now being studied more comprehensively are radiation, absolute vacuum, meteoroid impingement, and space temperature spectrums. Related probabilities and effects are being investigated. These efforts are costly and, although a large amount of information has been published, effects on specific equipments and missions frequently must be determined by extrapolation from available information and knowledgeable engineering judgments. When environmental effects are known, the level to which the equipment may be exposed may be in question. For example, relatively small temperature differences at a critical temperature level can have large effects on the performance of electronic parts, but even though the effects are recognized, there might be undetected hot spots or thermal gradients which could degrade reliability.

Materials properties may change unexpectedly, particularly when unfamiliar materials must be employed in new applications. Unforeseen changes in crystal structures, unique chemical changes, and so on may occur in the materials and may become manifest only after tens or hundreds of hours of use. Similarly, other design unknowns could adversely affect equipment capability and longevity.

Even when the undesirable conditions are known to the designer there may be no ready means of avoiding them. State-of-the-art materials and components may not have the necessary properties or characteristics but must be used because nothing better is available. Of course, in such cases research and development programs should be undertaken to correct the basic deficiencies. Also, the complexity may be so great that high failure rates are a natural consequence but cannot be avoided because system performance conditions or weight limitations preclude the use of redundant elements. The designer does the best he can to reduce the probability of failure. Where possible, maintainability is designed into the system so that when failures do occur, the deficiencies can be readily detected and appropriate maintenance or replacement actions can be accomplished.

Associated with design weaknesses are deficiencies in test programs, equipment, and operations. These, too, frequently arise from unknown and unanticipated factors. Funds may not be available to carry out a complete test program designed not only to verify performance capability but also to investigate areas of potential unreliability. Out-of-calibration instruments and maladjustments can also contribute to premature wear-out or failure of the test item. Failures of test equipment can lead to system failures either as secondary failures resulting from the imposition of excessive stresses or through fatigue by subjecting the system to an excessive number of cycles. Malfunctions of test valves, temperature controls, pressure regulators, etc., can impose stresses on the test item for which it was not designed. A simple example is the failure of a lubrication-cooling system which resulted in overheating the bearings of the test item and in subsequent seizure. In another instance a hydraulically operated dynamic brake was actuated because of low line pressure caused by a load elsewhere at the test facility, and this resulted in emergency shutdown of the test. Such shutdowns lead to excessive stresses on the test item and adversely affect longevity.

Poor communications is another area responsible for a large portion of unreliability. Incomplete or misinterpreted instructions from design engineering to manufacturing can result in undetected deviations of the hardware from the design intent; unclear directions to test engineering can result in incorrect test operations which damage the test item and cause premature failure. Field operating and maintenance manuals and instructions may be ambiguous, incomplete, or even nonexistent, leading to mishandling, improper maintenance, and early failure.

Required burn-in to weed out potential early failures may be bypassed in order to meet schedules or to make the equipment operational at an early date. Or, the life of the equipment itself may not be long enough (due to the state-of-the-art components) to avoid actual wear-out within the required operating period. Finally, the reliability requirement itself may be too high to permit a realistic probability of compliance within the budgeting, scheduling, and other program constraints.

1.5 CLASSES OF FAILURES

The causes of failure can be categorized into three basic types. It is recognized that there may be more than one contributing cause to a particular failure and that, in some cases, there may be no completely clear-cut distinction between some of the causes. Nevertheless, the use of the three classes facilitates discussion of the various activities within the discipline of reliability mathematics without invalidating the conclusions. We can then investigate problems of prediction, apportionment, and assessment and present methods for arriving at solutions to these problems which are both practical from the standpoint of ease of utilization and realistic in the sense that the solutions are valid and useful.

The three classes of failure are infant mortalities, random failures, and wear-out. A fourth possible class is out-of-tolerance failures which can occur at any time during the system operating life. The associated failure distributions for out-of-tolerance failures are the same as those for wear-out failures, and the mathematical methods used for evaluation are similar. Consequently, although these failures will later be considered separately, they are not classed here independently.

Infant mortalities, or early failures, are those which occur due to some fault in the part or assembly resulting from a design, manufacturing, or inspection deficiency. The failures which occur during the research and development phase of a program are usually infant mortalities. Corrective action is taken, the failure causes are eliminated, and verification tests are made. However, in some instances, certain failure mechanisms cannot be designed out of a part economically. It is less costly overall to produce many parts and sort out potentially defective units ("weak sisters" as they are sometimes called) by burn-in tests or other 100 percent inspections. Such practice is common in the electronics industry where many kinds of parts are routinely subjected to burn-in operation. The part characteristics are monitored during the 100-hr or other burn-in period and those displaying degradation or failure are deleted from the lot. Systems and subsystems are commonly "debugged" by operating them for a period of time prior to delivery to ensure the detection and elimination of early failures or potential failures. This is a must for subsystems of missiles and satellites which are essentially non-repairable during use, and it is highly desirable for repairable equipment such as commercial power distribution systems and radars whose reliable operation must not be jeopardized.

Because burn-in is a desirable and in some cases a necessary procedure and because infant mortalities can be eliminated, infant mortality is not a mode of failure which is considered in the discussion of reliability mathematics. If it does exist, the reliability of the component or system is compromised to such an extent that reliability calculations are not only invalid and misleading but are necessarily low because of the high incidence of failure. A review of the causes of failures and of their history will disclose when infant mortalities are present, and will aid in indicating the corrective measures (design modification, operator training, improved inspection, or burn-in) that are required to eliminate these early failures.

The remaining two classes of failures, random and wear-out (which class includes out-of-tolerance failures), are discussed in succeeding chapters in conjunction with applicable statistical distributions and mathematics. The negative exponential distribution is used for characteristics with randomly occurring failures, and the normal (gaussian) and associated distributions are used for wear-out and other characteristics whose distributions are "bell-shaped." Other, related distributions pertain to redundant systems. The relationship of the three classes of failure is shown pictorially in Fig. 1.1. This figure is frequently referred to as the "bathtub curve."

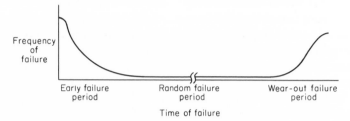

FIG. 1.1   The relative rates of failure during a component's or system's life history.

The heights of the early failure and wear-out failure periods do not necessarily have exactly the same relationship depicted in the figure. The frequency of early failures, for example, might be considerably lower for a mechanical component in the production phase. Similarly, the relative height of the curve in the random failure period might be quite different. However, the general relationship is as shown.

1.6 METHODS OF IMPROVING RELIABILITY

With all the potential areas of unreliability, we might well ask whether it is possible to attain the required reliability levels, particularly when they are very high. Fortunately, the picture is not as bleak as it appears. Problems are found and corrected and the corrective action is verified. Design weaknesses are eliminated, improved materials and parts are developed, adequate procedures and manuals are prepared. In addition there are several other methods of improving the reliability of a component or system. Many of these involve the application of numerical procedures. Comparison of alternate candidate designs, use of large safety margins, employment of redundancy, and overstress testing all utilize quantitative analyses. Assessment and correction of design deficiencies and maintenance scheduling also require the use of statistical techniques. Other methods such as failure analysis and burn-in, although not directly involving mathematical methods, can be most effective when the statistical methods are used as auxiliary aids.

In the comparison of alternate designs, two or more candidate configurations which will provide the required performance characteristics are prepared by the appropriate design groups. These are reviewed by members of reliability, quality assurance, stress, manufacturing engineering, and other organizations to determine potential weaknesses and to select the most promising candidate. One aspect of the review by the reliability function is the quantitative prediction of the reliability of each design. Such a prediction is shown in detail in Chap. 10. Without the quantitative comparison, the inclination with respect to reliability in the referenced application would have been to select that design which actually would have turned out to be somewhat less reliable than the others.

The use of large safety margins is widespread in the architectural profession and the construction industry. Margins of five and ten to one are common and can usually be achieved relatively inexpensively by using a larger beam, deeper foundation, etc. Weight is of little consideration. However, in such industries as electronics, aviation, and aerospace, weight and performance become major problems. The amount of additional fuel for a space mission is many times greater than the weight of the additional payload—each pound of payload costs dearly. Safety margins of as little as 10 percent (1.1 to 1) are encountered. If such small margins are to be used, knowledge of the distributions of stresses and strengths and their relationships is essential, and statistical methods are of major importance in the evaluation.

Redundancy (the availability of more than one means of performing a function when fewer are required) is a principal method of increasing reliability. Redundancy can involve the use of two or more identical components or assemblies so that, when one fails, others are available to perform the task; or it can include different means of accomplishing the function. A spare tire exemplifies the use of redundant parts; the manually operated hand sextant used for navigation of a spacecraft in the event of failure of the automatic controls is an example of the latter method. In both of these examples, the redundant component (the spare tire or the sextant) is used only when the primary system fails. This usage is known as *sequential* redundancy. Other redundant systems are operated together, all operable (unfailed) systems performing the function all the time. This type is called *active parallel* redundancy. The use of multiple resistive elements in a load bank, permitting one or more elements to fail without affecting system performance, is one example of active parallel redundancy. The use of four engines on an aircraft is another.

The type of redundancy is dictated primarily by system performance considerations. Sequential redundancy provides theoretically higher reliability than does active parallel redundancy *if* the failure detection and switching functions are extremely reliable. Otherwise, active parallel redundancy is preferred from a reliability standpoint. Both types result in greatly improved system reliability as compared with a non-redundant system. Calculations of reliabilities of redundant systems can become quite complex, and mathematical methods are presented in later chapters.

Overstress or accelerated testing can be used both to evaluate the capability of a component to meet reliability requirements and to provide quicker means of detecting potential weaknesses and failure modes. The relationship of failures and failure rates at accelerated conditions to those at nominal rated and actual operating conditions should be known from historical data if accelerated testing is to be most effective. Failure rates with respect to applied voltages and operating temperatures of capacitors and resistors, for example, are well known, and the relationships can be used for evaluating units of a new lot, type, or manufacturer. One frequently used relationship is the approximate doubling of the failure rate for each $10°C$ rise in temperature. Thus, a failure rate at $125°C$ is approximately thirty-two times the failure rate at $75°C$, provided that the unit can operate at the higher temperature. Units from a new lot can be operated at the higher temperature to more quickly evaluate the reliability. Since these components are usually very reliable, even higher temperatures are used in conjunction with overvoltage in order to determine failure rates in a reasonable time. Overstress testing of new types is common practice, and it is used to detect potential modes of failure. Another frequent application of accelerated tests is in conjunction with burn-in tests whereby new equipment items are subjected to initial operations in excess of the stresses that will be encountered in their application to weed out potential failures or "weak sisters." It is important, of course, that burn-in or other accelerated or overstress testing does not introduce new modes of failure that would not normally be experienced and that it does not degrade the test units if they are to be used in deliverable equipment.

In addition to prediction methods used to compare alternate candidate designs, reliability assessments of actual equipment are valuable for determining those areas where effort should be concentrated to achieve the greatest improvement. Those components which contribute most to the probability of system failure are revealed and can then be evaluated with respect to the various means of improving their reliabilities, e.g., larger safety margins, redundancy, etc. Statistical techniques are directly employed in assessing reliability, utilizing data from actual tests of the equipment. Chapters 13 and 14 describe the most frequently used assessment methods and include criteria for selection of the most applicable methods as well as detailed examples of their use.

Correction of design deficiencies is the most direct method of improving reliability. Engineering analyses are employed to detect weaknesses and to select means of eliminating them. Stress analysis, materials compatibility evaluation, and other disciplines contribute to this effort. In addition to performing auxiliary studies in support of this effort, the reliability organization can supply statistically designed test plans that will enable economic selection of optimum values of input characteristics and provide mathematical measures of significance of changes in the design. The statistical design of experiments, although not directly related to reliability (and not covered in this text), can be an area of great economic contribution.

When statistical techniques are applied to the scheduling of maintenance activities, the required level of reliability can be obtained in the most efficient manner. Maintainability is a major subject and includes such statistical measures as mean time between failures, mean time to repair, etc. The scope of this aspect of system effectiveness, of which reliability is another major facet, is quite broad and is the subject of other texts.

We will study both the mathematical methods and techniques used to measure and improve the reliability of components and systems and the criteria for their selection. Before proceeding with a detailed analysis of methods, some basic statistical concepts are presented. These will provide greater insight into the reasons for the validity and usefulness of the methods and will facilitate selection of the best techniques for particular reliability problems.

ADDITIONAL READINGS

ARINC Research Corporation: "Reliability Engineering," Prentice-Hall, Inc., Englewood Cliffs, N.J., 1964.

Bazovsky, I.: "Reliability Theory and Practice," Prentice-Hall, Inc., Englewood Cliffs, N.J., 1961.

Bureau of Naval Weapons, Chief of: "Handbook: Reliability Engineering," U.S. Government Printing Office, 1964.

Calabro, S. R.: "Reliability Principles and Practices," McGraw-Hill Book Company, New York, 1962.

Chorafas, D. N.: "Statistical Processes and Reliability Engineering," D. Van Nostrand Company, Inc., Princeton, N.J., 1960.

Dummer, G. W., and N. B. Griffin: "Electronics Reliability: Calculation and Design," Pergamon Press, Ltd., London, 1966.

Ireson, W. G.: "Reliability Handbook," McGraw-Hill Book Company, New York, 1966.

Lloyd, D. K., and M. Lípow: "Reliability: Management, Methods, and Mathematics," Prentice-Hall, Inc., Englewood Cliffs, N.J., 1962.

Shooman, M. L.: "Probabilistic Reliability: An Engineering Approach," McGraw-Hill Book Company, New York, 1968.

2

FUNDAMENTAL
STATISTICAL CONCEPTS

2.1 PROBABILITY

The initial set of concepts relating to reliability mathematics are those associated with probability. An understanding of probability is essential because all applications of reliability statistics depend on some aspect of probability, either implied in the formulation of the reliability activity or directly used in the calculations. The discussion which follows considers probability in terms of *discrete* events—events that make up a set of isolated values. If a die is tossed, assuming that it cannot stand on an edge or corner, the only possible results are a 1, 2, 3, 4, 5, or 6. A value of 2.34176, for example, is an impossible occurrence. However, the reader should bear in mind that most distributions are *continuous,* and any value within the limits of the distribution may occur. If the heights of physically normal, human adults are measured, values anywhere from somewhat under 5 ft to 7 or 7½ ft can be observed. Similarly, characteristics and reliabilities of components and systems have distributions which are continuous rather than discrete. Because the use of distributions of discrete events simplifies the discussion of probability and because many of these concepts are applicable to continuous functions, this section treats discrete cases.

The probability of an event is the ratio of the number of times it is expected to occur to the total number of trials. It can be expressed quantitatively as a fraction, percent, or decimal. When a perfect die is tossed, the probability of any particular face being up is 1/6. Because of chance, one number may appear more frequently than another when the number of trials is small. In the long run, after many tosses are made, the ratio of the number of times a number appears to the total number of tosses *approaches* 1/6. In later chapters we shall examine how much variation from a theoretical probability can be expected due to chance as a function of the number of trials that are made.

The range of possible values of probability is zero to one. If there is no chance of an event happening, for example, 2.34176 on a die, its probability is zero. If an event must happen each time a trial is made, its probability is one. It follows that, when a trial is made, the sum of all possible events must equal one. Thus, when a die is tossed, the combined probabilities of a 1, 2, 3, 4, 5, or 6 must add up to unity because there can be no other outcome.

Suppose that there are 10 balls in a box numbered consecutively starting from 1. If they are thoroughly mixed, the probability of picking at random any particular ball, say number 2, is 1 out of 10 or 1/10. Similarly, the probability of picking number 5 is also 1/10. The probability of picking either number 2 or number 5 is 1/10 + 1/10 = 2/10 or 1/5. This is true because these events are *mutually exclusive*; that is, the two events cannot occur simultaneously. In a group of mutually exclusive events, the probability of occurrence of *any* of several events is the sum of their individual probabilities. This is known as the *rule of addition.*

The probability of picking any particular ball remains 1/10 as long as each time a ball is picked it is returned to the box and all balls are thoroughly mixed. As long as this occurs, the probability of picking a particular ball a second time is *independent* of the results of the first selection. The *product rule* states that the probability of two or more independent events *all* occurring is the product of their individual probabilities of occurrence. Thus, the probability of picking *a particular* ball twice in succession under these conditions is 1/10 × 1/10 or 1 chance in 100.

However, the probability of picking *any* ball twice in succession is still 1/10. The probability of picking any ball the first time is unity since it is stated that a ball is picked. Then, the probability of picking that particular ball at the second selection is 1/10. The product of unity and 1/10 is, of course, 1/10. It is important to note the way in which a problem is stated in order to avoid making a significant error.

The probability of picking ball number 2 on the first trial or the second trial, at first guess, appears to be 1/10 + 1/10 or 1/5. However, the true probability of picking number 2 *at least once in two trials* is only 19/100. If the first pick is successful, i.e., if number 2 is picked, a second trial need not be made since its outcome adds nothing to the probability. In other words, a second selection is made only if number 2 is *not* picked the first time. The probability of picking number 2 is thus the probability of picking it the first time plus the probability of picking it the second time *if* the first trial was unsuccessful. The answer is $1/10 + (1/10 \times 9/10) = 19/100$. The difference between the two answers is $1/100$, which is the probability of picking number 2 twice in succession. The value of 19/100 may also be obtained by computing the probability of being unsuccessful and subtracting this probability from unity. The chance of not picking number 2 on the first trial is 9/10, and the chance of not picking it on the second trial is also 9/10. The probability of not picking it either time is $9/10 \times 9/10 = 81/100$. Subtracting this from 1 gives the result 19/100. When the performance or outcome of an event (e.g., the second trial) depends on the result of the first trial, we say that *conditional probabilities* are involved. Again, the particular way in which the problem is stated affects the answer. If a second pick is made regardless of what happens on the first pick, i.e., the two trials are independent, then the *average number of times* that ball 2 is picked in each pair of trials is indeed 1/5.

The various cases and their probabilities are summarized in Table 2.1. Cases A, B, and C constitute success (at least one 2) and since they are mutually exclusive the total probability of success is the sum of their individual probabilities = 19/100. If a second pick is made only if the first pick is unsuccessful, the probability of *at least* one 2 per case is the same (19/100). The *average number* of 2s if a second pick is always made is found by multiplying the probability of each case by the corresponding number of 2s and summing, as shown in the right-hand column. If a second pick is not always made, the average number of 2s per case decreases to 19/100 because some cases will consist of only one trial.

TABLE 2.1 Probability of picking at least one 2, and average number of 2s

Case	First trial		Second trial		Total case probability	Number of 2s picked	Product of total case probability and number of 2s
	Outcome	Probability	Outcome	Probability			
A	Success	1/10	Success	1/10	1/100	2	2/100
B	Success	1/10	Failure	9/10	9/100	1	9/100
C	Failure	9/10	Success	1/10	9/100	1	9/100
D	Failure	9/10	Failure	9/10	81/100	0	0
					Total = 1		Ave. no. of 2s = 20/100 = 1/5

2.1.1 Permutations

When the balls are not replaced after each selection, the probabilities change with each successive trial. If number 2 is picked the first time and not put back, then the probability of picking number 2 the second time is zero while the probability of picking any other ball increases from 1/10 to 1/9. It follows that if a particular ball is not picked either the first or the second time, the probability of choosing it at the third trial increases to 1/8.

The probability of picking number 2 on the first trial and number 5 on the second is 1/10 × 1/9 or 1/90. Similarly, the probability of picking number 5 first and then number 2 is also 1/90. When the order in which the balls are picked is considered, the term *permutation* is used. If there are n items and r are picked without replacement, the number of permutations is given by Eq. (2.1)

$$\text{Number of permutations} \; = \; \frac{n\,!}{(n\,-\,r)\,!} \tag{2.1}$$

The symbol "!" is read "factorial"; n! represents the product of all positive integers from 1 through n: $n! = n \times (n-1) \times (n-2) \times \cdots \times 2 \times 1$. Thus, the number of permutations of 10 balls taken 2 at a time becomes 10!/8 = $(10 \times 9 \times 8 \times \cdots \times 2 \times 1)/(8 \times 7 \times 6 \times \cdots \times 2 \times 1) = 10 \times 9 = 90$. The probability of picking a particular permutation is 1/90, which agrees with the previous calculations. The number of different ways in which all 10 balls can be picked, using the permutations formula $n!(n-r)!$, is 10!/0!. By definition, the value of 0! is unity. The number of permutations, then, becomes $10 \times 9 \times \cdots \times 2 \times 1 = 3{,}628{,}800$, and the probability of picking all 10 balls in any one specific sequence is 1/3,628,800 or less than 3 in 10 million. The symbolic notation for permutations is $P_r^{\,n}$ and is interpreted: the number of permutations of n things taken r at a time.

2.1.2 Combinations

If it makes no difference whether number 2 or number 5 is picked first as long as these two balls are selected, the probability can be found in two different ways. First, the probability of a permutation can be multiplied by the number of possible successful permutations: 1/90 × 2 = 1/45. This is equivalent to adding the individual probabilities of the successful permutations because they are mutually exclusive.

Second, the probability can be calculated from the number of combinations. The number of combinations is the number of ways in which some items can be picked from a larger group without regard to sequence. In this example, there are 45 different combinations of 2 balls picked from a total of 10 balls. These are enumerated in Table 2.2. The combinations formula can

TABLE 2.2   The 45 combinations of 2 balls picked from a total of 10

1 and 2	1 and 7	2 and 4	2 and 9	3 and 7	4 and 6	5 and 6	6 and 7	7 and 9
1 and 3	1 and 8	2 and 5	2 and 10	3 and 8	4 and 7	5 and 7	6 and 8	7 and 10
1 and 4	1 and 9	2 and 6	3 and 4	3 and 9	4 and 8	5 and 8	6 and 9	8 and 9
1 and 5	1 and 10	2 and 7	3 and 5	3 and 10	4 and 9	5 and 9	6 and 10	8 and 10
1 and 6	2 and 3	2 and 8	3 and 6	4 and 5	4 and 10	5 and 10	7 and 8	9 and 10

be used to simplify the enumeration. The number of combinations of n things taken r at a time is given by Eq. (2.2)

$$\text{Number of combinations} = \frac{n!}{r!\,(n-r)!} \tag{2.2}$$

For 10 balls taken 2 at a time this equation becomes $10!/(2! \times 8!)$ which is $10 \times 9 \times 8 \times 7 \times \cdots \times 2 \times 1)/(2 \times 1 \times 8 \times 7 \times \cdots \times 2 \times 1)$. This reduces to $(10 \times 9)/(2 \times 1) = 45$. Since there is only one combination of the second and fifth balls, the probability of picking this combination is $1/45$. The number of combinations of n things taken together is obviously one, and the combinations formula confirms this: $n!/r!(n-r)! = n!/(n! \times 0!) = n!/n! = 1$. The notation for combinations is $C_r^{\,n}$ and is interpreted: the number of combinations of n things taken r at a time.

Comparison of the permutations and combinations formulas

$$P_r^n = \frac{n!}{(n-r)!} \qquad C_r^n = \frac{n!}{r!\,(n-r)!}$$

shows that the number of combinations is equal to the number of permutations divided by r factorial. This is logical because $r!$ is the number of different ways in which r things can be picked, that is, the number of possible sequences of r things; and the number of combinations does not consider sequence.

The combinations formula is also used to compute probabilities of more complex events. The number of combinations of favorable events is divided by the total number of possible combinations to determine the overall probability. It is important to note that both the combinations formula, Eq. (2.2), and the somewhat simpler permutations formula, Eq. (2.1), consider chance selection of units *without replacement* so that the number of available choices decreases for each successive selection. Also, as mentioned earlier, the probability of picking a particular unit changes as each selection is made. Three examples will illustrate the calculations and point out the difference between conditional and independent probabilities.*

*The examples show the use of the *hypergeometric* distribution, also known as the *law of compound probabilities*.

1. If there are eight white balls and six black balls in a box and four balls are chosen at random, the probability of picking two of each color can be computed as follows: The number of combinations of eight things taken two at a time, C_2^8, is $8!/(2! \times 6!) = (8 \times 7 \times 6 \times \cdots \times 1)/(2 \times 1 \times 6 \times \cdots \times 1) = 28$; and the number of combinations of six things taken two at a time is $6!/(2! \times 4!) = 15$. These products are multiplied together to find the number of combinations of two white and two black balls picked from this box: $28 \times 15 = 420$. The total number of combinations of fourteen things taken four at a time is $14!/(4! \times 10!) = 1,001$. Therefore, the probability of choosing exactly two white and two black balls from this box is $420/1,001 \approx 0.42$. The reader can verify this result by enumerating all possible combinations in the manner indicated in Table 2.2 and designating particular units as white or black.

2. If there are four red and four green balls, the probability of picking two of each is $(C_2^4 \times C_2^4)/C_4^8 = (6 \times 6)/70 = 0.514$, which is slightly more than one-half. On the other hand, suppose that a family has four children and assume that the probabilities of a boy or girl for a single birth are equal. The probability of two boys and two girls is only 0.375.* The reasons for the apparently conflicting results derive from the effects of an event in the two different cases. In the case of the colored balls, as each ball is picked, the percent of red and green balls remaining changes, and the probability of picking a particular color changes for each successive selection. This is conditional probability. In the case of the children, the probability of a boy or a girl is not affected by the sex of the preceding children. This is independent probability.

3. The probability of picking all four aces in five cards from a standard 52-card deck is very, very small. The total number of combinations of five cards, C_5^{52}, is $52!/(5! \times 47!) = 2,598,960$. Since there are four aces and 48 other cards, the number of combinations of four aces and one other card is $C_4^4 \times C_1^{48} = 4!/(4! \times 0!) \times 48!/(1! \times 47!) = 1 \times 48 = 48$. The probability is $48/2,598,960 = 0.00001847$. The probability of picking all four cards of any one value is 13 times as high or 0.0002401, which is still a very small number. The probability of having two hands in succession with four of a kind is $(0.0002401)^2 = 0.0000000576$. If such an unlikely event occurred, we would suspect that the cards (or the dealers) were not unbiased.

---

*The probabilities of various combinations of boys and girls are as follows: four boys: $(1/2)^4(1/2)^0 = 1/16$; three boys and one girl: $4(1/2)^3(1/2)^1 = 1/4$; two boys and two girls: $6(1/2)^2(1/2)^2 = 3/8$; one boy and three girls: $1/4$; and four girls: $1/16$. Independent probability, when there are two possible outcomes (in reliability the outcomes are success or failure), is discussed more completely in Sec. 3.2 on the binomial distribution.

2.2 MEASURES OF CENTRAL TENDENCY

Sections 2.2 and 2.3 discuss descriptive measures of distributions. They can be used with either discrete or continuous functions, although they are usually thought of as being associated with the latter. There are a number of measures of central tendency, only one of which—the arithmetic mean—is used extensively in reliability. Two others—the median and the mode—have more limited application. Other measures such as the geometric mean and the harmonic mean are used very infrequently.

2.2.1 The Mean

The arithmetic mean, commonly called the *mean* or the *average,* is defined as the quotient of the sum of the values of individual measurements divided by the number of measurements. If there are five measurements X_1, X_2, X_3, X_4, and X_5, the mean or average is equal to the sum of the measurement divided by five, $\overline{X} = {}^{!}(X_1 + X_2 + X_3 + X_4 + X_5)/5$. The symbol $\overline{X}$ is usually read "X bar" and designates the arithmetic mean of the samples. The general expression for the mean of a group of measurements is given by

$$\overline{X} = \frac{\sum\limits_{i=1}^{n} X_i}{n} \tag{2.3}$$

where the symbol $\Sigma_{i=1}^{n}$ is read "the sum of all measurements from 1 through n."

The mean can be interpreted physically as the point of balance or center of gravity of the individual values; the first moment about the mean is zero. If, in the example, the five individual values are 2, 4, 7, 8, and 9, the sum of the values is 30 and their arithmetic mean is 30/5 = 6. We can imagine a weightless board marked off into equal segments, as in Fig. 2.1. Equal weights are placed at the second, fourth, seventh, eighth, and ninth marks. The board will balance at the sixth mark because the algebraic sum of the distances of the weights from 6 is zero.

FIG. 2.1 Physical representation of arithmetic mean.

2.2.2 The Median

The second measure of central tendency is the *median.* The median is defined as that value such that half the measurements are smaller and half are larger.

It is also defined as the 50th percentile of the distribution.* In the preceding set of five values, the median value is 7. When there is an odd number of measurements, the median value is one of the measurements. When there is an even number, the median is halfway between the two middle measurements unless these two measurements have the same value, in which case the median is also the same. The median finds greater usage in income and similar statistics than it does in reliability, although the concept is used occasionally and an understanding is required for some applications. One goodness-of-fit test, for example, compares sample frequencies with theoretical frequencies at various percentiles including the 50th.

2.2.3 The Mode

The *mode* is that value which has the greatest frequency. Its use is rather limited, although one method of normalizing skewed distributions described in Chap. 5 utilizes this parameter. Most distributions have only one mode and the existence of more than one often indicates a change in some input condition during the data generation or the presence of more than one population source, e.g., two different machines making the same part. (The definitions of mean and median automatically limit each of these measures of central tendency to one value.)

2.3 MEASURES OF DISPERSION

Dispersion or variation is an inherent characteristic of all populations. It is only the degree of variation that differs. Some parameters such as length of life have a large scatter while the variations of other parameters may be so small that extremely precise measurements are needed to detect them. If no variations existed, theoretically it would be possible to produce components and systems which never failed and hence had perfect reliability. Even most failures resulting from wear-out can ultimately be traced to minute variations in materials or finishes. A principal need for reliability activities stems from the existence of variations, and many of the mathematical functions are concerned with the detection of variations, their measurement, and the relationships between them.

2.3.1 The Range

The simplest measure of variability is the *range*. This is defined as the difference between the largest and smallest measured values. While the range is widely used in quality control, more accurate measures are needed for most

---

*A percentile is that value below which the specified percent of a distribution of values will fall. Thus, 10 percent of the values will fall below the 10th percentile, 50 percent will fall below the 50th percentile, etc.

reliability activity. The range is used occasionally for establishing acceptance criteria based on reliability requirements when the sample sizes are very small and other measures lose their accuracy.

2.3.2 The Standard Deviation

The most common measure is the *standard deviation* σ (the Greek letter sigma). It is found by taking the square root of the average sum of squares of deviations of all individual values from the arithmetic mean, as given by Eq. (2.4).

$$\sigma = \sqrt{\frac{\Sigma(X_i - \overline{X})^2}{n}} \tag{2.4}$$

For this reason, it is sometimes called the root-mean-square deviation. Equation (2.4a) provides an alternate, equivalent formula

$$\sigma = \sqrt{\frac{\Sigma(X_i^2)}{n} - \overline{X}^2} \tag{2.4a}$$

From the definition of moments* it is seen that the standard deviation is the square root of the second moment about the mean. This moment itself is called the *variance*. Because the units of variance are square units, the standard deviation is preferred since it is measured in the same units as the data. It is a direct measure of variability. Variations of two distributions can be compared by directly comparing their standard deviations. If the standard deviation of one distribution is double that of the other, then on the average the variations of the first are twice as great as those of the second. In fact, the only requirement for a direct comparison is that the units of measure be the same. The means do not have to be equal and in most cases (for example, stress versus strain) they are not.

To illustrate the computation, consider 10 filters whose lifetimes (based on pressure-drop measurements) were 17.5, 10.4, 12.8, 14.4, 20.0, 11.9, 16.3, 14.6, 21.7, and 15.4 days. It is convenient although not necessary to arrange these in order of increasing lifetimes, as shown in Table 2.3.

The standard deviation from Eq. (2.4) is

$$\sigma = \sqrt{\frac{\Sigma(X_i - \overline{X})^2}{n}} = \sqrt{\frac{111.62}{10}} = 3.34 \; days$$

*The kth moment about the mean is $(1/n)\Sigma(X_i - \overline{X})^k$ and the kth moment about the origin is $(1/n)\Sigma(X_i^k)$. The mean, therefore, is the first moment about the origin.

TABLE 2.3   Lifetimes of 10 filters

Individual times	Differences from mean	Squares of differences
10.4	– 5.1	26.01
11.9	– 3.6	12.96
12.8	– 2.7	7.29
14.4	– 1.1	1.21
14.6	– 0.9	0.81
15.4	– 0.1	0.01
16.3	0.8	0.64
17.5	2.0	4.00
20.0	4.5	20.25
21.7	6.2	38.44
$\Sigma = 155.0$		$\Sigma = 111.62$
$\overline{X} = \;\; 15.50$		

(the same unit of measure as the data and the mean). It is of interest to note that the values close to the mean contribute very little to the sum of squares while a value which is far from the mean affects the standard deviation significantly.

2.3.3   Populations and Samples

Up to this point, no distinction has been made between the means and standard deviations of samples as contrasted with those of complete populations. The symbol $\overline{X}$ has been used to denote the mean in either case. However, it is standard practice to designate only the sample mean as $\overline{X}$, and to use the Greek letter μ (mu) to denote the population mean. Mu is found in the same way as $\overline{X}$:

$$\mu \;=\; \frac{X_1 + X_2 + \cdots + X_n}{n}$$

except that n represents all units in the population rather than just the sample units. Thus, μ is the exact value of the population mean while $\overline{X}$ is an estimate of the population mean based on some particular sample. The accuracy of the estimate $\overline{X}$ improves as the sample size increases. The accuracy can be given in terms of the *standard deviation of the estimate* $\sigma_{\overline{X}}$, which is related to the standard deviation of the population and to the sample size as follows:

$$\sigma_{\overline{X}} \;=\; \frac{\sigma}{\sqrt{n}} \tag{2.5}$$

When the sample size is quadrupled, the variability of the sample means is reduced by a factor of 2, and $\overline{X}$ becomes a correspondingly better estimate of μ.

The Greek letter σ has been used for the standard deviation, and the formula for σ was given by Eq. (2.4), repeated here:

$$\sigma = \sqrt{\frac{\Sigma(X_i - \overline{X})^2}{n}} \tag{2.4}$$

As in the case of the mean, a distinction is made between the standard deviation of the population and the standard deviation of the sample. The population standard deviation is designated by σ. (Accepted practice is to use Greek letters for population parameters and English letters for statistics of samples.*) It is computed as shown above and exactly defines the variability of the population.

The sample standard deviation is denoted by s and is calculated by using $n - 1$ rather than n in the denominator of the expression where $n - 1$ represents the number of units in the sample minus one:

$$s = \sqrt{\frac{\Sigma(X_i - \overline{X})^2}{n - 1}} \tag{2.6}$$

An explanation of the use of $n - 1$ rather than n in Eq. (2.6) can be found in Sec. 1.36 of Freeman.† It is sufficient for purposes of reliability (and for most other statistical functions) to remember to use $n - 1$ rather than n when computing the standard deviation of a sample. Note that as the sample size increases, $\sqrt{n - 1}$ approaches $\sqrt{n}$. Also, the accuracy of the estimate of the standard deviation improves as does the accuracy of the estimate of the mean.

2.3.4 Other Descriptive Measures

Two other descriptive measures of a population are worthy of mention: *skewness* or "lopsidedness" and *kurtosis* or "peakedness." Skewness, defined by Eq. (2.7), describes the degree to which the distribution is unsymmetrical.

$$\text{Skewness} = \frac{(1/n)\,\Sigma(X_i - \overline{X})^3}{\sigma^3} \tag{2.7}$$

---

*A measurable characteristic of a population is called a *parameter*; a value computed from a sample is called a *statistic.*

†H. A. Freeman, "Industrial Statistics," John Wiley & Sons, Inc., New York, 1942.

A skewness of zero indicates a symmetrical distribution, but it is possible (though unlikely) for an unsymmetrical distribution to yield a calculated skewness value of zero. By defining it in this manner, the measure is a pure number, independent of both the scale of units and the origin. A negative value indicates a distribution with the tail to the left (Fig. 2.2a), while a positive value indicates a right-tailed distribution (Fig. 2.2b). The terms "negatively skewed" and "skewed to the left" are used for the first case, and "postively skewed" and "skewed to the right" are used for the latter. The data in Table 2.3 indicate a distribution with a slight positive skewness.

(a)   (b)

FIG. 2.2 Skewed distributions. (a) negative: skewness < 0.
(b) positive: skewness > 0.

A simpler method of obtaining a measure of skewness is to divide the difference between the mean and the mode (or median) by the standard deviation as in Eq. (2.8):

$$\text{Skewness} = \frac{\text{mean} - (\text{mode or median})}{\sigma \text{ or } s} \qquad (2.8)$$

Since skewness is used infrequently, there are no ready comparisons of the usefulness of the two methods. It is recommended that, if a measure of skewness is required, the simpler method be used. Because it is easier to make predictions and assessments for normally* distributed variables than for skewed distributions, methods have been developed for converting data into normal or near-normal distributions. Some of these are discussed in Chap. 5.

The other descriptive measure is kurtosis (peakedness), defined by

$$\text{Kurtosis} = \frac{(1/n) \Sigma (X_i - \overline{X})^4}{\sigma^4} \qquad (2.9)$$

The normal distribution has a kurtosis value of 3. If two distributions have the same standard deviation, the one with the higher kurtosis value will generally be more peaked. Because kurtosis is rarely used in reliability, it will not be discussed further.

2.4 SAMPLING AND CONFIDENCE

In the case of discrete events we have indicated that, when samples are picked

---

*The normal or gaussian distribution is discussed in Sec. 3.5.

at random, the statistics (parameters) which are calculated from the samples may differ considerably from those of the parent population. In the example of a family with four children, the probability of three children of one sex and only one of the other was 50 percent, while the probability of two of each was only 37½ percent. Since we know that there are about the same number of males as females, we would not make the error in our estimate of assuming a 3:1 ratio if we chose only one four-child family and that family happened to have this ratio of children. Similarly, with four red and four green balls, we would expect that if we picked two at random, on an average we would get one of each. However, we would not be too surprised if we picked two balls of the same color.

In a like manner, we could take actual measurements of dimensions or performance characteristics of samples and calculate the means and standard deviations. We would expect these statistics to be representative of the parameters of the population from which the samples were taken, but differences of the sample statistic values from values of the population parameters would not be unexpected, particularly if the sample sizes were small. The value of any other statistic, e.g., mode or skewness, calculated from a sample would also be expected to differ from the value of that parameter in the parent population.

When we know the exact nature of the distribution and the values of the parameters which describe it, we can predict the distribution of samples picked at random. Not only are we able to calculate the probability of the sample having about the same parameter values, but we can also determine mathematically the probability of the sample statistics differing from the population parameters by whatever amount we specify. That is, we can compute the probability of having four children of the same sex, of picking four red balls, of randomly selecting a sample with a mean of 6.0 when the population mean is 8.0, or of finding a sample standard deviation of 5.0 days when the standard deviation of the population is 3.3 days.

In reliability, as in quality control, test operations, and other functions associated with the evaluation of a product or process, we are usually concerned with the inverse problem. Rather than calculating the probability of obtaining samples with particular characteristics or statistic values from a known population, we wish to estimate the characteristics of the population from a set of sample data. We recognize that samples do not always present true pictures, that is, that they may not be representative of the population. Small samples can vary considerably from one to the next. As the sample size increases, there will be smaller and smaller differences among samples taken from the same population.

While we cannot be sure that a sample is representative, we can associate a *degree* of assurance with the sample results. This degree of assurance is called *confidence,* which can be defined as the *level of certainty which can be associated with conclusions based on the results of sampling.* This level of certainty can be determined for any of the characteristics which have been presented or with which we are familiar: the percent defective of a total population, the probability that the mean of a distribution lies within a given set of limits, or the probability that the true reliability is at least a certain level. The confidence will depend on the nature of the data, the parameters being estimated, the levels of reliability or quality that are chosen, and the amount of sample data available. We shall be directly involved with confidence throughout a large portion of this text, and particularly in Chaps. 3, 4, 13, and 14.

ADDITIONAL READINGS

ARINC Research Corporation: "Reliability Engineering," Prentice-Hall, Inc., Englewood Cliffs, N.J., 1964.

Bazovsky, I.: "Reliability Theory and Practice," Prentice-Hall, Inc., Englewood Cliffs, N.J., 1961.

Bureau of Naval Weapons, Chief of: "Handbook: Reliability Engineering," U.S. Government Printing Office, 1964.

Calabro, S. R.: "Reliability Principles and Practices," McGraw-Hill Book Company, New York, 1962.

Chorafas, D. N.: "Statistical Processes and Reliability Engineering," D. Van Nostrand Company, Inc., Princeton, N.J., 1960.

Dixon, W. J., and F. J. Massey, Jr.: "Introduction to Statistical Analysis," 3d ed., McGraw-Hill Book Company, New York, 1969.

Dummer, G. W., and N. B. Griffin: "Electronics Reliability: Calculation and Design," Pergamon Press, Ltd., London, 1966.

Freeman, H. A.: "Industrial Statistics," John Wiley & Sons, Inc., New York, 1942.

Grant, E. L.: "Statistical Quality Control," 3d ed., McGraw-Hill Book Company, New York, 1964.

Hoel, P. G.: "Introduction to Mathematical Statistics," John Wiley & Sons, Inc., New York, 1947.

Ireson, W. G.: "Reliability Handbook," McGraw-Hill Book Company, New York, 1966.

Lloyd, D. K., and M. Lipow: "Reliability: Management, Methods, and Mathematics," Prentice-Hall, Inc., Englewood Cliffs, N.J., 1962.

Mood, A. M.: "Introduction to the Theory of Statistics," McGraw-Hill Book Company, New York, 1950.

Munroe, M. E.: "The Theory of Probability," McGraw-Hill Book Company, New York, 1951.

Shooman, M. L.: "Probabilistic Reliability: An Engineering Approach," McGraw-Hill Book Company, New York, 1968.

3

DISTRIBUTIONS USED IN
RELIABILITY

3.1 INTRODUCTION

This chapter describes briefly the distributions which are used most frequently in reliability. Their basic characteristics are explained in conjunction with their applications to reliability analyses. The distribution used for discrete events or probabilities—the binomial—is discussed first because it is most closely associated with the aspects of probability discussed in Chap. 2. The Poisson and negative exponential are then explored because, as we will see, they follow from the binomial.

The normal distribution, which characterizes many natural and man-made phenomena, is presented next, along with an associated distribution—the Student's t—used for obtaining confidence limits when averages are estimated. The chi-square distribution is then introduced because, although it is used to analyze discrete values, it is the foundation of other distributions used to evaluate variability. These distributions are then presented.

The most complex distribution commonly found in reliability is the Weibull, discussed only briefly in this chapter. A more thorough discussion is deferred until Chap. 14 where a detailed example is worked out. The final topic is the Tchebycheff inequality, a sort of "last resort" which is used only when a very minimum of information is available.

Familiarity with these distributions and their applications will be of great value as we investigate reliability principles and means of evaluating achieved reliability and of attaining reliability goals.

3.2 THE BINOMIAL DISTRIBUTION

The binomial distribution is one of the simplest distributions, although it is used for computing reliabilities of complex redundant systems. It is the one usually associated with probability problems and games of chance. The distribution is $(a + b)^n$. This can be expanded to

$$a^n + na^{n-1}b + \frac{n(n-1)a^{n-2}b^2}{2!} +$$

$$\frac{n(n-1)(n-2)a^{n-3}b^3}{3!} + \cdots + b^n$$

It is applicable when there are only two possible states or outcomes of an event, i.e., when a component or system either has or does not have a particular characteristic, either conforms to requirements or is discrepant, either operates or fails, and so on. If we substitute p and q for a and b, where p is the probability of an event happening (a component being good) and q is the probability of the event not happening, which is the alternate event (the component being defective), then $p + q = 1$, and $(p + q)^n$ also equals 1. It follows that the expansion of $(p + q)^n$ must also equal unity:

$$p^n + np^{n-1}q + \frac{n(n-1)p^{n-2}q^2}{2!} + \frac{n(n-1)(n-2)p^{n-3}q^3}{3!} + \cdots$$

$$+ q^n = 1 \qquad (3.1)$$

The mathematical expression for the rth term of the expansion is given by Eq. (3.2). In *this* equation, the initial term p^n is called the zero term ($r = 0$).

$$r\text{th term} = \frac{n!}{r!(n-r)!} p^{n-r}q^r \qquad (3.2)$$

where n is the total number of events (units) and r is the number of unfavorable events (defectives).

We can next examine the meaning of each term in the series on the left side of Eq. (3.1). Suppose that there are n identical units of a component, that the probability of a unit being good is p, and that the probability of its being defective is $q (= 1 - p)$. If there is only one unit ($n = 1$), then the probability of its being good is simply p. If there are two units ($n = 2$), then the probability of both being good is $p \times p = p^2$; if there are three units, the probability of all three being good is p^3. Consequently, if there are n units, the probability of all n being good is p^n. Therefore, the first term in the series (renumbering from 1 rather than from 0) is the probability of all units being good.

Next, suppose there are two units A and B, one good and one defective. There are two ways that this can occur: A good and B defective, with probability $p \times q$, and A defective and B good, with probability $q \times p (= p \times q)$. Since these are mutually exclusive and constitute all possible combinations of one good unit and one defective unit, the total probability is $2pq$. If there are three units $A, B,$ and $C,$ two good and one defective, there are three possible combinations: A and B good and C defective, A and C good and B defective, and B and C good and A defective. The probability of each combination is $p \times p \times q = p^2 q$. Again, since each combination is mutually exclusive and together they constitute all possible combinations, the probability of two good and one defective unit is $3p^2 q$. Similarly, if there are n units, the probability of $n - 1$ good units and one defective one is $np^{n-1}q$. Thus, the second term of the binomial expansion series is the probability of exactly one defective unit, all other units being good.

By extending these derivations to cover each succeeding term, we find that the third term is the probability of exactly two defective units, the fourth term is the probability of exactly three defectives, and so on. There are $n + 1$ terms in the expansion, and the last term is the probability of all units being defective.

The binomial is used to calculate the probability of system success for *active* redundant configurations when the redundant components are identical. It is equally applicable for components whose reliabilities are time dependent and for equipment which is called on for a single operation (popularly termed "one-shot" items), such as flash bulbs, missile stage separation devices, gas generators, and other single-usage components. It also applies to reliability calculations of such single-usage equipment as missile engines, short-life batteries, etc., which are required to operate for one time period and are not reused. The two main requirements are (1) knowledge of the reliability of a single unit with respect to the mission, and (2) the relative constancy of this reliability after failure of other units. For example, if there are two units operating in parallel and only one is required, then if one fails, the reliability (or failure rate) of the second should not be affected. Strictly speaking, after one or more units have failed, the failure rates of the remaining units do change because their loads might be higher, the environment might be different, e.g., the surrounding temperature might decrease, etc. For practical purposes, it is assumed that the failure rates are not affected. While not 100 percent accurate, the assumption is not so inaccurate as to negate the usefulness of the reliability calculations.

The reliability of a group of redundant items depends not only on the reliability of each individual item and on the number of items in redundant configuration, but also on how many are required to work to achieve system success. If all are required, then the first term of the binomial series represents system success. In this case there is really no redundancy. However, if all but one are required (one failure permitted), then success is achieved if either no failures occur or exactly one failure occurs. The system reliability is then the sum of the first two terms of the series. If two failures are permitted, then the sum of the first three terms represents the probability of system success. In general, if r failures are permitted, system success is the sum of the first $r + 1$ terms. Table C.1* provides reliabilities of redundant systems when there are n redundant items and r may fail. R_i is the reliability of an individual unit.

The coefficients of the terms in the binomial expansion can be found as indicated by the multiplier in Eq. (3.2) or by using Pascal's triangle, Fig. 3.1. Note that each number in the triangle is the sum of the nearest two numbers in the preceding line.

```
Sample size, n                        I
    I                        I        I
    2                   I    2    I
    3              I    3    3    I
    4         I    4    6    4    I
    5    I    5   10   10    5    I
    6  I  6   15   20   15    6    I
    7 I  7  21   35   35   21    7    I
                        etc.
```

FIG. 3.1 Pascal's triangle: coefficients of terms in the binomial expansion.

*C.1 refers to the first table in Appendix C. All references to tables in the Appendixes will identify the Appendix (A, B, or C) by uppercase letter and the particular table by number. When confidence levels are involved or separate tables are provided for different sample sizes, the table number will be followed by a lowercase letter.

3.3 THE POISSON DISTRIBUTION

To maintain consistency, we now consider q as the probability of the alternate event (in this case, failure) occurring during a given time interval, and n as the number of such intervals. If the product nq is held constant as n becomes infinite, the binomial distribution $(p + q)^n$ approaches a limit. This limit is known as the Poisson distribution. For each term in the binomial expansion

$$p^n + np^{n-1}q + \frac{n(n-1)p^{n-2}q^2}{2!} + \cdots$$

in Eq. (3.1), there is a corresponding term in the Poisson distribution in Eq. (3.3).

$$e^{-c} + ce^{-c} + \frac{c^2}{2!}e^{-c} + \cdots = 1 \tag{3.3}$$

where c is the product nq and e is the base of natural logarithms = 2.71828. The expression for the rth term is given by Eq. (3.4) wherein the initial term e^{-c} is again called the zero term:

$$r\text{th term} = \frac{c^r}{r!}e^{-c} \tag{3.4}$$

Since each term in the Poisson distribution directly corresponds to a term in the binomial when n becomes infinite, the meaning of each term in the Poisson series can be readily determined. As in the binomial, the probability of no failures occurring is given by the first term e^{-c}, the probability of exactly one failure is given by the second term ce^{-c}, and so on. We have seen that the sum of the terms in the binomial is equal to unity; similarly, the sum of the terms of the Poisson is also equal to unity. The distinction is that the binomial normally has a finite number of terms, that is, n is a real positive integer. Because the Poisson is the limit of the binomial as n becomes infinite, the Poisson has an unlimited number of terms. However, we will see shortly that this fact does not interfere with its use in reliability calculations.

Since the product of nq (= c) is a constant, the number or probability of failure is constant. This occurs in reliability when a component or system has a constant failure rate and the time interval is also constant. We can replace c with λt where λ (the Greek letter lambda) is the *failure rate* and t is time, so that the product λt again represents number or probability of failure as did the product nq. In order for a system to maintain a constant failure rate, the number of operating components must be kept constant. (Other considerations are discussed in later chapters.) If one component fails, it either must be repaired and restored to its original condition or must be replaced with an identical component. This manner of system operation is known as *sequential redundancy,* as contrasted with active redundancy associated with the binomial distribution.

The Poisson distribution is used to calculate the reliability of sequentially redundant systems. Equation (3.3) is rewritten as shown by Eq. (3.5):

$$e^{-\lambda t} + \lambda t e^{-\lambda t} + \frac{(\lambda t)^2}{2!} e^{-\lambda t} + \cdots = 1 \tag{3.5}$$

The first term defines the probability of no component failures, the second term defines the probability of one component failure, etc., as stated above; and there are an infinite number of terms in the expansion. However, a system has a finite number of possible repairs or replacements. Thus, the probability of system success can be determined from a finite number of terms in the expansion. For example, if a system has five units, with one operating and four on standby to be used as replacements in case of failure, then system success is achieved if not more than four failures are experienced. The first five terms of the Poisson expansion give the probabilities of zero, one, two, three, and four component failures, and their sum is thus the probability of system success. The use of both the binomial and Poisson distributions in reliability is discussed at length later in the book, where the Poisson distribution shown in Eq. (3.5) is usually written as

$$e^{-\lambda t} \left(1 + \lambda t + \frac{(\lambda t)^2}{2!} + \cdots \right) = 1 \tag{3.6}$$

The first term of the Poisson, $e^{-\lambda t}$, is very important because it denotes the probability of no failures—the reliability—of a single component or non-redundant system which has a constant failure rate. This term, known as the negative exponential distribution, is discussed next.

3.4 THE NEGATIVE EXPONENTIAL

Many components, particularly electronic ones, are operated during the period of time when failures occur randomly due only to "chance causes."*

---

*It might be proposed that, in a very strict sense, there is no such thing as a "chance" cause and that the causes of all failures can ultimately be traced, perhaps requiring an investigation into the physical, chemical, and metallurgical properties of the component and its materials. This detail is unnecessary for prediction purposes, and since these defects occur infrequently and necessarily at random, the failures are classed as "chance" or random failures. This practical approach is perfectly satisfactory for our purposes and the general practice of classifying these failures as random will be followed.

During this period there is no one failure mechanism which is predominantly responsible for failure. Failures occurring as the result of "infant mortality" have been eliminated by corrective actions relating to the design, the operating procedures, or the inspection methods, or by burn-in. Similarly, the wear-out time has not been approached and failures due to wear-out or performance degradation beyond tolerance limits do not have a significant effect on the overall failure rate. The number of failures occurring during any time interval appears to be related only to the total number of units operating, the length of the interval, and the test conditions. If sufficient numbers are tested and if the interval is long enough (without approaching either the infant mortality or the wear-out period), a failure rate will be observed which tends toward some relative constant value.

For any given component type, the failure-rate value will depend on the operating and external environmental conditions (voltage, temperature, pressure, vibration level, etc.) and will be characteristic of this set of conditions. Of course, when the conditions change, the failure rate will correspondingly change. The effects of these changes in conditions on failure rates are discussed more fully in Chap. 9. We are concerned here with defining a distribution which can characterize this constant failure rate and thus can be used for a large portion of reliability-prediction activity.

The negative exponential is the applicable distribution. Engineers and technicians familiar with electrical components have had experience with this distribution as it applies to dc voltage decay in a capacitor and to similar phenomena. The distribution is $y = e^{-x}$ and is shown in Fig. 3.2. As we have seen, in reliability applications, y represents reliability and x is the product of the failure rate λ and time t: $R = e^{-\lambda t}$. For any constant failure rate the value of reliability depends only on time. The limits of the reliability value are 1.0 at time equal to zero, and 0.0 as time becomes infinite. This distribution is called a

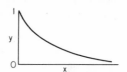

FIG. 3.2 The negative exponential distribution.

"one-parameter distribution" as is the Poisson distribution since, once the failure rate is set, the reliability function is completely defined.* Figure 3.3 depicts the distribution as a function of time (or λt, with λ = a constant). For a high failure rate (large λ), the reliability curve (a) descends steeply and rapidly approaches zero. When the failure rate is low (small λ), the descent is more gradual (c) and approaches zero more slowly. The general shape of the curve remains the same, but it is compressed or stretched out along the time

*In contrast, the binomial distribution is a two-parameter distribution because values of the probability p (or q) and the exponent n must both be established to describe this distribution. The normal distribution, Sec. 3.5, is also a two-parameter distribution because the mean μ and standard deviation σ are required to define it.

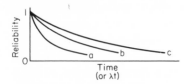

FIG. 3.3 Reliability as a function of time when λ is constant. (a) Large λ; (b) intermediate λ; (c) small λ.

axis depending on the value of λ. Also, in a series system which has a constant failure rate, the mean time to failure m is the reciprocal of the failure rate ($m = 1/\lambda$ or $\lambda = 1/m$) and the equation $R = e^{-t/m}$ is interchangeable with $R = e^{-\lambda t}$. Table 3.1 gives values of reliability for a few values of failure rate and time and provides a general idea of the relationship. A more complete table is found in Appendix B.

This distribution leads to the simplest calculations when either component or system reliability is computed. Although this is desirable, it should not be the determining factor when a distribution is chosen to depict the mathematical reliability function. The choice of distributions must still be based on the technical considerations associated with the equipment. Fortunately, as indicated above, many, if not most, items of equipment are operated in the time interval after the infant-mortality or burn-in period has passed and before wear-out becomes a significant factor affecting the failure rate. In other words, the equipment is used during the interval when its failure rate is relatively constant, and therefore the negative exponential is truly representative of the reliability as well as being easy to use.

An important corollary to the applicability of the negative exponential distribution is the fact that reliability is independent of starting time when there is no redundancy.* If an item of equipment has a failure rate λ at time

*Redundancy considerations are discussed in Chaps. 8 and 10 on mathematical models and system reliability prediction.

TABLE 3.1 Reliability as a function of failure rate and time using the negative exponential equation: $R = e^{-\lambda t} = e^{-t/m}$

Failure rate, %/1,000 hr	Mean time to failure, hr	Time, 1,000 hr								
		0.1	0.5	1.0	5	10	50	100	500	1,000
0.01	1×10^7	0.99999	0.99995	0.9999	0.9995	0.999	0.995	0.990	0.951	0.905
0.02	5×10^6	0.99998	0.99990	0.9998	0.9990	0.998	0.990	0.980	0.905	0.819
0.05	2×10^6	0.99995	0.99975	0.9995	0.9975	0.995	0.975	0.951	0.779	0.607
0.10	1×10^6	0.9999	0.9995	0.999	0.995	0.990	0.951	0.905	0.607	0.368
0.20	500,000	0.9998	0.9990	0.998	0.990	0.980	0.905	0.819	0.368	0.135
0.50	200,000	0.9995	0.9975	0.995	0.975	0.951	0.779	0.607	0.082	0.0067
1.0	100,000	0.999	0.995	0.990	0.951	0.905	0.607	0.368	0.0067	0.00005
2.0	50,000	0.998	0.990	0.980	0.905	0.819	0.368	0.135	0.00005	0.00000
5.0	20,000	0.995	0.975	0.951	0.779	0.607	0.082	0.0067	0.00000	0.00000
10.0	10,000	0.990	0.951	0.905	0.607	0.368	0.0067	0.00005	0.00000	0.00000

zero, its reliability for the period of time t is $e^{-\lambda t}$. If, at the end of this time period, it is still operable, its reliability for the next time period of equal length is still $e^{-\lambda t}$. This holds true as long as the equipment is operated during the period of constant failure rate.

It follows that the reliability of the equipment for any time period from time t_1 to time t_2 is $\exp[-\lambda(t_2 - t_1)]$* provided that the equipment was operable at time t_1. If all that is known is that it was operable at some prior time t_0, then the probability of its being operable at time t_2 (its reliability) is the product of $\exp[-\lambda(t_1 - t_0)]$ which is its reliability for time t_1 and $\exp[-\lambda(t_2 - t_1)]$. This is equal to $\exp[-\lambda(t_1 - t_0) - \lambda(t_2 - t_1)]$, or $\exp[-\lambda(t_2 - t_0)]$. This, of course, is the same result that would be obtained if we had calculated the reliability for the time period t_0 to t_2.

3.5 THE GAUSSIAN (NORMAL) DISTRIBUTION

The normal or gaussian† distribution (the "bell-shaped" curve) is perhaps the most familiar. Many natural phenomena have this distribution. Also, many characteristics of manufactured equipment are distributed normally, and when the university professor states that he "grades on the curve," he is referring to the normal distribution. The equation of this distribution is

$$Y = \frac{1}{\sigma\sqrt{2\pi}} e^{-(X_i - \mu)^2/2\sigma^2} \qquad (3.7)$$

In this equation, X_i is the abscissa or the individual value along the horizontal axis, and Y is the ordinate or the height of the curve at the corresponding value of X. The center (mean) and spread (standard deviation) of the distribution are the two parameters which define it. These are, respectively, μ and σ. The mean μ is the arithmetic average, and the standard deviation σ is the measure of the distribution's variability. Sigma is defined as $[\Sigma(X_i - \mu)^2/N]^{1/2}$, where N is the number of individuals in the population. The larger the value of sigma, the greater is the distribution's spread. Once μ and σ are established, the distribution is uniquely defined, and for any given distribution μ and σ are constant. Figures 3.4 and 3.5 compare some normal distributions with different values of μ and σ. The area under each curve is unity.

The normal distribution differs from the binomial, and from other discrete distributions, in an important respect. We recall that the binomial is a distribution of only two discrete values; that is, regardless of the relative

---

*Note that exp $(x) = e^x$ and that this form will be used whenever it enhances legibility.

†The distribution is named after the German mathematician, Karl Gauss.

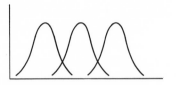

FIG. 3.4 Normal curves with same mean and different standard deviations.

FIG. 3.5 Normal curves with same standard deviation and different means.

probabilities, a unit either has a particular property or characteristic or it does not, e.g., a unit may be either good or bad, it cannot be 0.5372198 good. On the other hand, the normal distribution is continuous and although the probability of a unit or characteristic having any particular value is very small, any value is possible.*

Because the normal distribution is symmetrical, it follows that exactly half of the distribution is on each side of the mean. It is not quite so clear that, regardless of the value of σ, the same percentage or fraction of the distribution will lie between the mean and any particular number of standard deviations $z\sigma$ or between $z\sigma$ and $\pm \infty$.† For example, if $z = +1$, then 34.13 percent of the distribution will be between the mean and the value represented by $+ 1\sigma$. Also, 15.87 percent of the distribution will have values greater than the value represented by $+ 1\sigma$ and 84.13 percent will have smaller values. For example, in Fig. 3.6, if the mean $\mu = 50$ and $\sigma = 5$, then 34.13 percent of the distribution will be between 50 and 55, 15.87 percent will be greater than 55, and 84.13 percent will be less than 55. If, instead of 5, σ were 12, and the mean remained at 50, then $+ 1\sigma$ would be equal to 62 and the percentages just given would be with respect to a value of 62 rather than 55.

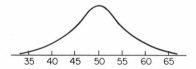

FIG. 3.6 Normal distribution with mean $\mu = 50$ and standard deviation $\sigma = 5$.

Figure 3.7 shows the percent of the normal distribution between different values of z (different numbers of σ). Table B.2 provides percentages between $- z$ and $+ z$ standard deviations.

*Theoretically, because the number of possible values is infinite, the probability of any particular value is zero. Because it is not usually possible to measure to a great number of places of accuracy, we assign probabilities to the smallest range of values which we can discriminate.

†This can be proved by integrating the value of Y for $- \infty < X \leq \mu + z\sigma$ for any value of z. When $z = 0$, the result is 0.5.

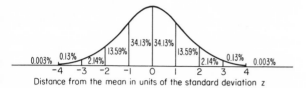

Distance from the mean in units of the standard deviation z

FIG. 3.7   Percent of distribution between integer numbers of standard deviations of a normal distribution.

We are constantly working with samples rather than with entire populations, and frequently the samples are quite small. This is necessarily true for one or more of a number of reasons. The entire population may be unavailable: it may not even exist yet, the decision to produce it depending on the acceptability of the sample units. Evaluation, inspection, or testing may be too costly or time-consuming to be practical for an entire population. The evaluation might be destructive in nature so that if the entire population were checked, no units would remain to perform the required task. Consequently, the relationship of a sample to the population from which it was taken is of interest. Of even more importance is the definition or description of a population based on information which we have obtained from a usually limited sample.

If a population has a normal distribution with mean μ and standard deviation σ, then samples taken at random from this population will tend to have a normal distribution also. As the size of the sample increases, its distribution becomes more nearly normal, and the sample average (designated $\overline{X}$, read "X bar") will be nearer the population mean. However, since the sample units are randomly selected, $\overline{X}$ will probably not be exactly equal to μ, but it is the best estimate.

The standard deviation of the sample s measures the variation of individual units in the sample just as σ measures the variability of the entire population. The computation of s is very similar to the computation of σ: $s = [\Sigma (X_i - \overline{X})^2/(n - 1)]^{1/2}$. The difference is that $\overline{X}$ is used instead of μ, and the denominator of the fraction is one less than the sample size n. The value of s, computed using $n - 1$ rather than n, is the best (unbiased) estimate of σ. This is due to the nature of its distribution. (For a more detailed discussion, the reader may wish to refer to a text on mathematical statistics.) We use $\overline{X}$ and s to estimate the values of μ and σ.

Often the means and standard deviations of samples are compared or the possible range of the mean or variance of a population is estimated from a sample. Some of these calculations are considered in later chapters. In order to make them, we must know how sample parameters themselves are distributed. When the population distribution is normal, the distribution of sample averages $\overline{X}s$ is also normal. Even when the population is not normal, the distribution of $\bar{X}$s is very close to normal.* While the normality of sample averages is desirable for quality control work and is especially useful for control chart activities, it can lead to errors in reliability calculations unless it is recognized that the original population may not be normal.

Of equal importance is the variation of sample averages. The standard deviation of sample averages $\sigma_{\bar{X}}$ is equal to the standard deviation of the population divided by the square root of the sample size: $\sigma_{\bar{X}} = \sigma/\sqrt{n}$. If s is used to estimate σ, then $\sigma_{\bar{X}} = s/\sqrt{n}$. Here, n is used in the denominator even though $n - 1$ is used in calculating the value of s.

There are two principal applications of the normal distribution in reliability. The first is concerned with the relationship of variable characteristics of equipment to specifications. This involves the probability of the characteristics meeting performance requirements. The second relates to wear-out life of components and the distribution of component failures due to extended periods of operation. Actually, wear-out could be considered as another variable characteristic which affects the component's reliability. However, since performance characteristics are usually thought of as being effective as soon as the component is operated or at least within a short time thereafter while wear-out is generally a relatively long-term phenomenon, the two types of characteristics will be considered separately.

A component most often is required to operate within certain prescribed limits. An ac generator, for example, has to provide a specified frequency within a certain number of cycles per second; a check valve must open between specified pressure levels; and an explosive device must detonate when the applied voltage is within certain values. All these applications involve the relationship of a variable operating characteristic to a set of limits. Sometimes only one limit, either an upper or a lower one, is defined, such as the current limit for a fuse. The distribution of these characteristics can be represented by a normal distribution, as discussed above, and the values of μ and σ can be estimated. The relationship of the distribution to specification limits can then be established. In this manner the probability of meeting requirements—the reliability—is determined.

We recall that, after the period of random failures, the component or assembly is subject to wear-out. The distribution of wear-out failures is often considered to be normal although it would be unusual for it to be exactly normal. The wear-out failure distribution is frequently sufficiently close to normal that the use of this distribution for predicting or assessing reliability is valid and provides useful information. In the case of wear-out failures, we are concerned with only one limit, the required mission lifetime, and the

---

*See E. L. Grant, "Statistical Quality Control," 3d ed., p. 75, McGraw-Hill Book Company, New York, 1964, or W. A. Shewhart, "Economic Control of the Quality of Manufactured Product," p. 182, D. Van Nostrand Company, Inc., Princeton, N.J., 1931.

distribution of wear-out failures is compared with the minimum required life to evaluate the probability of successful completion.

3.6 STUDENT'S t DISTRIBUTION

In the discussion of the normal distribution, it was stated that when sample units are taken from a total population, the mean and standard deviation of the sample may vary from the true mean and standard deviation of the population. This variation is known as sampling error. When only a few sample units are picked, the sampling error may be quite large. For example, when only three units are chosen from a population which has a normal distribution, it is possible that all three could be smaller than the mean, and the average of the three would then be less than the mean. Also, the standard deviation calculated for the three units could be considerably greater or smaller than the standard deviation of the population. If we were to estimate these parameters for the population from a small sample, our estimates could be considerably in error.

We could reduce the amount of our potential error by taking more and more units in the sample. The greater the number of units, the smaller the error because it becomes less and less likely that all units will be on one side of the mean and none on the other. Eventually, as the sample becomes so large that it includes the entire population, the sampling error goes to zero.

However, it is seldom possible to measure every unit or even a large number of units from a population to find its mean. Instead, the population mean must be estimated from a relatively small sample. Since this gives rise to an error, we would like to be able to state that the true mean lies somewhere between certain values or that the true mean is at least as great as some specified value; we would also like to be able to state how certain (confident) we are of our estimates. The Student's t distribution enables us to do this when there is reasonable assurance that the population is normally distributed.

This distribution accounts for the error resulting from taking a finite sample. It provides the capability for attaining confidence in the estimate of the mean by relating the amount of error in the estimate to the confidence level. The more absolute error that can be accepted in the estimate, the more sure we are that the true value of the mean or percentage is within the set limits. Conversely, if we state that the true value is within very restricted limits (corresponding to a close estimate), we are less sure that we are correct and the associated confidence level goes down. As indicated above, as the sample size is increased, our confidence that the true mean is within a set of limits of a given width also increases; or, if we wished to retain the same confidence level, then the width of the limits narrows; i.e., the error is reduced. Stated more concisely, an increase in sample size results in an increase in confidence or in a decrease in error or in some combination of both.

The relationship between error magnitude, confidence level, and sample size as related to the population mean is shown pictorially in Fig. 3.8. Interpretation of the figure is as follows. For a sample size of 2, we are 50 percent sure (confident) that the true mean μ lies within $\pm 1.0\sigma_{\overline{X}}$ ($= 1.0$ $s/\sqrt{n}$)* of the estimated mean $\overline{X}$. We are 75 percent confident that μ is not less than $\overline{X} - 1.0\sigma_{\overline{X}}$. We are also 75 percent confident that it is not greater than $\overline{X} + 1.0\sigma_{\overline{X}}$. Using the same curve, we are 70 percent sure that μ is within $\overline{X} \pm 1.96\sigma_{\overline{X}}$ and 90 percent sure that it is within $\overline{X} \pm 6.31\sigma_{\overline{X}}$. If the sample size is greater than 2, we would determine our error (in terms of $\sigma_{\overline{X}}$) and confidence from another of the curves.

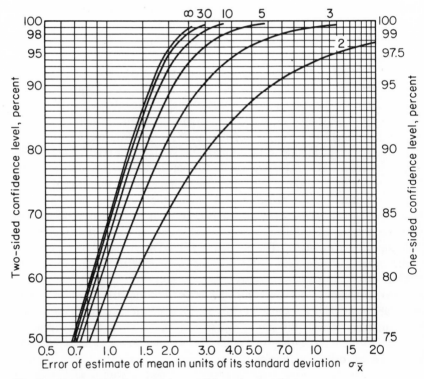

FIG. 3.8 Relationship of sample size, confidence level, and error in standard deviations.

*We recall from Sec. 3.5 that the standard deviation of sample averages, σ_X, is equal to the standard deviation of the population divided by the square root of the sample size.

Table B.3 provides more detailed data about this relationship. Sample sizes are listed in the left-hand column in terms of number of degrees of freedom (d.f.). The number of degrees, denoted by the Greek letter ν (nu), is one less than the sample size: d.f. (or ν) = $n - 1$. The concept of degrees of freedom is discussed more fully in Chap. 4. Confidence (probability) values are given as headings of the other columns. Error magnitudes in terms of units of standard deviation are given in the body of the table. Three brief examples will serve to illustrate the use of the table.

1. Sample size = 10 (d.f. = 9); Confidence level = 0.90, two-sided

   $t = 1.83$

   This means that the probability is 0.90 that the population mean μ is between ± 1.83 standard deviations from the sample average $\overline{X}$. Also, the probability is 0.05 that μ is smaller than the value corresponding to − 1.83 standard deviations from the sample average and 0.05 that it is greater than the value corresponding to $\overline{X}$ + 1.83 standard deviations.

2. Sample size = 4; Confidence level = 0.95, two-sided

   $t = 3.18$

   The probability is 0.95 that μ is between ± 3.18 standard deviations from $\overline{X}$.

3. Sample size = 30; Confidence level = 0.995, one-sided

   t = plus *or* minus 2.76.

   The probability is 0.995 that μ is greater than $\overline{X}$ − 2.76 standard deviations. It is also 0.995 that μ is less than $\overline{X}$ + 2.76 standard deviations. The probability of μ being within ± 2.76 standard deviations of $\overline{X}$ is 0.99.

Note that as the sample size increases, the values of t in the table approach the values of z in the normal table.

Another directly related use of the t distribution is the comparison of a sample average $\overline{X}$ with a known or assumed population mean μ to determine whether the sample could have come from the population. Such comparisons are known as tests of hypotheses because a hypothesis is formulated regarding the relationship of $\overline{X}$ to μ and a statistical test is made to confirm or disprove the hypothesis. The t table provides numerics for valid comparisons even if the original distributions are not normal if the sample sizes are large. Such tests are discussed in the next chapter.

3.7 CHI-SQUARE DISTRIBUTION

The chi-square (χ^2) distribution is another important distribution used in reliability. Although the distribution itself is continuous, it is used like the binomial in conjunction with discrete values. While the binomial distribution is limited to only two possible outcomes of an event, e.g., good or bad, the

χ^2 can consider many possible outcomes. However, each outcome is a discrete occurrence, and no other possibilities exist. For example, there are six and only six possibilities when a die is rolled (assuming that the die does not rest on edge), and the result of a roll must be one of these six.

Frequently, a characteristic can be and is measured on a continuous scale, and any value within the measurement precision is possible. However, the measured values are then placed in assigned categories, and the data are treated as discrete values with the number of possible values equal to the number of categories. For example, if the heights of adult males were being measured, any result between the largest and smallest measurements is possible, within the capability of the measuring equipment. For purposes of classification, the measurements would be grouped into convenient categories, perhaps to the nearest inch. The relative quantities in each of the categories or classifications can then be analyzed.

The χ^2 distribution is used to compare probabilities of observed frequencies with theoretical frequencies of an assumed distribution. Goodness-of-fit tests, described in Chap. 5, utilize the χ^2 distribution to determine whether a particular theoretical distribution adequately represents the data, i.e., to determine if the observed data could reasonably be expected to come from the distribution. In a goodness-of-fit test, the value of χ^2 is found from Eq. (3.8).

$$\chi^2 = \sum_1^n \frac{(f_{oi} - f_{ti})^2}{f_{ti}} \tag{3.8}$$

where f_{oi} is the observed frequency in category i, f_{ti} is the theoretical frequency of the same category, and n is the number of categories.

The table of χ^2 values, Table B.4, is arranged in a manner similar to the table of values of the t distribution. The side columns list the number of degrees of freedom. Probabilities or percentages are listed as headings of the remaining columns and the χ^2 values are given in the body of the table. Since detailed examples are given in Chap. 5, we shall consider here only the use of the table.

Suppose that, as the result of an analysis, a χ^2 value of 6.57 is obtained and that there are 10 categories. There are 9 degrees of freedom (d.f. $= n - 1$) associated with the data. The table value for 9 degrees of freedom and 70 percent probability is 6.39 and for 50 percent probability it is 8.34. This means that a χ^2 value as high as 6.57 will occur almost 70 percent of the time due to chance, and we conclude that the theoretical distribution represents the data. On the other hand, the probability of a value as high as 20 is only about 3.5 percent and if this value had been obtained from the data we would conclude that the distribution did not represent the data.

Note that, unlike the t distribution, as the number of degrees of freedom of the χ^2 distribution increases, the table values also increase. This is because the χ^2 distribution sums the differences of all categories and the more categories there are the larger the sum will be. The t distribution uses averages and the larger the number of samples the closer will be the sample average to the true average.

3.8 CHI-SQUARE/DEGREES-OF-FREEDOM

Closely related to the chi-square distribution is the chi-square/degrees-of-freedom (χ^2/d.f.) distribution. Values of this distribution are found by dividing the chi-square value by the number of degrees of freedom, hence the name χ^2/d.f. This distribution is utilized as frequently as is the χ^2 distribution and is used to compare the value of a sample standard deviation or variance s^2 with a known or assumed variance σ^2 of a normally distributed population. It is also used to determine confidence regions or limits for σ^2 from the sample variance s^2. Thus, the relationship of the χ^2/d.f. distribution to the variance of a normal distribution is the same as the relationship of the t distribution to the mean.

When the χ^2/d.f. distribution is used to compare variances, the probability that a sample variance of a particular magnitude, s^2, comes from a population whose variance is σ^2 is determined. If the probability is high, the conclusion is reached that there is no statistical evidence that the s^2 is different from σ^2. If the probability is low, then the conclusion is that the sample variance is different from the population variance and that the sample did not come from the population.

The χ^2/d.f. value is found from the ratio of s^2 to σ^2. If a statistical analysis results in a χ^2/d.f. value of 2.00 and the associated number of degrees of freedom is 14 (corresponding to $n = 15$), then the probability that s^2 is equal to σ^2 is small—there is less than a 2 percent probability that a value as high as 2.00 would occur by chance (from Table B.5). If the 2.00 value had resulted from a sample of only 5 (d.f. = 4), then the probability would be about 8 percent that s^2 equals σ^2 and the conclusion of a difference would be much less definite.

Confidence limits for σ^2 can be derived from the χ^2/d.f. table just as confidence limits for μ can be obtained from the t table. The numerics in the χ^2/d.f. table are for ratios of s^2/σ^2. The procedure is to find two multipliers k_1 and k_2 for s^2 such that σ^2 will be between $k_1 s^2$ and $k_2 s^2$ a given percentage of the time. For example, if we wanted a 90 percent probability that σ^2 would be between $k_1 s^2$ and $k_2 s^2$, we would find values of k_1 and k_2 such that 5 percent of the time σ^2 would be less than $k_1 s^2$ and 5 percent of the time it would be greater than $k_2 s^2$. The steps are as follows:

$$c_2 < \frac{s^2}{\sigma^2} < c_1$$

$$\frac{1}{c_2} > \frac{\sigma^2}{s^2} > \frac{1}{c_1}$$

$$\frac{1}{c_1} < \frac{\sigma^2}{s^2} < \frac{1}{c_2}$$

$$\frac{s^2}{c_1} < \sigma^2 < \frac{s^2}{c_2}$$

We now let $k_1 = 1/c_1$ and $k_2 = 1/c_2$ and obtain

$$k_1 s^2 < \sigma^2 < k_2 s^2$$

Table B.5 is used to find values of c_1 and c_2 such that the desired probabilities are obtained. For 90 percent probability and a sample size of 15 (d.f. = 14), $c_1 = 1.692$ and $c_2 = 0.469$. The corresponding values of k_1 and k_2 are 0.591 and 2.13, respectively.

If it were desired to find a value $k_2 s^2$ such that the probability was 90 percent that σ^2 was not greater, then a one-sided confidence limit would be found. For $n = 15$ and 90 percent probability, the value of c_2 is 0.556 and $k_2 = 1.80$. The probability of σ^2 exceeding $1.80 \ s^2$ would be 10 percent. If desired, a k_1 value could be similarly found such that the probability of σ^2 being less than $k_1 s^2$ would be 10 percent.

3.9   THE F DISTRIBUTION

The F distribution is similar to the χ^2/d.f. distribution in that it compares ratios of two variances. Whereas the χ^2/d.f. compares a sample variance s^2 with a known or assumed population variance σ^2, the F distribution compares the two sample variances s_1^2 and s_2^2 with each other. Comparisons of two sample variances are made in conjunction with tests of hypotheses, discussed in Chap. 4, and with analyses of variance of data from complete sets of tests in which several variances are being evaluated. This latter use is associated with the design and analysis of experiments. The F test can help to determine which individual input characteristics have significant effects on the output being studied and which input characteristics interact with each other.

Since the F distribution compares pairs of sample variances, both variances have associated degrees of freedom. Table B.6 is a table of F distribution values with various numbers of degrees of freedom for both the numerator and denominator. Two examples will illustrate its use.

1. If the variances from two samples of ten units each are being compared and the resulting ratio of the two variances, s_1^2/s_2^2, is 3.50, there is less than 5 percent probability that this high a ratio is due to chance. This conclusion is derived from the fact that the 0.95 probability value for 9 and 9 degrees of freedom is only 3.18 which means that, if the two population variances were the same, 95 percent of the ratios of variances of two samples taken from these populations would be less than 3.18. Similarly, there is only a 5 percent probability that the ratio would be as small as 0.314. Consequently, the probability is 0.90 that the ratios of two sample variances which come from populations of equal variance will be between 0.314 and 3.18, and a value outside this range will occur only 10 percent of the time. In this example, there were the same number of degrees of freedom associated with both variances. When the numbers of degrees of freedom are different as in the second example, more care must be taken to be sure that the correct sets of table values are used.

2. Suppose that the variance of a sample of five pieces is 2.80 and that the variance of another sample of 16 pieces is 2.10. The ratio of the variance is 1.33. The number of degrees of freedom for the numerator ν_1 is 4 and for the denominator ν_2 is 15. For 4 and 15 degrees of freedom, the probability of a value of 1.51 or less is 75 percent, and therefore the probability of a value of 1.33 is greater than 25 percent.

   If the variances have been inverted so that ν_1 was 15 and ν_2 was 4 and the ratio was 0.75, the F table value for 15 and 4 degrees of freedom would be used for comparison. The probability of a value as small or smaller than 0.664 occurring by chance is 25 percent, and therefore the probability of a ratio of 0.75 is greater than 25 percent. This confirms the original comparison.

   Note that 0.664 is the reciprocal of 1.51. In order to find a confidence limit α for the ratios of two variances from the reciprocal limit $1 - \alpha$, the number of degrees of freedom must be interchanged and the ratio must be inverted as shown by Eq. (3.9).

$$F_{\alpha, \nu_2, \nu_1} = \frac{1}{F_{1-\alpha, \nu_1, \nu_2}} \tag{3.9}$$

Equation (3.9) can be used to find some table values which are not provided. For instance, if the 0.95 value for $\nu_1 = 10$ and $\nu_2 = 50$ were desired, the reciprocal of the 0.05 value for $\nu_1 = 50$ and $\nu_2 = 10$ can be used:

$$F_{0.95,\, 10,\, 50} = \frac{1}{F_{0.05,\, 50,\, 10}} = \frac{1}{0.493} = 2.03$$

3.10 WEIBULL DISTRIBUTIONS

The Weibull is the most complex of the distributions generally used in reliability analyses. It was developed by W. Weibull of Sweden and used for problems involving fatigue lives of materials. While the negative exponential distribution is described by only one parameter λ (or m), and the normal distribution is described by two parameters μ and σ, three parameters are required to uniquely define a particular Weibull distribution. These three parameters are the scale parameter α, the shape parameter β, and the location parameter γ, and they are related by the Weibull density function, Eq. (3.10). The corresponding reliability function is given by Eq. (3.11).

$$f(t) = \frac{\beta (t - \gamma)^{\beta - 1}}{\alpha}\, e^{-\frac{(t - \gamma)^{\beta}}{\alpha}} \tag{3.10}$$

$$R(t) = e^{-\frac{(t - \gamma)^{\beta}}{\alpha}} \tag{3.11}$$

Figures 3.9 to 3.11 illustrate the effects of different values of these parameters on the form of the distribution, and they can be compared with Figs. 3.3 to 3.5, which depicted different normal and negative exponential

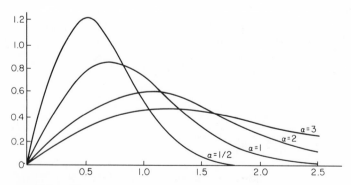

FIG. 3.9   Weibull distributions with different values of α; $\beta = 2$; $\gamma = 0$.

distributions.* In the first of the three figures, β and γ are held constant and four members of a family of Weibull distributions with various values of α are shown. It can be seen that the general shape of the distribution remains the same as does the initial zero point, but the distribution is compressed or stretched out along the abscissa or time axis. The shape of the distribution would be different for a different value of β, but all members of the family would have a comparable shape for the same β. Similarly, for a different value of γ, the initial zero point would change but all members of the family would have the new zero.

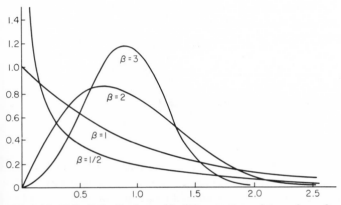

FIG. 3.10 Weibull distributions with different values of β; $\alpha = 1$; $\gamma = 0$.

In the second figure, β is varied while α and γ are held constant. Here, the shape of the distribution changes while the scale and initial points do not. Because the shape changes, it is more difficult to envision the constancy of the scale, as compared with the ease of recognizing the general constant shape maintained in the first figure.

The last figure depicts a family of Weibull distributions with the same shape and scale but with different values of γ. When γ is negative, it indicates that some units are defective at the time of initial operation; i.e., they have failed in storage. When γ is positive, there is some period of time which is failure free. In the case of ball bearings, for example, it may be the length of time it takes for microscopic incipient flaws beneath the bearing surface to propagate to the surface and later cause fatigue failures.

Values of α, β, and γ can be found for any particular set of failure data, and corresponding reliability-confidence numerics can be determined. Although

---

*When the value of β is approximately 3 1/5, the Weibull distribution most closely approximates the normal. The negative exponential is a particular set of Weibull distributions with $\beta = 1$ and $\gamma = 0$. The value of α corresponds to m, the mean time to failure, which is the reciprocal of λ.

mathematical procedures are available, graphical solutions are preferred because of their comparative ease of application. Since the principal use of the Weibull distribution in reliability is in connection with assessments, and because even the graphical method involves considerable calculation, a detailed example is deferred until Chap. 14.

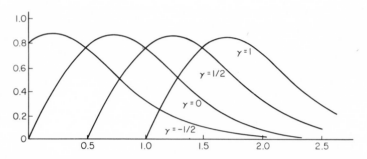

FIG. 3.11   Weibull distributions with different values of γ; $\alpha = 1$; $\beta = 2$.

3.11   TCHEBYCHEFF'S INEQUALITY

When nothing is known about a distribution other than its mean and standard deviation, the Tchebycheff inequality can be used to find a confidence region or limits. The inequality, Eq. (3.12), states that the portion of a distribution that is further than $k\sigma$ from the mean will not exceed $1/k^2$.

$$\text{Tchebycheff's inequality: } P(|X - \mu| > k\sigma) \leq 1/k^2 \qquad (3.12)$$

For example, not more than one-ninth, or about 0.11, of the distribution will be further from the mean than three sigma units. This compares with the normal distribution which has less than three-tenths of 1 percent (< 0.003) of its total further than three sigma units.

One refinement of this inequality can be used when the distribution is symmetrical and has only one mode. This refinement, known as the Camp-Meidell inequality, is given by Eq. (3.13), and reduces the portion from $1/k^2$ to $4/9k^2$.

$$\text{Camp-Meidell inequality: } P(|X - \mu| > k\sigma) \leq 4/9k^2 \qquad (3.13)$$

Using the latter inequality, the portion of a distribution further than three sigma units from the mean will not exceed $4/9 \times 1/9$ or $4/81$, which is less than 5 percent.

It is seen that neither the Tchebycheff nor the Camp-Meidell inequality can provide very high confidence or reliability. If we wanted 99 percent confidence, using the Tchebycheff inequality, the mean would have to be 10 sigma units away from the specification limit. Even with the Camp-Meidell the difference or safety margin would have to be at least 6 2/3 sigma units. If the normal distribution were applicable, a margin of only about 2 1/2 sigma units would be needed. There is seldom the luxury of a large difference (in sigma units) between the mean of a distribution and the specification requirements, and the Tchebycheff inequality is not a very useful tool in reliability mathematics.

ADDITIONAL READINGS

Barlow, R. E., and F. Proschan: "Statistical Theory of Reliability," John Wiley & Sons, Inc., New York, 1965.

Bazovsky, I.: "Reliability Theory and Practice," Prentice-Hall, Inc., Englewood Cliffs, N.J., 1961.

Bureau of Naval Weapons, Chief of: "Handbook: Reliability Engineering," U.S. Government Printing Office, 1964.

Calabro, S. R.: "Reliability Principles and Practices," McGraw-Hill Book Company, New York, 1962.

Chorafas, D. N.: "Statistical Processes and Reliability Engineering," D. Van Nostrand Company, Inc., Princeton, N.J., 1960.

Dixon, W. J., and F. J. Massey, Jr.: "Introduction to Statistical Analysis," 3d ed., McGraw-Hill Book Company, New York, 1969.

Grant, E. L.: "Statistical Quality Control," 3d ed., McGraw-Hill Book Company, New York, 1964.

Hiltz, P. A., and J. L. Gaffney: "Statistical Techniques for Reliability," North American Aviation, Inc., Downey, Calif., 1965.

Hoel, P. G.: "Introduction to Mathematical Statistics," John Wiley & Sons, Inc., New York, 1947.

Ireson, W. G.: "Reliability Handbook," McGraw-Hill Book Company, New York, 1966.

Johnson, L. G.: "The Statistical Treatment of Fatigue Experiments," American Elsevier Publishing Company, Inc., New York, 1964.

Lloyd, D. K., and M. Lipow: "Reliability: Management, Methods, and Mathematics," Prentice-Hall, Inc., Englewood Cliffs, N.J., 1962.

Mood, A. M.: "Introduction to the Theory of Statistics," McGraw-Hill Book Company, New York, 1950.

Munroe, M. E.: "The Theory of Probability," McGraw-Hill Book Company, New York, 1951.

Shewhart, W. A.: "Economic Control of Quality of Manufactured Product," D. Van Nostrand Company, Inc., Princeton, N.J., 1931.

Shooman, M. L.: "Probabilistic Reliability: An Engineering Approach," McGraw-Hill Book Company, New York, 1968.

Villars, D. S.: "Statistical Design and Analysis of Experiments for Development Research," Wm. C. Brown Company Publishers, Dubuque, Iowa, 1951.

Working, H.: "A Guide to Utilization of the Binomial and Poisson Distributions in Industrial Quality Control," Stanford University Press, Stanford, Calif., 1943.

4

TESTS OF HYPOTHESES

4.1 INTRODUCTION

There are many questions which arise concerning reliability data. One may wish to determine whether there is a difference between the averages or variabilities of two lots of parts or whether a new lot meets an established standard or is significantly better or worse. There may be a need to determine whether a particular theoretical distribution can be used to describe the observed distribution of a characteristic or if the value of one variable can be used to estimate the value of a second. The latter questions are considered in the next chapter, which covers some statistical procedures related to reliability and other engineering activities and which should be familiar to engineers and statisticians. This chapter treats testing hypotheses relating to the mean and standard deviation of one or more samples or populations. Confidence intervals associated with these parameters are also discussed since the decision to accept or reject a hypothesis depends on how certain one wishes to be that his decision is correct.

4.2 CONFIDENCE AND RISK

In Chap. 2 we stated that confidence was the degree of certainty that we can associate with a conclusion based on sampling. Correspondingly, the risk is the degree of uncertainty. The particular way in which the problem is formulated and in which the conclusion is stated affects the confidence and the associated risk. The ensuing paragraphs distinguish between two types of conclusions. The first, which is of immediate interest, pertains to the main subject of this chapter: tests of hypotheses concerning sample means and standard deviations. The second type of conclusion relates more directly to reliability numerics and has a different type of risk associated with it. The distinction between the two is important.

The confidence or probability of error can be set at whatever level one wishes. In deciding whether or not to change a manufacturing process, for example, one may wish to be 99 percent certain (confident) that the new process is better than the old before making the change. The probability of error—the risk of making the change when there is no difference between the two processes—would be 1 percent (0.01). In this type of decision, a test program is set up and the hypothesis that the *new process is not better than the present* is formulated. This is known as the *null hypothesis*. The hypothesis is rejected only if the probability is 1 percent or less that the test results which are obtained can result from chance.

The type of error associated with rejecting a hypothesis when it is true, i.e., saying that the new process is better when it really is not, is called an α or type 1 error. It is also known as the producer's risk because, in many sampling plans used by purchasers, the hypothesis is made that the lot is good unless the sampling results prove otherwise. Thus, when the sampling results are such that the hypothesis is rejected when it is really correct (and the lot is rejected when it is really good), the manufacturer suffers a loss because of product returned in error.

There is also the opposite type of error—that of accepting the hypothesis when it is wrong. This is known as a β or type 2 error. It is also known as the consumer's risk because, in sampling plans, it is the risk of accepting a lot as meeting requirements when it does not. In most tests of hypothesis of means and variances, the α risk is established and the β risk is unknown unless the true value of the population parameter being examined is known. But if the value of the parameter were known, there would be no reason to test the hypothesis; hence the β risk, in these tests of hypotheses, is generally unknown. For a given sample size, the smaller the α risk that is established, the larger will be the β risk. The only way to reduce both risks simultaneously or to reduce one without increasing the other is to test a larger sample.

The hypothesis usually used in reliability numerics holds that the reliability is as good or better than some specified level. This is the opposite of the null hypothesis. While the null hypothesis states that two means, standard deviations, processes, etc., are *equal,* the statement used in reliability is that the equipment meets (*is better than*) a specified level of reliability. The associated risk in this case is the β risk—the risk of accepting the conclusion (that the reliability meets a specified level) when it is wrong (when the reliability does not meet this level). Reliability-confidence relationships are discussed at length in Chaps. 13 and 14.

4.3 DEGREES OF FREEDOM

Before we discuss the hypotheses tests themselves, the concept of degrees of freedom, introduced in Sec. 3.6, should be explained more fully. The number of degrees of freedom, designated by the Greek letter ν (nu), is defined as the number of independent values of a variable. To state this in another way, it is equal to the number of values of a variable minus the number of constraints. If there are n values of a variable from which one constant or parameter has been calculated, then only $n - 1$ values of the variable remain independent or unfixed. That is, after $n - 1$ values are found, the last unit must have that value which results in the calculated value of the parameter. When determining the value of the standard deviation s from a sample, if $\overline{X}$ is computed from n different X's, then after $n - 1$ X's have been found independently, the final X is no longer independent but must assume that value which yields the calculated value of $\overline{X}$. Again, the number of independent values is called the number of degrees of freedom.

If one parameter has been calculated from n values of a variable, then there are $n - 1$ degrees of freedom when a second parameter is calculated. If two parameters have been calculated, for example, α and β for a Weibull distribution, then there are $n - 2$ degrees of freedom remaining in calculating the third parameter, for example, γ. Similarly, if three parameters have been calculated, only $n - 3$ degrees of freedom remain when the fourth is calculated, and so on. As we shall see, the number of degrees of freedom is important in the analysis of data because the significance of the results depends on the number of degrees of freedom in the data.

4.4 HYPOTHESES FOR MEANS AND FOR VARIANCES

The two main types of hypotheses being considered here are those pertaining to means and those pertaining to variances or standard deviations. Different statistical distributions are used to evaluate the significance of test results for these two basic types. Normal (gaussian) and Student's t distributions are used for means, while chi-square/degrees-of-freedom (χ^2/d.f.) and F distributions are used for variances. The particular one of each pair that is used depends on the nature of the data, i.e., whether parameter values are known or whether they are estimated. When differences in means are evaluated, if the standard deviations are known, the normal is used; if one or both of the standard deviations are estimated from a sample, the t distribution is used. Similarly, when differences in standard deviations are evaluated, if one is known the χ^2/d.f. is used; if both are estimated from the data the F distribution is used. These statistical distributions are also used for evaluating other types of test results. For example, the F distribution is used in the analyses of variances of data from statistically designed experiments. Discussion of their utilization will be limited here to those hypotheses stated above and to directly related questions.

4.5 HYPOTHESES RELATING TO MEANS

The first group of hypotheses that will be considered relate to the means of samples and populations. Comparisons of sample and universe means are discussed first, and examples of both one-sided and two-sided cases are presented. These are followed by comparisons of two sample means, and by discussion of other alternatives and of confidence intervals.

4.5.1 Comparisons of Means When the Variance Is Known

The first and simplest case is that of comparing the mean of a new distribution with the mean of an existing one when the universe standard deviation is known or assumed. Such a comparison can arise when the average value of a characteristic of all parts from a new machine is being compared with the average value of parts obtained from existing equipment. Comparison with existing standards of the octane rating of samples from a new batch of fuel is another example. The normal distribution is used for these comparisons. It provides exact probabilities when the universe is normally distributed and is approximately correct for large-size samples even when the universe distribution is not normal.

The standard deviation of means of samples, $\sigma_{\bar{X}}$, is utilized in the analysis. It is equal to the standard deviation of individuals divided by the square root of the number of units used to calculate the mean, as given by

$$\sigma_{\bar{X}} = \frac{\sigma}{\sqrt{n}} \tag{4.1}$$

The difference in the average of the new lot or batch in terms of $\sigma_{\bar{X}}$ is computed and the significance of this difference is determined. If the amount of difference could occur only very rarely due to chance, e.g., less than 1 percent of the time, the difference is said to be very significant, and the conclusion is made that there is a difference in the average of the new material. If the amount of difference could occur frequently due to chance alone, then there is no basis for rejecting the hypothesis that the new material has the same average. It may, in fact, be different, but the statistical evidence is not sufficient to warrant this conclusion with a reasonable level of certainty. A value of 5 percent is frequently used as the criterion of significance. One percent is sometimes used if the consequences of an α type error are serious and those of a β type error are not.

Suppose that the mean octane rating of a type of fuel is 98.0. One sample from each of the 25 drums of a new batch is tested and a mean value of 97.7 is obtained. The standard deviation σ is known from past experience to be 0.8. The question is whether the new batch is at least as good as previous lots or whether the 0.3 difference is significantly lower. From Eq. (4.1) the standard deviation of the average, $\sigma_{\overline{X}}$, is equal to the standard deviation of the individuals divided by the square root of 25. $\sigma_{\overline{X}} = \sigma/\sqrt{25} = 0.8/5.0 = 0.16$. Three-tenths (0.3) is thus equal to a difference of 1.875 standard deviations. The question here is not whether the new batch is *different* but whether it is *worse*. Consequently, a statistical test is made to determine if a value of 1.875 standard deviations *lower* is significant. Using Table B.2, the probability of a 1.875 standard deviation difference in one direction due to chance is 3.0 percent.* If a probability of 5 percent or less is considered significant, the mean of the new batch is judged to be lower than the average.

When differences in only one direction are involved, a *one-sided* test is made. If differences in both directions are considered, a *two-sided* test is made. If the new batch of fuel were being evaluated for difference *regardless of direction,* a difference as large as 1.875 sigma units could occur 6.1 percent of the time. If the value of 5 percent were retained as the criterion of significance, the new lot would be judged as not being significantly different. Note that when the direction of difference is not considered, the probability of any particular amount of difference between means is twice as high as the probability of the difference in one particular direction.

When a machine is being set up and it is desired that the average of a dimension be maintained at the same value that existed on previous items, a two-sided test is made. Assume that the dimension has had a mean of 3.278 in. and a standard deviation of 0.002 in. and that it is normally distributed. Ten pieces from the new setup are measured and the following values (in inches) are obtained:

---

*From the table, the proportion of the normal distribution between $- 1.875$ and $+ 1.875$ standard deviations is 0.9392. The proportion beyond $\pm 1.875\ \sigma$ is $1.0000 - 0.9392 = 0.0608$. Since the normal distribution is symmetrical, one-half of 0.0608 is less than $- 1.875\ \sigma$. This is 0.0304 or 3.0 percent.

3.281	3.278
3.276	3.282
3.278	3.279
3.280	3.280
3.279	3.277
	Total 32.790
	Average 3.279

$\sigma_{\overline{X}} = \sigma/\sqrt{n} = 0.002/\sqrt{10} = 0.00063$
Difference $= 3.279 - 3.278 = 0.001$
Difference in sigma units $= \dfrac{0.001}{0.00063} = 1.59$

The probability of a difference of 1.59 $\sigma_{\overline{X}}$, from Table B.2, is 0.11. Since this is greater than 10 percent and certainly greater than 5 percent, the hypothesis that there is no difference in the mean of the new setup is accepted. If the same average value had been obtained from 25 pieces, $\sigma_{\overline{X}}$ would be $0.002/\sqrt{25} = 0.0004$, and the difference in sigma units would be $0.001/0.0004 = 2.5$. The probability of a difference as large as 2.5 $\sigma_{\overline{X}}$ due to chance is only about 1¼ percent, and the hypothesis of no difference in the mean would then be rejected.

This use of the normal distribution for comparison of means, i.e., when the standard deviation is known, is denoted as the z test. The standard notation used for designating the application of this test is given by

$$z = \frac{\overline{X} - \mu}{\sigma/\sqrt{n}} \tag{4.2}$$

The Greek letter μ (mu) denotes the known or assumed mean of the universe and $\overline{X}$ is the average of the sample. The standard deviation of the average, $\sigma_{\overline{X}}$, is $\sigma/\sqrt{n}$ as stated in Eq. (4.1). The value of z, which is the measure of the difference between the sample average and the universe average, is thus obtained in units of the standard deviation of sample averages.* It can be compared directly with a table of the normal distribution which gives probabilities of differences in terms of units of standard deviation.

4.5.2 Comparisons of Means When the Variance Is Estimated

Even though the distribution may be normal or near normal, in most reliability work as well as in the other technical activities, the value of σ is not

*This is called the *normalized* value. This term is often found in statistical literature and frequently refers to any data taken or derived from a normal distribution which are measured in terms of sigma units.

known and must be estimated from the sample data. Recalling that, on the average, s is the best estimate of σ, s can be substituted for σ in Eqs. (4.1) and (4.2). However, the distribution of s is not the same as the distribution of σ, and s is actually less than σ more than half the time. Consequently, z is no longer the most accurate parameter to use for decision making and is replaced by t. The t test is comparable to the z test with t replacing z and s replacing σ:

$$t = \frac{\overline{X} - \mu}{s/\sqrt{n}} \qquad (4.3)$$

Table B.3 gives the probabilities of exceeding values of t for different degrees of freedom. The number of degrees of freedom is one less than the sample size. It can be seen from a comparison of Table B.3 with Table B.2 that, as the sample size increases, values of t become closer and closer to values of z for corresponding probabilities.

As a typical example of the use of the t test, the mean burst pressure of one type of liquid storage tank was calculated to be 549 psi. The burst-pressure distribution is approximately normal but the standard deviation is unknown and must be estimated from the test data. Five units of a new lot are tested to failure and the rupture pressures are recorded to the nearest 5 psi. It is desired to determine whether the new lot has a significantly *lower* rupture pressure than previous lots. Data are given in Table 4.1.

TABLE 4.1 Burst-pressure data from new lot of storage tanks

X_i	$X_i - \overline{X}$	$(X_i - \overline{X})^2$
545 psi	2	4
530	-13	169
550	7	49
545	2	4
545	2	4
Total 2,715 psi		$\Sigma = 230$

$\overline{X} = $ 543 psi

$$s = \sqrt{\frac{230}{4}} = 7.58 \text{ psi}$$

$$t = \frac{543 - 549}{7.58/\sqrt{5}} = \frac{-6}{3.39} = -1.77*$$

*Greater accuracy is not warranted because only two-place accuracy was obtained from the original data.

Using Table B.3 for a t distribution with 4 degrees of freedom, an absolute value as large as 1.77 in one direction will be obtained about 8 percent of the time due to chance, and the conclusion could be reached that the average difference of 6 psi is not significant. If the same average and standard deviation had been obtained from a sample of 15 or more pieces (14 or more degrees of freedom), the probability of this large a difference in one direction would be less than 5 percent and the results would be considered significant.

Just as there are one-sided and two-sided z tests, there are also one-sided and two-sided tests using the t distribution. The last example illustrated the one-sided case. If the test had been performed to determine whether there was *any* difference in the average burst pressure of the new lot rather than to determine if it was lower, a two-sided test would have been made. With 4 degrees of freedom, an absolute value as large as 1.77 sigma units in any direction would occur due to chance about 16 percent of the time and would not be considered statistically significant.

4.5.3 Comparison of Two Sample Means

The last group of hypotheses involving means which will be considered here involves the comparison of means of two samples. When two samples are being compared it is necessary that both averages, both variabilities, and both sample sizes be taken into account. This is accomplished by modifying Eqs. (4.2) and (4.3) as follows:

$$z = \frac{\overline{X}_1 - \overline{X}_2}{\sigma\sqrt{1/n_1 + 1/n_2}} \qquad (4.4)$$

Equation (4.4) is used to compare the means of two samples when both sigmas are known and are equal (as is usually assumed).* In almost all cases involving two samples, however, the values of sigma are not known and Eq. (4.5) should be used (again assuming that both standard deviations are equal).

$$t = \frac{\overline{X}_1 - \overline{X}_2}{s_p\sqrt{1/n_1 + 1/n_2}} \qquad (4.5)$$

*When both sigmas are known but are unequal, the formula

$$z = \frac{(\overline{X}_1 - \overline{X}_2) - (\mu_1 - \mu_2)}{\sqrt{\sigma_1^{\,2}/n_1 + \sigma_2^{\,2}/n_2}}$$

is used. When the population means are equal, $\mu_1 - \mu_2 = 0$.

In Eq. (4.5) s_p is a pooled estimate of the standard deviation and is found from Eq. (4.6).*

$$s_p = \sqrt{\frac{\Sigma(X_{1i} - \overline{X}_1)^2 + \Sigma(X_{2i} - \overline{X}_2)^2}{n_1 + n_2 - 2}}$$

$$= \sqrt{\frac{(n_1 - 1) s_1^2 + (n_2 - 1) s_2^2}{n_1 + n_2 - 2}} \tag{4.6}$$

The subscripts 1 and 2 refer to the sample groups from which the readings were taken; there are $n_1 + n_2 - 2$ degrees of freedom.

Both one-sided and two-sided tests are used for these comparisons in the same manner as used in previous examples. When the comparison is made to determine *only* if $\overline{X}_1$ is smaller (or larger) than $\overline{X}_2$, a one-sided test is used. If the comparison is to determine a possible difference between the two averages, *regardless of direction,* a two-sided test is used. As stated earlier, for any given value of z or t in terms of sigma units, the probability of exceeding a given absolute value is twice as great when direction is not considered (two-sided test) as when the probability in only one direction is of concern (one-sided test).

An example of a one-sided test for differences in means would be the comparison of a new, more costly boiler design with an existing one when both mean values are computed from samples. If the average amount of fuel consumed by boilers of the new design were to be compared with the average amount consumed by boilers of the existing design for the same amount of heat, the two averages would be compared using Eq. (4.5) (since the value of s_p would be computed from the data). The hypothesis of equality would be

---

*With unknown and unequal standard deviations, the formula

$$t = \frac{(\overline{X}_1 - \overline{X}_2) - (\mu_1 - \mu_2)}{\sqrt{s_1^2/n_1 + s_2^2/n_2}}$$

is used. There are f degrees of freedom, where

$$f = \frac{(s_1^2/n_1 + s_2^2/n_2)^2}{[(s_1^2/n_1)^2/(n_1 + 1)] + [(s_2^2/n_2)^2/(n_2 + 1)]} - 2$$

Since the value of f will not usually be an integer, the closest value in Table B.3 may be used. For precise significance levels, interpolation is necessary. Again, with equal population means, $\mu_1 - \mu_2 = 0$.

rejected only if the average fuel consumption of the new design were significantly *lower.*

A two-sided test might be used to determine which of two alternative processes yielded the smaller percent defective (or higher reliability). The test for significance would be valid if the percent defective were a function of some normally distributed variable, e.g., tensile strength, minimum power, wear-out life, etc. The test would compare means of the variable obtained from the two processes; if a significant difference existed, that process which produced the higher-quality product would be chosen, all other factors being equal.

4.5.4 Other Alternatives

Up to this point, two alternatives have been discussed. The hypothesis of equality was either accepted when there was not sufficient statistical evidence to disprove it or it was rejected when the probability was remote that the observed difference between the means occurred due to chance. When the hypothesis was accepted, the means were assumed to be equal. When it was rejected the means were judged to be different, but the question as to how different remains to be answered. If there is no evidence to the contrary and there are no technical reasons for assuming any particular values, the best estimates of the means are the observed averages and the best estimate of the difference is the observed difference.

This approach seems to leave no middle ground. Either the means are the same or they vary by the observed difference. We realize, however, that while $\overline{X}$ is the best estimate of μ, it is *only* an estimate. The true average may vary from the estimate even though the amount of this variation may not be as large as the difference between the observed averages or between one observed average and a previously established value of μ.* Figure 4.1 illustrates this relationship. In this figure the difference between $\overline{X}$ and μ is statistically significant so that $\overline{X}$ is assumed to come from a different population. The true mean of this other population may differ from the estimate $\overline{X}$ and may be somewhere within the range indicated by the dotted lines. At the same time, the averages of samples of size n taken from the original population may vary from μ by an amount indicated by the dashed lines. Note that the distributions of the two averages need not necessarily overlap. The relationship will depend on the true difference in means, on the variabilities of the two populations, and on the sample sizes.

In the example, there is a small probability of obtaining a sample average from the new population that is so low that it falls in the shaded area. Since there is also a small probability of the average of a sample taken from the

---

*When the hypothesis of equality is rejected, the true difference may actually be larger than the observed difference.

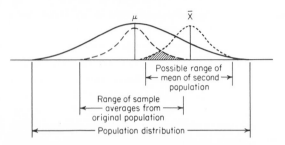

FIG. 4.1   Relationships among population and sample means when populations are different.

original population falling in this area, there is a resultant probability of accepting the hypothesis that the means are equal when they are not. This is the β or type 2 error. As indicated, the probability of this error depends on the amount of overlap which, in turn, is dependent on the relationship of the two distributions. It also depends on the α risk that is taken and on the sample size. Note also that only a portion of the shaded area represents the β error since only some of the averages in this area would be judged as not being significantly different.

It follows from this discussion that tests of hypothesis other than equality can be made. From technical or other considerations, a particular difference in means might be expected. For example, the tensile strength of a particular steel would be expected to be lower at higher temperatures, while the leakage current of a transistor would increase. If a new chemical composition of the steel were being tried with the objective of obtaining a smaller decrease in strength at high temperatures, a hypothesis could be formulated that the difference in tensile strength at ambient and at a specified elevated temperature is equal to the previously experienced decrease. If the hypothesis were proved wrong and the decrease in tensile strength was significantly less than expected, the new composition would be adopted.

We shall let $\overline{X}_a$ and $\overline{X}_e$ represent the average tensile strengths of the new compositions at ambient and at elevated temperatures, respectively, and d the previously experienced decrease (d is equal to $\mu_1 - \mu_2$ in the equations in the footnotes in Sec. 4.5.3). A z test using the previous values of the standard deviation and difference in tensile strength is then appropriate for evaluating the hypothesis. The test is defined by

$$z = \frac{\overline{X}_a - (\overline{X}_e + d)}{\sigma \sqrt{1/n_1 + 1/n_2}} \tag{4.7}$$

Assume that $\overline{X}_a = 75$, $\overline{X}_e = 61$, $d = 16$, $\sigma = 2$, and $n_1 = n_2 = 8$. All values are in 1,000 psi. Substituting in Eq. (4.7), we have

$$z = \frac{75 - (61 + 16)}{2\sqrt{1/8 + 1/8}} = -2.0$$

Because we are interested only in whether the difference is less than previously experienced, that is, if $\overline{X}_a < (\overline{X}_e + d)$, a one-sided test is used. Using a table of the normal distribution, a value as low as -2.0 sigma units is obtained only about 2¼ percent of the time due to chance, and hence is considered significant. Consequently, the new composition would be adopted.

This z test could have been expressed as a test of differences rather than as a test of means, as in Eq. (4.8). This is in the same form as the equations in the previous footnotes, where $d = \mu_1 - \mu_2$.

$$z = \frac{(\overline{X}_a - \overline{X}_e) - d}{\sigma\sqrt{1/n_1 + 1/n_2}} \tag{4.8}$$

This second form of the test is mathematically equivalent to Eq. (4.7) and yields identical results.

4.5.5 Confidence Intervals for Means

We have seen that the average of a sample will vary within some range, the variation of the sample averages from Eq. (4.1) being $1/\sqrt{n}$ of the population variation. While we may not wish to compare the sample average with some other average, it may be desirable to estimate within what range we can expect the true mean of the population from which the sample was taken to fall. This is basically the same question that has been discussed before but it is now considered from another viewpoint. With respect to reliability we might wish to know whether the mean value of a population meets some minimum (or maximum) requirement, e.g., whether the average yield of some chemical process is satisfactory. A number of sample units can be taken, their average and standard deviation computed, and the results compared with the requirement. If the standard deviation is known or assumed from previous experience, the z test is applied; if s is calculated from the sample data, the t test should be used.

Suppose that 10 sample tests are made from a new chemical process and that the yields are as given in Table 4.2, which also shows the calculation of s. Units of measure are not given but may be pounds, kilowatts, calories, or whatever units are suitable to the process.

TABLE 4.2 Chemical process yield

X_i	$(X_i - \overline{X})$	$(X_i - \overline{X})^2$
196	1	1
193	- 2	4
197	2	4
195	0	0
195	0	0
192	- 3	9
194	- 1	1
198	3	9
196	1	1
194	- 1	1
Total 1,950		$\Sigma = 30$
Average 195.0		$s = \sqrt{30/9} = 1.826$

The standard deviation of averages, $\sigma_{\overline{X}}$, is equal to $s/\sqrt{n} = 1.826/\sqrt{10} = 0.577$.

Assuming that the distribution of yields is normal or fairly close to normal, using the t distribution we can calculate confidence intervals for μ from the data. The true value of μ will be between $\overline{X} - K\sigma_{\overline{X}}$ and $\overline{X} + K\sigma_{\overline{X}}$ a certain proportion of the time.* In this example, if we wished to form a 95 percent *confidence interval*, we would choose a K for 9 degrees of freedom such that 2½ percent of the time the true mean would be less than $\overline{X} - K\sigma_{\overline{X}}$ and 2½ percent of the time it would be greater than $\overline{X} + K\sigma_{\overline{X}}$. This value of K, from Table B.3, is 2.26. Thus, the probability is 0.95 that the true mean will be between $195.0 \pm 2.26 \times 0.577$, (193.7 and 196.3).

A one-sided test is made if only a minimum (or maximum) mean value is desired. If we wish to determine the minimum mean value such that we were 95 percent confident that the true mean was at least that high, we would choose a value of K such that the probability was 5 percent that μ is less than $\overline{X} - K\sigma_{\overline{X}}$. For 9 degrees of freedom, this K is 1.83. The probability is 0.95 that the true mean is equal to or greater than $195.0 - 1.83 \times 0.557$ (193.9).

4.6 HYPOTHESES RELATING TO VARIABILITY

Tests for the significance of differences in standard deviations can be made in the same manner as tests of differences in means. Such tests for variations

---

*The use of $K\sigma_{\overline{X}}$ rather than $z\sigma_{\overline{X}}$ is preferred in reliability because the standard deviation of averages is most often computed from s rather than from σ.

utilize the variance (the square of the standard deviation), and the comparison tables are prepared accordingly. While the nature of the distributions of s^2 and σ^2 is inherently different from that of z and t (the standard deviation can never be less than zero while z and t distributions are symmetrical about μ and $\overline{X}$), the distributions are equally well known for normally distributed variables, and differences between observed and theoretical values can be evaluated using the appropriate tables.

4.6.1 Comparison of a Single Variance with a Known Standard

When a hypothesis is made regarding the comparison of a single variance or standard deviation with a known or assumed variance, the chi-square/degrees-of-freedom (χ^2/d.f.) test is used. When both values are determined from observed test data, the F test is used to compare the two variances or standard deviations. Since in both the χ^2/d.f. and F tests at least one of the values of the standard deviation is estimated, application of degrees of freedom is required. In the χ^2/d.f. test only one value of d.f. is needed, and the χ^2/d.f. table (Table B.5), which provides comparison numerics for many levels of probability and for degrees of freedom ranging from 1 to ∞, is used.* The F table is somewhat more complex because both variance values are observed, and degrees of freedom apply to both the numerator and denominator of the ratio being tested for significance. Consequently, the F table (Table B.6) is considerably larger, and values are provided for sets of three variables: levels of probability, degrees of freedom in the numerator, and degrees of freedom in the denominator.

Several examples will illustrate the use of these tests, and will cover both one-sided and two-sided cases. The relationship between one-sided and two-sided tests for differences in variance is the same as the relationship for differences in means. When a test is made to determine whether one variance is larger (or smaller) than another, a one-sided test is used. When the test is for difference regardless of direction, a two-sided test is applicable.

A one-sided test was used to compare the variance of an automatic lathe (used for turning the diameters of automobile engine pistons) with previous standards. Effects of modifications to the machine were being evaluated in an effort to establish a more uniform process. A modification was made and 25 pistons were machined. The variance in diameters, s^2, for the sample run of 25 pistons was computed to be 0.00066. The population variance from previous production runs prior to the modification had been 0.00040. If the

---

*The relationship between the χ^2/d.f. and χ^2 tables is quite simple. Values in the χ^2/d.f. table can be obtained from the χ^2 table merely by dividing by the number of degrees of freedom. For this reason some texts omit the χ^2/d.f. table and leave it up to the user to perform the appropriate division.

modification had an adverse effect, an opposite modification would be made in the hopes of reducing the variance, but if the change was not significant, a different modification would be tried. A χ^2/d.f. test, Eq. (4.9), was made to determine the significance.

$$\chi^2/\text{d.f.} = \frac{s^2}{\sigma^2} \tag{4.9}$$

The χ^2/d.f. value was 0.00066/0.00040 = 1.65. The χ^2/d.f. table shows that for 24 degrees of freedom (d.f. = $n - 1$) a value as high as 1.65 will occur less than 2½ percent of the time due to chance. (If the variances are the same, 97½ percent of the ratios will be less than 1.64.) Therefore, the variance of 0.00066 was judged to be significantly higher than the previous variance and an opposite modification was indicated.

As a second example, suppose that a new type of pressure transducer is being evaluated and compared with transducers of current design. If both types can be calibrated to the same degree of accuracy, the comparison between the two types will be with respect to their variation when checked against a standard pressure. The present type of transducer is known to have a standard deviation of 1.50 psi—equivalent to a variance of 2.25 psi^2 (variance = σ^2). Ten independent measurements are taken with the new type of transducer and its variance is estimated by calculating s^2 from the measurements. It is important that the measurements be independent so that the variation or error in one measurement does not influence the next. The value of s^2 is computed from $s^2 = \Sigma(X_i - \overline{X})^2/(n - 1)$ where each X_i is the *difference* between the observed and true values (the error) and $\overline{X}$ is the *average of the differences* (the average error), not the average value. Table 4.3 provides the data.

We wish to compare the estimate of the variance of the new type of transducer (1.38) with the known variance of the present type (2.25) to see whether the new type is significantly better. A χ^2/d.f. test is employed, using Eq. (4.9): χ^2/d.f. = 1.38/2.25 = 0.613. Using Table B.5 for 9 degrees of freedom, a value as low as 0.613 could occur more than 20 percent of the time due to chance. Therefore, the evidence is not sufficient to conclude that the population variance of the new type is less than that of the present transducers. (However, since the sample variance of the new type was less, we might wish to test additional units before making a final decision.)

There are relatively few instances in industry where some variability is desired. Usually, the objective of a study, process modification, or other change is to reduce the amount of variation in order to achieve a more uniform product. One activity where variation is desirable is the personnel selection process whereby industrial workers are assigned to jobs for which they are best suited. Aptitude and dexterity tests which provide a range of

TABLE 4.3 Calculation of average measurement error $\overline{X}$ and its variance. True value = 50 psi

Observed measurement	X_i: error (absolute difference from true value	$(X_i - \overline{X})$	$(X_i - \overline{X})^2$
52	2	0.4	0.16
54	4	2.4	5.76
49	1	- 0.6	0.36
50	0	- 1.6	2.56
47	3	1.4	1.96
51	1	- 0.6	0.36
49	1	- 0.6	0.36
48	2	0.4	0.16
49	1	- 0.6	0.36
51	1	- 0.6	0.36
	$\Sigma = 16$		$\Sigma = 12.40$
Average error $\overline{X} = 1.6$ psi			$s^2 = 12.40/9 = 1.38$

results are desired provided that there is good correlation between test results and job performance.

Suppose that the standard deviation of test results of a dexterity test is 5.0 points ($\sigma^2 = 25$), and that this permits adequate selection and provides good correlation. The cost of giving the test and evaluating the results is high, and a less expensive test would be adopted if the test results had the same variance and adequate correlation. Fifty applicants are given the new test and the test scores have a standard deviation of 3.9. The variance, therefore, is 15.2. The χ^2/d.f. test ratio s^2/σ^2 is 15.2/25 = 0.61. A two-sided test will be used.

The procedure for a two-sided test is to choose a level of significance $1 - \alpha$ and establish a region such that the ratio s^2/σ^2 will fall in the region $1 - \alpha$ percent of the time. The region is made symmetrical in the sense that $\chi^2_{\frac{1}{2}}$ /d.f. $< s^2/\sigma^2 < \chi^2_{1-\frac{1}{2}\alpha}$/d.f., so that the percentage below the lower limit and the percentage above the upper limit are each equal to $\frac{1}{2}\alpha$. If the test ratio falls outside the region, i.e., *either* below $\chi^2_{\frac{1}{2}\alpha}$/d.f. or above $\chi^2_{1-\frac{1}{2}\alpha}/$ d.f., the results are considered significant. Using the χ^2/d.f. table for 49 degrees of freedom, by interpolation s^2/σ^2 will be between 0.64 and 1.43 about 95 percent of the time due to chance (and below 0.64 or above 1.43 only 5 percent of the time). Therefore, a value as low as 0.61 is considered significant, and the standard deviation of the new test is judged to be different from that of the standard test.*

*It should be pointed out that, while a sample size of 50 is relatively large when evaluating a homogeneous process, lot, or production run, it is really quite small when used for evaluating a group of people with unknown characteristics and possibly wide variations among them.

4.6.2   Comparisons of Variances of Two Samples

Many comparisons of variances are concerned with variances of two samples and require the use of an F distribution to determine the significance of the differences. These comparisons occur in the same types of situations that have already been discussed such as process variabilities, octane or power ratings, machine capabilities, strength characteristics, and so on. The only difference is that both variances are estimated from sample data. One-sided tests will be considered first.

If the variability of the crush strength of a new type of shock mount is being compared with the variability of the present type, both variances are first estimated by crushing a number of samples and then an F test is performed. Suppose that the new type has some advantageous characteristics and will be used unless there is significant evidence that its crush strength variance is greater than the variance of the present type. The variance of the new type, s_1^2, is estimated from tests on 10 samples to be 12 lb^2, while the present type has an estimated variance s_2^2 from 16 samples of only 9 lb^2. The F test, defined by Eq. (4.10), is used.

$$F_{\nu_1, \nu_2} = \frac{s_1^2}{s_2^2} \tag{4.10}$$

where ν_1 and ν_2 are the degrees of freedom of the numerator and denominator, respectively. In this example, there are 9 and 15 degrees of freedom and the F ratio = $12/9 = 1.33$. For 9 and 15 degrees of freedom, the probability of a ratio as high as 1.33 due to chance, from Table B.6, is greater than 25 percent (the 0.75 value in the table is 1.46). Hence, the conclusion is reached that there is no significant evidence that the new type of shock mount has greater variability than the present type.

An equivalent comparison could be made by determining if the second variance (present type) is significantly less than the first (new type). The crush strength variance of the present type of shock mount would be evaluated to see if it is significantly *smaller* than that of the new type. Using the same set of values ($s_1^2 = 12$, $s_2^2 = 9$) the F test would compare s_2^2 with s_1^2 as follows: $F = 9/12 = 0.75$, but the numbers of degrees of freedom are now 15 and 9. Because the comparison is to determine whether s_2^2 is significantly *smaller*, the value of 0.75 is compared with small values of probability in the F table.* The F table value for 15 and 9 degrees of freedom, $F_{15,9}$, for 25 percent probability is 0.69. Since 0.75 exceeds 0.69, there is more than a 25 percent probability of a value as small as 0.75 occurring due to chance, and the same conclusion of no significant difference is reached.

*The relationship between the two F table values is given by Eq. (3.9) in Sec. 3.9.

The F table in Appendix B gives probabilities and corresponding numerical values such that there is the stated probability that the ratio of two variances will be *less* than the numerical table value due to chance. When an F test is made to see if the s^2 in the numerator is significantly greater than the s^2 in the denominator, the ratio must exceed the F table value associated with a high probability, there being only a small likelihood of exceeding the table value due to chance. When the test is made to see if the numerator s^2 is significantly less than the denominator s^2, the ratio must be less than the F table value associated with a low probability.

When a test is made to determine whether the variance (or standard deviation) of one population or group is significantly different from the variance of another regardless of which may be larger or smaller, a two-sided F test is used. The variance of each is computed, and the ratio of the two variances is calculated and compared with the appropriate F table values. The standard procedure is to decide on the level of significance $1 - \alpha$, and to establish a region such that the probability is $\frac{1}{2}\alpha$ that the F ratio will be below the lower limit and $\frac{1}{2}\alpha$ that it will be above the upper limit due to chance. This follows the same procedure that was used for the $\chi^2/$d.f. test when only one variance was estimated from sample data except that, because both variances are estimated, an F test is used rather than a $\chi^2/$d.f. test. If we wished to use a 90 percent significance level, then $\alpha = 0.10$, and F table values would be selected such that the ratio of variances would be less than the lower limit only 5 percent of the time and greater than the upper limit 5 percent of the time. If the sample results were such that the F ratio did fall below the lower or above the upper limit, we would say that there is a significant difference in the two variances.

Suppose that a canning company is comparing the variation in net filled weights of cans from two different filling machines. Ten cans are taken from each machine and the net contents are weighed and recorded as shown in Table 4.4. All measurements are in ounces.

Lower and upper 5 percent points for 9 and 9 degrees of freedom, from Table B.6, are 0.31 and 3.18, respectively. The value of 0.47 is within these limits and we therefore conclude that, at the 90 percent level of significance, there is no evidence of a difference in variance between the two filling machines.

If each machine has many individually controlled filling heads, as is common in this industry, then a sample size of 10 is not large enough to warrant drawing conclusions regardless of the significance of the F test. There could easily be differences from head to head within a machine, and the sampling procedure either would have to include a large enough number of cans to ensure that cans from all or at least most of the filling heads were included, or preferably would have to compare variances among heads as well as variances between machines. A similar situation occurred in an automobile

TABLE 4.4 Net filled weights of cans from two filling machines

Machine 1			Machine 2		
Net weight	$(X_i - \overline{X})$	$(X - \overline{X})^2$	Net weight	$(X_i - \overline{X})$	$(X_i - \overline{X})^2$
15.54	- 0.02	0.0004	15.57	0.01	0.0001
15.55	- 0.01	0.0001	15.54	- 0.02	0.0004
15.57	0.01	0.0001	15.58	0.02	0.0004
15.55	- 0.01	0.0001	15.56	0.00	0.0000
15.56	0.00	0.0000	15.53	- 0.03	0.0009
15.57	0.01	0.0001	15.59	0.03	0.0009
15.58	0.02	0.0004	15.56	0.00	0.0000
15.55	- 0.01	0.0001	15.55	- 0.01	0.0001
15.56	0.00	0.0000	15.57	0.01	0.0001
15.57	0.01	0.0001	15.55	- 0.01	0.0001
$\Sigma = \overline{155.60}$		$\Sigma = \overline{0.0014}$	$\Sigma = \overline{155.60}$		$\Sigma = \overline{0.0030}$

$\overline{X}_1 = 15.560$ $\overline{X}_2 = 15.560$

$$s_1^2 = \frac{0.0014}{9} = 0.000156 \qquad s_2^2 = \frac{0.0030}{9} = 0.000333$$

$$F = \frac{0.000156}{0.000333} = 0.47$$

engine manufacturing plant which vividly illustrates this point. A four-head machine was used to bore the inside diameter of wrist pin bushings after they were installed in pistons. Initial measurements indicated that the variance of the machine was too great and that the abnormally high percentage of rejects was caused by this variance. The company was about to purchase a new machine whose variance was guaranteed to be considerably less. The reliability engineer first made a more detailed study of the machine and found that each of the four heads of the machine had very little variance, but the settings were not all the same and this gave the appearance of a large variance. The boring machine was reworked at relatively little expense, and when care was taken to adjust all four heads to the same setting, the number of rejects was reduced to well below the acceptable level. In this instance, a net savings in excess of $50,000 was achieved by the simple application of proper analytical procedures, and this savings did not include the savings in scrap and rework costs.

4.6.3 Confidence Intervals for Variances

Just as confidence intervals can be established for means, they can also be established for variances. In fact, the last example on the net filled weight of cans, in effect, required the determination of a confidence interval and then a comparison of the F ratio with the confidence limits. The confidence interval numeric defines the probability of the true variance falling within the corresponding upper and lower confidence limits. By using the appropriate χ^2/d.f. or F table, confidence intervals can be set up for any desired levels of probability. For example, if a confidence interval of only 80 percent is desired, then the 10 percent and 90 percent table probability values are used. If a very large confidence interval, for example, 99 percent, is wanted, then the 0.5 percent and 99.5 percent probability values are utilized.

4.7 CONCLUDING REMARKS

The tests of hypotheses which have been discussed utilize tables which provide probabilities for samples with normal or near-normal distributions. When the sample size is greater than 30, the distribution of sample means is approximately normal even for very non-normal distributions. Also, many distributions which are not normal can be normalized by the simple transformation of a parameter, e.g., the logarithm may be normally distributed. Some transformations are discussed in Chap. 5.

Furthermore, as indicated in the opening paragraph, tests for the goodness of fit of distributions to sample data and for correlation between variables can also be formulated as tests of hypotheses. Such tests are independent of the nature of the distribution and are called non-parametric tests because the comparison is between distributions rather than between parameters of distributions. Consequently, the need for assuming normality is obviated and these tests can be applied to any set of data.

Finally, many reliability predictions and assessments involve the estimation of both means and standard deviations (or variances) and the comparison of resulting distributions with specification requirements. The hypotheses in this chapter dealt with either one or the other of these parameters and presented appropriate statistical analysis methods, but they did not consider both of them simultaneously. Chapter 14 on reliability assessment methods describes in detail the evaluation of reliability when both means and variances are estimated, and it defines the use of Table C.4, which consists of a set of curves providing reliability (probability)—confidence relationships for a large range of sample sizes and estimated parameter values.

ADDITIONAL READINGS

ARINC Research Corporation: "Reliability Engineering," Prentice-Hall, Inc., Englewood Cliffs, N.J., 1964.

Chorafas, D. N.: "Statistical Processes and Reliability Engineering," D. Van Nostrand Company, Inc., Princeton, N.J., 1960.

Dixon, W. J., and F. J. Massey, Jr.: "Introduction to Statistical Analysis," 3d ed., McGraw-Hill Book Company, New York, 1969.

Fisher, Sir Ronald A.: "The Design of Experiments," Oliver & Boyd Ltd., London, 1935.

—— and Frank Yates: "Statistical Tables for Biological, Agricultural and Medical Research," 4th ed., Hafner Publishing Company, Inc., New York, 1953.

Freeman, H. A.: "Industrial Statistics," John Wiley & Sons, Inc., New York, 1942.

Hiltz, P. A., and J. L. Gaffney: "Statistical Techniques for Reliability," North American Aviation, Inc., Downey, Calif., 1965.

Hoel, P. G.: "Introduction to Mathematical Statistics," John Wiley & Sons, Inc., New York, 1947.

Ireson, W. G.: "Reliability Handbook," McGraw-Hill Book Company, New York, 1966.

Mood, A. M.: "Introduction to the Theory of Statistics," McGraw-Hill Book Company, New York, 1950.

Simon, Gen. L. E.: "Engineer's Manual of Statistical Methods," John Wiley & Sons, Inc., New York, 1941.

Snedecor, G. W.: "Statistical Methods," 5th ed., Iowa State College Press, Ames, 1957.

Villars, D. S.: "Statistical Design and Analysis of Experiments for Development Research," Wm. C. Brown Company, Publishers, Dubuque, Iowa, 1951.

5

RELATED STATISTICAL
PROCEDURES

5.1 GOODNESS OF FIT

When reliability predictions and assessments are performed and when other statistical procedures are followed, an assumption is made about the nature of the distribution that fits the data and sometimes about the values of the parameters which describe it. If a normal distribution is assumed, its mean and standard deviation are calculated from the data. Confidence levels and associated reliabilities are derived from the calculations based on this assumption of normality. Similarly, when the assumed distribution is the negative exponential, i.e., a constant failure rate, the value of λ and the resulting confidence—reliability relationships are determined for the assumed distribution. The same is true for any set of data and any distribution which is chosen to represent the data. In a test of a hypothesis, values of the parameters are assumed and a test is made to determine the validity of the hypothesis and thereby assess the choice of the assumed parameter values. We have seen examples of tests for differences in means and standard deviations, and the significance of these differences has been examined.

These procedures are generally quite accurate. There are usually sound engineering reasons for the choice of the distribution or the parameter values. Historical data substantiate the nature of the failures, their frequency, and their pattern of distribution. Prior knowledge often exists of the effects of a change of input characteristics on output values. It is seldom if ever necessary to resort to that most pessimistic method, the Tchebycheff inequality, to evaluate a set of data. However, even when there is a high degree of certainty concerning the choice of a distribution and the values of its parameters, we may still wish to verify the choices.

In later chapters we shall see that occasionally assumptions are made regarding a distribution when there may be little prior statistical information. In these cases it is even more desirable that there be verification that the distribution fits the data. Comparisons of test data with an assumed distribution are accomplished by a *goodness-of-fit* test. Such a test, as its name implies, evaluates how well the data fit the assumed distribution or, to be more exact, determines if there is evidence of disagreement. One may choose a particular degree of assurance that the data did not come from the assumed distribution (i.e., that the distribution does not fit the data) which he feels is necessary to reject the assumption. He can then compare the results of the goodness-of-fit test with a mathematically determined requirement. If desired, the degree to which the data do not match the distribution can be determined first and then the decision can be made as to whether there is sufficient disagreement to reject the assumed distribution. Note that, in either case, the goodness-of-fit test is for *disagreement* rather than for agreement of the distribution with the data. The data are assumed to fit the distribution unless there is sufficient evidence to disprove the assumption.

One potential disadvantage of goodness-of-fit tests is that, even though an assumed distribution fits the data, there may be other distributions which fit the data equally well or perhaps even better. Another distribution may be of a different type, e.g., normal rather than exponential, or it may be the same type of distribution but with different parameter values—a different mean or standard deviation. For example, in one case, a negative exponential distribution of failures was assumed for the 10 major components in a system. However, as testing time was accumulated and more failures occurred, it became evident that the failures of one particular component were resulting from wear-out rather than from chance causes and were normally rather than exponentially distributed. While there was no significant evidence from the first few failures that the failure rate was not constant, i.e., that the failure distribution did not follow the decreasing exponential, nevertheless a normal distribution would have fit the data better. It was not until additional failures had been experienced that the evidence strongly suggested the change in distributions. Consequently, the reader is cautioned, when employing a goodness-of-fit test, to remember that the test only rejects the assumed distribution when there is definite evidence that the distribution is incorrect; it does not ensure that the selected distribution is the best one when there is not sufficient evidence to reject it.

There are two tests which are the most frequently used for goodness-of-fit analyses. The first test, the *chi-square test* (χ^2), compares the observed frequencies of different events with their theoretically expected frequencies and checks actual differences against the difference that would occur due to chance alone. It is restricted to relatively large samples, i.e., greater than 25. The second test, the *Kolmogorov-Smirnov test,* compares cumulative frequencies so that each individual unit value can be evaluated. While this test can be used for data from a large number of samples, it is more useful for evaluating the appropriateness of distributions for data from small samples.

5.1.1   The Chi-square Test

In the chi-square test, differences between *observed* and *theoretically expected frequencies* are squared and divided by the expected frequency. The quotients are then added as shown by Eq. (5.1). The numerator terms are squared because, if they were not, positive and negative values would tend to cancel each other.

$$\chi^2 = \sum_{i=1}^{k} \left[\frac{(f_{oi} - f_{ei})^2}{f_{ei}} \right] \tag{5.1}$$

where f_{oi} is the observed frequency in each group or class, f_{ei} is the expected frequency, and the summation covers k groups or classes of data. Two examples will serve to illustrate the test.

If an unbiased die is rolled, the probability of a particular face turning up on any one roll is $1/6$. If 150 rolls are made, the expected frequency of each number is $1/6 \times 150$, or 25. From experience we would not expect each of the numbers to appear on the top face exactly 25 times, but we would expect that the variations would not be too large. If one or two faces appeared much more frequently than the others we would tend to say that the die is untrue or *biased.* The chi-square test is used to determine whether the amount of difference could reasonably be caused by chance. Note that we are assuming a particular distribution based on knowledge of the behavior of an unbiased die.

Suppose that the numbers 1 through 6 appeared 33, 23, 24, 23, 28, and 19 times, respectively. We wish to know whether a frequency as high as 33 or as low as 19 could be due to chance alone. Calculation of the χ^2 comparison statistic is as follows:

$$\chi^2 = \frac{(33 - 25)^2}{25} + \frac{(23 - 25)^2}{25} + \frac{(24 - 25)^2}{25} + \frac{(23 - 25)^2}{25}$$

$$+ \frac{(28 - 25)^2}{25} + \frac{(19 - 25)^2}{25}$$

$$= \frac{64}{25} + \frac{4}{25} + \frac{1}{25} + \frac{4}{25} + \frac{9}{25} + \frac{36}{25} = \frac{118}{25}$$

$$= 2.56 + 0.16 + 0.04 + 0.16 + 0.36 + 1.44 = 4.72$$

The value of 4.72 is compared with the values in a χ^2 table to determine its significance. The significance of the difference—in this case 4.72—depends on the number of degrees of freedom associated with the data. If none of the parameters of the distribution is estimated from the data, the number of degrees of freedom ν is one less than the number of groups or classes: $\nu = k - 1$.* One additional degree of freedom is lost for each parameter estimated from the data. Since none of the parameters was estimated (the value of 25 was assumed), and there were six possible outcomes (six classes) for each roll of the die, there are 5 degrees of freedom. From Table B.4, for 5 degrees of freedom, there is about a 45 percent probability that this high a value of chi-square is due to chance. Therefore, the assumption that the die is unbiased, i.e., that the probability of each face turning up is 1/6, is accepted.

Assume that there was some prior knowledge that the die was untrue and it was believed that the probability of rolling a 1 was 20 percent greater than normal and that of rolling a 6 was 20 percent less than normal. The expected frequencies in 150 rolls would then be 30, 25, 25, 25, 25, and 20 for the numbers 1 through 6, respectively. The χ^2 value would be

$$\chi^2 = \frac{(33 - 30)^2}{30} + \frac{(23 - 25)^2}{25} + \frac{(24 - 25)^2}{25} + \frac{(23 - 25)^2}{25}$$

$$+ \frac{(28 - 25)^2}{25} + \frac{(19 - 20)^2}{20}$$

$$= \frac{9}{30} + \frac{4}{25} + \frac{1}{25} + \frac{4}{25} + \frac{9}{25} + \frac{1}{20}$$

$$= 0.30 + 0.16 + 0.04 + 0.16 + 0.36 + 0.05 = 1.07$$

The probability of a value of 1.07 or larger due to chance is greater than 95

---

*The assumption of the particular theoretical distribution and number of groups results in the loss of 1 degree of freedom.

percent. Thus, this distribution fits the data even better than does the one based on the assumption of an unbiased die, even though there was no statistically significant reason for rejecting the first assumption.

A second, more applicable example concerns a characteristic whose failure rate was assumed to be constant. A large number of units (250) were tested until all failed, and the failures were grouped into 10 classes or time periods. Those failing in the first 100 hr were placed in the first period, those failing in the second 100 were placed in the second, and so on. The mean time to failure (MTTF) was approximately 300 hr. The frequencies for the various time periods are given in Table 5.1.

TABLE 5.1 Frequencies of failures

Time interval	Number of failures	Time interval	Number of failures
0-100 hr	39	500- 600 hr	22
100-200 hr	58	600- 700 hr	11
200-300 hr	47	700- 800 hr	6
300-400 hr	33	800- 900 hr	7
400-500 hr	25	900-1,000 hr	2
			250

A table of the negative exponential distribution is used to find the expected failure frequencies for each time period. Since the MTTF is 300 hr, 100 hr constitutes one-third of the MTTF. Table 5.2 shows the probability of survival at the end of each 100-hr period (each 1/3 MTTF), and the

TABLE 5.2 Number of expected failures in 250 units

End of time period	Probability of survival*	Probability of failure	Expected failures	End of time period	Probability of survival	Probability of failure	Expected failures
1/3 MTTF (100 hr)	0.7165	0.2835	71	2 MTTF (600 hr)	0.1353	0.0536	13 1/2
2/3 MTTF (200 hr)	0.5134	0.2031	50 3/4	2 1/3 MTTF (700 hr)	0.0970	0.0383	9 1/2
1 MTTF (300 hr)	0.3679	0.1455	36 1/4	2 2/3 MTTF (800 hr)	0.0695	0.0275	6 3/4
1 1/3 MTTF (400 hr)	0.2636	0.1043	26	3 MTTF (900 hr)	0.0498	0.0197	5
1 2/3 MTTF (500 hr)	0.1889	0.0747	18 3/4	3 1/3 MTTF (1,000 hr)	0.0357	0.0141	3 1/2
				> 3 1/3 MTTF (> 1,000 hr)		0.0357	9

*Probability of survival $= R = e^{-\lambda t} = e^{-t/m}$ where $m =$ MTTF $= 1/\lambda$.

probability of failure during the period is found by subtraction. The expected number of failures is the product of the probability of failure times the sample size, 250.

Expected frequencies, actual frequencies, differences, squares of differences, and chi-square values are provided in Table 5.3. The last term in the table, $(0 - 9)^2/9$, is added to account for the balance of the expected distribution. Note that if the last time period is divided into two or more periods, the sum of the resulting terms would still be 9 because the observed frequencies are all zero. There are 11 classes of data, and one parameter—the MTTF—was estimated from the observed values. Consequently, there are 9 degrees of freedom: $v = k - 1$ — number of estimated parameters. The chi-square value for 9 degrees of freedom and 0.1 percent probability is 27.88, which is less than the calculated chi-square value of 38.72. Therefore, there is less than 1 chance in 1,000 that the data came from this distribution. A Weibull distribution, discussed in detail in Chap. 14, might be more suitable. The reader may wish to return to this example later and test the fit of other distributions.

TABLE 5.3   Chi-square test for 250 failures

a	b	c	d	e
Expected frequencies	Observed frequencies	Difference	Squares of differences	Quotient (d/a)
71	39	32	1,024.00	14.42
50 3/4	58	7 1/4	52.56	1.04
36 1/4	47	10 3/4	115.56	3.19
26	33	7	49.00	1.88
18 3/4	25	6 1/4	39.06	2.08
13 1/2	22	8 1/2	72.25	5.35
9 1/2	11	1 1/2	2.25	0.24
6 3/4	6	3/4	0.56	0.08
5	7	2	4.00	0.80
3 1/2	2	1 1/2	2.25	0.64
9	0	9	81.00	9.00
				$\Sigma = 38.72$

5.1.2   The Kolmogorov-Smirnov Test

As stated earlier, the Kolmogorov-Smirnov (K-S) test compares observed *cumulative* frequencies with theoretical *cumulative* frequencies and therefore can be used for small samples. A third example illustrates its application. The value of a variable is measured on 15 units. The 15 observations, in rank order, are 11.1, 13.1, 13.8, 14.3, 14.7, 14.8, 15.0, 15.1, 15.1, 15.4, 15.7,

16.3, 16.7, 17.5, and 18.2. The characteristic is believed to be normally distributed with a mean of 15 and a standard deviation of 2, and the K-S test will be used to verify this assumption.

The assumed distribution and the observed values are divided into convenient classes in a manner similar to that used for the χ^2 test. In the example, a class width of 1 (= 0.5σ) is used. However, rather than comparing the observed and expected frequencies class by class, cumulative values are used and the maximum difference between the observed cumulative frequency and the expected cumulative frequency is found. This maximum difference is then compared with a table of K-S critical values to determine whether the difference could reasonably be expected by chance or whether it is so large that the assumed distribution does not truly represent the data. Table 5.4 shows the cumulative observed frequencies, the cumulative expected frequencies for a normal distribution with $\mu = 15$ and $\sigma = 2$, and the absolute differences. If an observed value falls on the border of two classes, for example, 15.0, it is included in both cumulative totals.

The largest difference between observed and expected cumulative frequencies is 0.11. This is less than any value in Table 5.5 for a sample size of 15 and is equivalent to a very high probability. This means that the probability is close to one that, for a sample size of 15, a value as high as 0.11 will be observed due to chance; and there is no evidence that the data are not from this distribution.

If the K-S test were made for an assumed normal distribution with a mean of 16 and a standard deviation of 2, a maximum difference in cumulative frequencies would be 0.23. While this is higher than 0.11, there is still considerable probability (about 35 percent) of its occurring by chance, and this distribution would be accepted as reasonable. A normal distribution with

TABLE 5.4   Date for the Kolmogorov-Smirnov test

Value	Cumulative frequencies equal to or less than value in first row												
	9	10	11	12	13	14	15	16	17	18	19	20	21
Observed cumulative quantities	0	0	0	1	1	3	7	11	13	14	15	15	15
Observed cumulative frequencies	0.00	0.00	0.00	0.07	0.07	0.20	0.47	0.73	0.87	0.93	1.00	1.00	1.00
Expected cumulative frequencies	0.00	0.01	0.02	0.07	0.16	0.31	0.50	0.69	0.84	0.93	0.98	0.99	1.00
Absolute difference	0.00	0.01	0.02	0.00	0.09	0.11	0.03	0.04	0.03	0.00	0.02	0.01	0.00

$\mu = 15$ and $\sigma = 1\frac{1}{2}$ fits even better than the originally assumed distribution, and it gives a maximum difference in cumulative frequencies of only 0.05.

We have studied tests that may be made to examine the relationship between an assumed distribution and the data which it was selected to

TABLE 5.5 Kolmogorov-Smirnov numerics for goodness-of-fit test*

Sample size	Significance level (probability)											
	0.10 (0.90)	0.20 (0.80)	0.30 (0.70)	0.40 (0.60)	0.50 (0.50)	0.60 (0.40)	0.70 (0.30)	0.80 (0.20)	0.90 (0.10)	0.95 (0.05)	0.98 (0.02)	0.99 (0.01)
1	0.55	0.60	0.65	0.70	0.75	0.80	0.85	0.90	0.95	0.975	0.99	0.995
2	0.40	0.43	0.46	0.49	0.53	0.57	0.62	0.68	0.78	0.84	0.90	0.93
3	0.33	0.35	0.38	0.41	0.44	0.47	0.51	0.56	0.64	0.71	0.79	0.83
4	0.29	0.31	0.33	0.36	0.38	0.41	0.45	0.49	0.56	0.62	0.69	0.73
5	0.26	0.28	0.30	0.32	0.34	0.37	0.41	0.45	0.51	0.56	0.63	0.67
6	0.24	0.25	0.27	0.29	0.31	0.34	0.37	0.41	0.47	0.52	0.58	0.62
7	0.22	0.24	0.25	0.27	0.29	0.32	0.34	0.38	0.44	0.49	0.54	0.58
8	0.21	0.22	0.24	0.26	0.28	0.30	0.32	0.36	0.41	0.46	0.51	0.54
9	0.20	0.21	0.23	0.24	0.26	0.28	0.31	0.34	0.39	0.43	0.48	0.51
10	0.19	0.20	0.22	0.23	0.25	0.27	0.29	0.32	0.37	0.41	0.46	0.49
11	0.18	0.19	0.21	0.22	0.24	0.26	0.28	0.31	0.35	0.39	0.44	0.47
12	0.17	0.19	0.20	0.21	0.23	0.25	0.27	0.29	0.34	0.38	0.42	0.45
13	0.16	0.18	0.19	0.21	0.22	0.24	0.26	0.28	0.32	0.36	0.40	0.43
14	0.16	0.17	0.18	0.20	0.21	0.23	0.25	0.27	0.31	0.35	0.39	0.42
15	0.16	0.17	0.18	0.19	0.21	0.22	0.24	0.27	0.30	0.34	0.38	0.40
16	0.15	0.16	0.17	0.19	0.20	0.22	0.23	0.26	0.30	0.33	0.37	0.39
17	0.15	0.16	0.17	0.18	0.19	0.21	0.23	0.25	0.29	0.32	0.36	0.38
18	0.14	0.15	0.16	0.18	0.19	0.20	0.22	0.24	0.28	0.31	0.35	0.37
19	0.14	0.15	0.16	0.17	0.18	0.20	0.22	0.24	0.27	0.30	0.34	0.36
20	0.14	0.15	0.16	0.17	0.18	0.19	0.21	0.23	0.26	0.29	0.33	0.35
25	0.12	0.13	0.14	0.15	0.16	0.17	0.19	0.21	0.24	0.26	0.30	0.32
30	0.11	0.12	0.13	0.14	0.15	0.16	0.17	0.19	0.22	0.24	0.27	0.29
35	0.10	0.11	0.12	0.13	0.14	0.15	0.16	0.18	0.21	0.23	0.25	0.27
> 35	$\dfrac{0.61}{\sqrt{n}}$	$\dfrac{0.65}{\sqrt{n}}$	$\dfrac{0.70}{\sqrt{n}}$	$\dfrac{0.75}{\sqrt{n}}$	$\dfrac{0.81}{\sqrt{n}}$	$\dfrac{0.87}{\sqrt{n}}$	$\dfrac{0.95}{\sqrt{n}}$	$\dfrac{1.07}{\sqrt{n}}$	$\dfrac{1.22}{\sqrt{n}}$	$\dfrac{1.36}{\sqrt{n}}$	$\dfrac{1.48}{\sqrt{n}}$	$\dfrac{1.63}{\sqrt{n}}$

*Column headings not in parentheses give probabilities of maximum absolute differences between observed cumulative frequencies and theoretical cumulative frequencies being smaller than or equal to the table values. The probabilities of maximum absolute differences exceeding the table values are in parentheses.

Values for probabilities from (0.01) through (0.20) are adapted from L. H. Miller, Table of Percentage Points of Kolmogorov Statistics, *J. Am. Statist. Assoc.*, vol. 51, 1956, with the permission of the Association. Other values are derived from the expression $\sqrt{\log_e(1/a)/2n}$ by Smirnov which appeared in equation (3) of the article, modified by an error-correction factor developed by the author, who accepts responsibility for any error resulting from the modification.

represent. These tests provide means of determining when the assumed distribution differs significantly from the observed data. While a goodness-of-fit test cannot prove that a distribution which does not *disagree* with the data is the best choice, the fact that the variations between observed and theoretical frequencies can be attributed to chance provides some degree of assurance that the assumed distribution can be used and that the results of its use will be reasonably valid.

5.2   REGRESSION AND CORRELATION

We can now turn our attention to a somewhat related subject. Rather than finding a distribution that represents the data and then deriving further relationships and probabilities based on the distribution, we may wish to determine the relationship between two variables and to see how well a value of one variable can be used to predict the value of the second. The first part of this related subject—the relationship between two variables—is called *regression*; the second part—the accuracy of the relationship—is called *correlation*. The distinction between regression and correlation can perhaps best be shown pictorially. Figure 5.1 depicts the relationships between four different variables.

The value of the independent variable A is measured on the horizontal axis (abscissa). The values of three dependent variables B, C, and D are measured on the vertical axis (ordinate). It can be seen from the figure that there are very close relationships between the dependent variables and the independent variable A. In the first case, the observed values of B, shown by crosses, are linearly related directly to the values of A according to the relationship $B = 3/2\ A$. Observed values of the second variable, shown by circles, are inversely related to A: $C = 5 - A$. The relationship of observed values of D (triangles) is also quite good, but it is not linear; it can be described algebraically as $D = 1/8\ A^2$. In these three cases the values of the dependent variables can be determined quite accurately from the value of the independent variable within the range of the graph. (Extrapolation outside a graph range should not be made unless it is known that the observed relationship holds true throughout the extended range.)

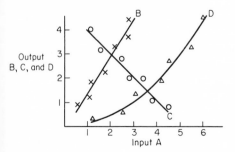

FIG. 5.1   Relationships between one independent and three dependent variables, all with high correlation.

In Fig. 5.1, we say that there is excellent, or at least very good, correlation between variables *B, C,* and *D* and variable *A.* However, the regression lines between the variables are quite different, one being linear and direct, the second also being linear but inverse, and the third being non-linear (in this case, second degree). Regression lines of higher order exist but will not be treated in this text.

Correlation is the measure of how well the data fit the curve, regardless of what type of regression line exists. Correlation analysis is often performed after the best fitting line has been found by regression methods although the coefficient of correlation can be found before finding the regression line. Figure 5.2 shows three additional regression lines, all linear, all direct, and all with the same slope. However, they have greatly different values of correlation. The data for line *E,* shown by squares, have a very high correlation, and the value of dependent variable *E* can be predicted quite accurately from the value of *A.* Values of dependent variable *F* (dots) have somewhat more variation from line *F,* and the correlation between *A* and *F* is less than that between *A* and *E.* It is still high enough to make reasonably accurate predictions of values of *F* from values of *A.* There is only slight correlation between *G* (crossed circles) and *A,* and values of *A* can be used to estimate values of *G* with very little accuracy.

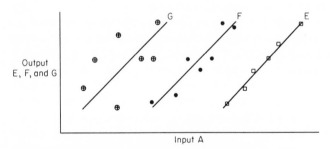

FIG. 5.2   Relationships between one independent and three dependent variables with different degrees of correlation.

5.2.1   Linear Regression

Simple linear regression represents a straight-line relationship between the dependent and independent variables and is the easiest to compute. In all cases in Figs. 5.1 and 5.2, except for variable *D,* the relationship is linear. A straight line can even be used to define the relationship between *D* and *A* within the range of the graph with only a slightly lower correlation.

The line of best fit, whether the relationship is linear or of a higher order, is defined as that line which minimizes the sum of the squares of the *vertical* differences between the regression line and the observed values of the dependent variable. If we let Y_o represent the observed value of the dependent variable and Y_R the value of the dependent variable found from the regression equation, those values of Y_R which result in the smallest value for the term $\Sigma(Y_{oi} - Y_{Ri})^2$ are on the line of best fit. Figure 5.3 illustrates the values whose sum of squares is being minimized.

The general equation for a straight line is

$$Y = a + bX \tag{5.2}$$

where a is the intercept on the Y (vertical) axis and b is the slope of the line.*
This value of Y is used for Y_R, thereby yielding values of Y on the regression line. In minimizing $\Sigma(Y_{oi} - bX_i - a)^2$, calculus is used to solve for the algebraic expressions of a and b which result in the least sum of squares.† The expressions are

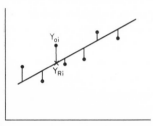

FIG. 5.3   Regression line show-ing vertical deviations whose sum of squares is minimized.

$$a = \overline{Y} - b\overline{X} \tag{5.3}$$

$$b = \frac{\Sigma[(X_i - \overline{X})(Y_i - \overline{Y})]}{\Sigma[(X_i - \overline{X})^2]} = \frac{\Sigma(x_i y_i)}{\Sigma(x_i^2)} \tag{5.4}$$

In Eqs. (5.3) and (5.4), the notations used throughout the book are followed: capital letters X and Y represent actual observed values; lowercase letters x and y are differences between observed values and means. When calculating $x_i y_i$ products, it is important that corresponding x and y values be used. (There are other, equivalent expressions for $\Sigma(x_i^2)$, for example, $(\Sigma X_i^2 - n\overline{X}^2)$, but because values of x_i must be computed in the numerator, computing the denominator as shown requires the fewest calculations.) The full equation for the line of best fit is found by substituting the value of a from Eq. (5.3) into Eq. (5.2). Y_R is substituted for Y because the specific line being defined is the regression line:

$$Y_R = bX + \overline{Y} - b\overline{X} \tag{5.5}$$

Final values of Y_R for a particular set of data are then determined by solving for b in Eq. (5.4).

---

*This is the standard notation used in statistical literature and is comparable to the equation $Y = mX + b$ used in other areas of mathematics.

†The algebraic values of a and b which minimize $\Sigma(Y_{oi} - bX_i - a)^2$ can be found by taking partial derivatives with respect to a and b and setting them equal to zero. Appendix A.1 includes a complete derivation.

5.2.2 Non-linear Regression

The methods described here consider only linear relationships. Non-linear relationships are also possible, as was indicated in Fig. 5.1. Square, square root, logarithmic, and exponential relationships as well as other non-linear functions are often encountered in engineering, and regression analyses may be required. Some computer programs are available for performing the mathematical operations, which are considerably more complex than those for linear relationships. Some of these programs even plot the analysis results. It has been found, however, that when the parameter values are accurately plotted on fine-grid graph paper, a curved line of best fit can be drawn with considerable precision by sighting. In fact, a computer program considers all data points, and a single abnormal point—perhaps the result of a test or measurement error—has a disproportionate influence on the numerical results. The sighting method will disclose any anomaly in the data, and engineering judgment may dictate the deletion of a particular value. It is therefore suggested that curved regression lines be plotted using visual methods, with the general restriction that, in most cases, the direction of curvature be kept constant (no points of inflection in the curve). The degree of curvature, however, may be changed smoothly to best fit the data.

5.2.3 Correlation

The degree of correlation is measured by the correlation coefficient. A value of unity represents perfect correlation;* a value of zero is no correlation. The correlation coefficient is denoted by the lowercase letter r and is expressed by Eq. (5.6).† A qualitative explanation of the equation follows:

$$r = \sqrt{1 - \frac{\Sigma(Y_i - Y_R)^2}{\Sigma(Y_i - \overline{Y})^2}} \qquad (5.6)$$

Since the correlation coefficient is a measure of how close the observed values of Y lie to the regression line, a deviation of zero should yield a correlation of unity. Therefore, the terms of deviation should be subtracted

---

*Most texts use both positive and negative values of correlation, depending on whether the slope b of the regression line is positive or negative. The significance of the coefficient considers the fact that both positive and negative values are possible. However, since the degree of correlation is the same regardless of the sign of the coefficient and since the coefficient is found by taking a square root, a positive value will be used here to avoid the confusion which sometimes arises from a negative value of correlation.

†In Eq. (5.6) and in the subsequent discussion, Y_i refers to the observed value and Y_R refers to the corresponding regression-line value.

from one but cannot exceed one. When the sum of squares of deviations from the regression line is made the numerator of a ratio in which the sum of squares of deviations from the mean is the denominator, the ratio varies from zero to plus one. The deviations are squared because, by definition, the unsquared algebraic sum of deviations from the mean is always zero, and the fraction would have a zero denominator if the terms were not squared. Because squares of differences are used, the square root of the resulting expression is used.

The expression for r in Eq. (5.6) can be easily visualized as a measure of agreement of the observed values to the regression-line values. However, its calculation involves considerable computation because the value of Y_R must be calculated for each value of X (from $Y_R = bX + \overline{Y} - b\overline{X}$), each difference term $(Y_i - Y_R)$ must be calculated, and so on. There are two equivalent expressions for the correlation coefficient, either of which requires fewer computations. These are given by Eqs. (5.7a) and (5.7b).*

$$r = \frac{\Sigma x_i y_i / n}{\sigma_X \sigma_Y} \tag{5.7a}$$

$$r = \frac{\Sigma X_i Y_i / n - \overline{X}\overline{Y}}{\sigma_X \sigma_Y} \tag{5.7b}$$

The terms of these expressions are more readily calculated because many of the computations are performed in the calculation of b in Eq. (5.4). Even the sigma values are readily available because all values of x_i and y_i are also calculated in Eq. (5.4).

A second advantage of these expressions is that the correlation coefficient can be computed without computing the regression line, and it is sometimes required only to determine whether there is correlation between two variables. The values of σ_X and σ_Y can be calculated either from general Eq. (2.4) or from Eq. (2.4a), both of which are repeated here for convenience.

$$\sigma_X = \sqrt{\frac{\Sigma x_i^2}{n}} = \sqrt{\frac{\Sigma (X_i - \overline{X})^2}{n}} \tag{2.4}$$

$$= \sqrt{\frac{\Sigma (X_i^2)}{n} - \overline{X}^2} \tag{2.4a}$$

*Calculations for Eqs. (5.7a) and (5.7b) may result in a negative value of r. The sign can be ignored and the absolute value used.

While the value of r is calculated more easily from Eq. (5.7a) or from Eq. (5.7b) than from Eq. (5.6), it is more difficult to visualize the latter expressions as measures of correlation because there is no reference to a regression line. Therefore, the equivalence of these three expressions is presented in Appendix A.2. Starting with Eq. (5.6), a step-by-step transformation is performed, first to expression (5.7a) and then to expression (5.7b). Each step is numbered, and the corresponding mathematical operation is explained after the complete transformation.

The question arises as to the significance of the correlation coefficient. As a general rule, if the coefficient when multiplied by the square root of the sample size is greater than 2.0, the correlation is said to be significant in the statistical sense.* Of course, the greater the value of r, the better the relationship is between the two variables, and the more accurate is the estimate of one from the value of the other.

A short example illustrates the application of regression and correlation analysis. In this case the relationship between the original value of tensile strength is compared with a final value after the application of a particular heat treatment to the test specimens. The effect of the heat treatment on this property of the test items is to be evaluated. Original tensile strength values were 63, 68, 59, 58, 65, 61, 63, and 59. Final values after heat treating, in the same sequence, were 72, 75, 70, 66, 71, 72, 69, and 69. All data are in 1,000 psi and are depicted graphically in Fig. 5.4. Table 5.6 provides the details of the analysis.

$$\text{The slope } b = \frac{\Sigma xy}{\Sigma x^2} = \frac{51.0}{82} = 0.62 .$$

$$Y_R = bX + \overline{Y} - b\overline{X} = 0.62X + 70.5 - (0.62 \times 62) = 0.62X + 32.0$$

$$r = \frac{\Sigma xy/n}{\sigma_X \sigma_Y} = \frac{51.0/8}{\sqrt{82/8} \times \sqrt{50.0/8}} = 0.80$$

Significance test $= 0.80\sqrt{8} = 2.3$; therefore, r is considered to be significant.

When distributions were fitted to data it was mentioned that, although there might be no statistical evidence that the distribution which was selected

---

*If there is no correlation between two variables, the true correlation coefficient is zero, and the *average* correlation calculated from samples will also be zero. However, some samples will show some degree of correlation, either positive or negative. The distribution of the correlation coefficient from a population with a zero correlation is approximately normal with a mean of zero and a standard deviation of $1/\sqrt{n}$. If the value of r from a sample of size n exceeds the 2σ value $(2/\sqrt{n})$ for a zero correlation, then there is less than 5 percent probability that the correlation value came from a population with zero r. Therefore, if the product of $r \times \sqrt{n}$ is greater than 2.0, the correlation coefficient is considered significant.

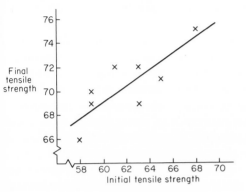

FIG. 5.4 Tensile strength values (1,000 psi).

did not fit the data, other distributions could also fit. Similarly, while the regression line is the line of best fit, other lines might also be satisfactory. In the example, a second-degree curve with a slight downward concavity might fit slightly better. Prior engineering knowledge could suggest that the expected increase in tensile strength might be a constant value or even a fixed percentage of the original value. The line $Y = X + 8.5$ (determined from the difference between $\bar{X}$ and $\bar{Y}$) describes an average increase in tensile strength of 8,500 psi, and has a correlation coefficient of 0.63 using Eqs. (5.2) and (5.6) for the calculations. A test for significance of this $45°$ line which has a somewhat lower value of r results in a value which is just below the 5 percent criterion of significance ($r \times \sqrt{n} = 1.8$) and which corresponds to a probability of no correlation of about 7 percent.

TABLE 5.6 Data for analysis of tensile strength

Original X	Final Y	$X - \bar{X}$ x	$Y - \bar{Y}$ y	x^2	xy	y^2
63	72	1	1 1/2	1	1.5	2.25
68	75	6	4 1/2	36	27.0	20.25
59	70	- 3	- 1/2	9	1.5	0.25
58	66	- 4	- 4 1/2	16	18.0	20.25
65	71	3	1/2	9	1.5	0.25
61	72	- 1	1 1/2	1	- 1.5	2.25
63	69	1	- 1 1/2	1	- 1.5	2.25
59	69	- 3	- 1 1/2	9	4.5	2.25
Total 496	564			82	51.0	50.0
Average 62	70 1/2					

5.3 NORMALIZATION

The final topic of this chapter is the transformation of data from a non-normal distribution to a gaussian one. While the normal distribution is frequently assumed to be descriptive of a set of variables data and does correctly describe the distribution of many characteristics found in nature and industry, there are variables characteristics which are not normally distributed. A goodness-of-fit test may indicate non-normality or conditions may exist which, from an engineering standpoint, are known to result in non-normal variability.

Since a great deal is known about the gaussian distribution and tables which provide probability-variation relationships are readily available, it is desirable that reliability analyses utilize this distribution when practical. Percentages of this distribution which are further than a given number of standard deviations from the mean can be found from Table B.2, so that, if the data are normally distributed, probabilities of particular values can be quickly determined. Therefore, when it is possible to transform non-normal data into a normal distribution by a simple change of the variable, this should be accomplished. If a skewed distribution can be evaluated using procedures applicable to a normal distribution, then these procedures should be used. Obviously, such distributions as the negative exponential (constant failure rate) cannot be normalized, nor would we wish to do so because analysis methods and tables are readily available for this distribution in its original form. However, many skewed distributions are easily transformed and valid transformations are routinely made.

5.3.1 Skewed Distributions

We will first examine a technique derived from experience for evaluating a skewed distribution. It combines both empirical and theoretical procedures. Because the technique does not depend on an exact mathematical description of the original distribution, a strict mathematical proof is not presented. However, like many other engineering and reliability methods, the usefulness and validity of results have been verified by actual application.

When a distribution has a single mode X_0 and has the general shape shown in Fig. 5.5, it is called a skewed distribution. It should be evaluated as such if measurements of at least 25 units were included when determining its shape. (If there are fewer units, there is ordinarily not sufficient evidence of non-normality, based either on observation or a goodness-of-fit test, to assume the existence of skewness.) A distribution with a right-hand tail as shown in the figure is said to be skewed to the right.

Suppose that the value of a characteristic is measured on 30 units and, when plotted, the measurements form the skewed distribution shown in Fig. 5.6. In the figure, a continuous curve is superimposed over a histogram of the

30 readings. Suppose, also, that the characteristic is required to have a minimum value of 45. A suggested procedure for this, and for any skewed distribution with a single mode, is to consider the entire distribution as the sum of two normal distributions, each having a mean value equal to the mode

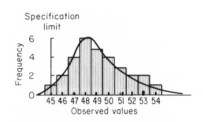

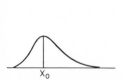

FIG. 5.5 A skewed distribu-tion with mode X_0.

FIG. 5.6 Histogram of 30 measure-ments, showing specification limit.

of the skewed distribution (in this case, 48). That portion of the distribution which is on the same side of the mode as the requirement, and including the mode, is evaluated as one-half of a normal distribution. In the example, the left side of the distribution including the modal value is analyzed. This includes the measured values of 45 through 48, inclusive. The standard deviation is computed by assuming that the remaining part of the assumed normal distribution except for the modal value itself is a mirror image, as shown in Fig. 5.7. Calculation of the standard deviation is shown in Table 5.7.

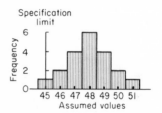

FIG. 5.7 Complete assumed normal distribution.

The difference between the mean, 48, and the minimum requirement, 45, is 3/1.49 = 2.01 standard deviations. Since the number of actual original values used in the assumed normal distribution was 13, this number is used when finding reliability-confidence numerics. Reliability-confidence values can then be determined using methods described in Chap. 14 and the appropriate curves in Table C.4.

However, the resulting reliability value must be modified to account for the skewness of the original distribution. This is accomplished as follows.

1. The unreliability Q is first computed: $Q = 1 - R$.
2. The fraction of the original skewed distribution that is related to the reliability calculation is determined. This is that portion which is between the mode and the specification limit, including one-half of the number of measurements at the mode itself. Any measurements beyond the limit are also included. In this example, these measurements include

TABLE 5.7   Calculation of standard deviation of assumed normal distribution

Value X_i	Frequency n	Difference $X_i - \bar{X}$	Difference squared $(X_i - \bar{X})^2$	Total of squares $n(X_i - \bar{X})^2$
45	1	- 3	9	9
46	2	- 2	4	8
47	4	- 1	1	4
48	6	0	0	0
49*	4	1	1	4
50*	2	2	4	8
51*	1	3	9	9
				$\Sigma = 42$

$$s = \sqrt{\frac{\Sigma(X_i - \bar{X})^2}{n - 1^\dagger}} = \sqrt{\frac{42}{19}} = 1.49$$

*Assumed values.

†The value of n is determined from the number of values used for the complete assumed normal distribution. Since values of 45 through 51 were used, $n = 20$ and $n - 1 = 19$.

values of 45, 46, 47, and half of those at 48, equal to $1 + 2 + 4 + \frac{1}{2}(6) = $ 10 measurements. Since there were 30 original measurements, 10 is one-third of the original number.

3. The value of Q from step 1 is multiplied by twice the fraction calculated in step 2. The new value is denoted $\hat{Q}$. In our example:

$$\hat{Q} = 2 \times \frac{1}{3} \times Q = \frac{2}{3}Q$$

4. The final reliability value, denoted as $\hat{R}$, is $1 - \hat{Q}$. For example, if the reliability were 0.94 and $Q = 0.06$ at the desired level of confidence, then $\hat{Q}$ in this case would be $2/3 \times 0.06 = 0.04$, and $\hat{R}$ would be 0.96.

5.3.2   Known Non-normal Functions

Sometimes there are variables which appear to have skewed distributions but which are some function of another variable which is normally distributed. The relationship between the two variables might be logarithmic, exponential, square, square root, or some other function derived from technical considerations of the variables. For example, windage (power) losses in a turbine are proportional to the fifth power of the speed. If the speed is regulated by a speed controller so that it is normally distributed, then the fifth root of the losses will be normally distributed also. Similarly, the temperature on one surface of a plate is a function of the radiant heat energy impinging on the opposite side of the plate. In this instance, the function is exponential: $T = ke^{-bx}$ where T is temperature and x is thickness. If the thickness is normally distributed, then the natural logarithm of the temperature will be too. If losses or temperatures were plotted directly, a skewed distribution would result. Because the mathematical functions are known, they can be applied to the data to transform them to variables which are normally distributed.

Suppose that some output Y is proportional to the cube of the input, X; for example, the volume of a sphere is proportional to the cube of its diameter. While it may be difficult to measure the diameter directly, it may be much easier to determine the volume, perhaps by water displacement. If we wished to evaluate the parameters of the diameter (its mean and standard deviation), we could determine displacement and statistically analyze the cube root. The volumetric determination of 25 spheres gives the following results. (Units are not given, but could be cc, in.^{3}, etc., as applicable.) A histogram of the data is shown in Fig. 5.8. Note that if the cell widths (divisions) of the histogram had been 50 instead of 40, there would be little visual evidence of significant skewness. This confirms the fact that the use of too few divisions can mask important information. The histogram also indicates that a minimum of 25 measurements are needed to disclose the presence of non-normality.

28	84	117	148	179
43	91	121	155	195
58	97	124	161	212
69	102	129	166	274
81	109	137	170	342

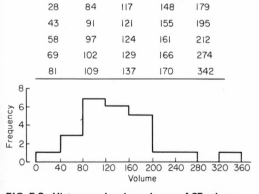

FIG. 5.8   Histogram showing volumes of 25 spheres.

When the cube roots of the measured values are used, a much more normal distribution results. The cube roots are as follows, and the corresponding histogram is given in Fig. 5.9, using about the same number of cells that was used in Fig. 5.8. Plus and minus signs are used as needed to indicate the particular cells in which borderline readings are included.

3.1	4.4	4.9	5.3	5.6
3.5+	4.5-	4.9	5.4	5.8
3.9	4.6	5.0-	5.4	6.0-
4.1	4.7	5.1	5.5-	6.5-
4.3	4.8	5.2	5.5+	7.0-

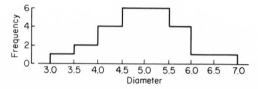

FIG. 5.9 Histogram of the diameters of 25 spheres.

The choice of the cube root was made because of prior knowledge of the mathematical function. If this information had not been available, the square root, logarithm, or other function might have been used and could have resulted in a normal-appearing distribution. This further confirms the fact that 25 values are a minimum number for determining normality, and points out that if the mathematical function is unknown, many more values will be needed. Computation of the mean and standard deviation is left as an exercise for the reader. A goodness-of-fit test may also be made, if desired.

After the transformation is made and the mean and standard deviation have been computed, standard methods applicable to normally distributed variables can be used to determine reliability-confidence relationships. The reliability numbers are directly applied to the original data without modification.

ADDITIONAL READINGS

Amstadter, B. L., and T. A. Siciliano: "Reliability Assessment Guides for Apollo Suppliers," North American-Rockwell Corporation, SID 64-1447A, Downey, Calif., 1965, IDEP no. 347.40.00.00-F1-28.

ARINC Research Corporation: "Reliability Engineering," Prentice-Hall, Inc., Englewood Cliffs, N.J., 1964.

Chorafas, D. N.: "Statistical Processes and Reliability Engineering," D. Van Nostrand Company, Inc., Princeton, N.J., 1960.

Dixon, W. J., and F. J. Massey, Jr.: "Introduction to Statistical Analysis," 3d ed., McGraw-Hill Book Company, New York, 1969.

Draper, N. R., and H. Smith: "Applied Regression Analysis," John Wiley & Sons, Inc., New York, 1968.

Fisher, Sir Ronald A., and Frank Yates: "Statistical Tables for Biological, Agricultural and Medical Research," 4th ed., Hafner Publishing Company, Inc., New York, 1953.

Freeman, H. A.: "Industrial Statistics," John Wiley & Sons, Inc., New York, 1942.

Hiltz, P. A., and J. L. Gaffney: "Statistical Techniques for Reliability," North American-Rockwell Corporation, Downey, Calif., 1965.

Hoel, P. G.: "Introduction to Mathematical Statistics," John Wiley & Sons, Inc., New York, 1947.

Ireson, W. G.: "Reliability Handbook," McGraw-Hill Book Company, New York, 1966.

Johnson, N. L., and F. C. Leone: "Statistics and Experimental Design in Engineering and the Physical Sciences," John Wiley & Sons, Inc., New York, 1964.

Lloyd, D. K., and M. Lipow: "Reliability: Management, Methods, and Mathematics," Prentice-Hall, Inc., Englewood Cliffs, N.J., 1962.

Miller, L. H.: Tables of Percentage Points of Kolmogorov Statistics, *J. Am. Statist. Assoc.,* vol. 51, 1956.

Mood, A. M.: "Introduction to the Theory of Statistics, " McGraw-Hill Book Company, New York, 1950.

Shewhart, W. A.: "Economic Control of Quality of Manufactured Product," D. Van Nostrand Company, Inc., Princeton, N.J., 1931.

Simon, Gen. L. E.: "Engineer's Manual of Statistical Methods," John Wiley & Sons, Inc., New York, 1941.

Snedecor, G. W.: "Statistical Methods," 5th ed., Iowa State College Press, Ames, 1957.

Villars, D. S.: "Statistical Design and Analysis of Experiments for Development Research," Wm. C. Brown Company Publishers, Dubuque, Iowa, 1951.

6

DATA CONSIDERATIONS

6.1 INTRODUCTION

The need for meaningful and applicable data has been discussed by many writers in almost all areas of human enterprise. Valid information is a primary requisite in all technologies and disciplines and particularly in the field of reliability mathematics and related phases of reliability. While the need is most obvious, of course, in those activities which are concerned with the collection, summarization, and analysis of data, it is equally applicable to activities where data evaluation is not a principal effort. To cite a rather self-evident example, an architect relies on handbooks, tables, and other data sources for information pertinent to the materials and designs which he employs. The accuracy and validity of his information has been verified many times and he accepts it without question. The engineer similarly depends on comparable sources of information related to his areas of activity. From a broad viewpoint, the entire history of human progress is based on acceptance of facts previously verified and documented. When these facts are wrong, advancement is delayed (the late discovery and development of the Western

Hemisphere is a notable example of this); when the facts are correct, they form the foundation for further advancement.

The acquisition of good data, then, is a necessary requisite for proper performance of reliability functions. Data which are accurate and complete enable better estimation of reliability, provide for timely and effective corrective action of design and test deficiencies, and lead to improvements in procedures, processes, and instructions which prevent failures. The problem of securing these data is not an easy one. Except for those systems which are fully automated, human effort is required to record and transmit information as well as to summarize and evaluate it. Even in those systems which provide for automatic data processing, interpretation is needed and investigation of discrepancies is necessary.

6.2 DATA PLANNING

For data to be useful not only must they be accurate and valid, but they must also be pertinent—oriented toward providing needed information. To obtain pertinent data the purposes for which the data will be used require definition, the sources of the data must be delineated, the means of evaluation should be known, and applicable criteria must be formulated with respect to the data's classification. Only when all these factors are considered and satisfied will the data be truly useful. Without useful data neither reliability nor any other activity can contribute toward the accomplishment of program objectives.

First, the purposes of the information should be defined. Inspectors, technicians, and other personnel cannot be expected to provide meaningful information if they do not know or understand what information is desired. The responsibility for specifying what data are needed lies with the engineers and analysts. To do this they must first have a clear picture of the purposes for which the data are being collected. These purposes need not all be immediate (historical information is often obtained even though it may not be used in the foreseeable future), but they should be stated.

Second, data sources must be specified to ensure that the data are unbiased and are accurately reported. If failure information is being gathered, not only failures but also successes must be recorded in an impartial manner. All applicable sources which would provide pertinent data should be used. "Applicable" means that selectivity should be applied to cost, availability, and ease of data collection; "pertinent" refers to the collection of data without bias—preference is given to data sources only on the basis of the validity of the data and not because they support or disprove a hypothesis of reliability or unreliability. If one were summarizing the failure histories of two or more components or subsystems which were being considered for the same function, the fact that one component or subsystem had experienced more failures than another may not be significant if the first had operated for a greater number of hours or had operated in a more severe environment.

Third, evaluation procedures should be determined before the data collection is undertaken. In fact, the most successful economic programs are those in which these procedures are established before the test program itself is initiated. Efficiency and effectiveness in data collection are directly related to the prior knowledge of the evaluation methods that will be employed to analyze the information. It would be ineffective to collect only attributes (go/no go) data, for example, if the analysis required variables information. Likewise, considerable time and money would be wasted if actual readings were taken and recorded and only the total numbers of successes and failures were to be used in the analysis. Consequently, a thorough study of the objectives of the overall data program and of the means of analysis to be employed must precede at least the data collection and, preferably, the test program itself. The study should result in specific determination of the analysis methods.

Finally, there should be criteria for the data classification. "What constitutes failure?" "What defines appropriate test conditions?" "When is a test applicable?" and "What is a suspended test?" are among the relevant questions that require answers *before* the data are collected. One actual example relates to a fluid pump where the system engineers did not define performance criteria and the design engineers would not. Performance anomalies occurred which resulted in system failure when the pump was installed but the cognizant engineers refused to recognize the occurrence of a component failure because there were no criteria of acceptable performance. If analyses are to be meaningful and if unbiased evaluations are to be made, the criteria must be defined in advance; their formulation cannot be delayed until the tests have been run and the data have been collected and summarized.

6.3 DATA COLLECTION

The steps of defining purposes, formulating classification criteria, and so on which have just been discussed are only preliminary and do not in themselves ensure that the data will be meaningful. Operating, test, and maintenance personnel must be encouraged, oftentimes prodded, to provide information. In this they are no different from ourselves, for few of us enjoy putting into writing the history of events, conditions, parameter values, and other details so necessary for adequate data evaluation. There is no magic formula for obtaining data. A constant monitoring and auditing is needed, for as soon as vigilance is relaxed, data reporting correspondingly decreases. This reluctance to write down information is not necessarily or even usually deliberate; it merely reflects our normal human dislike of writing down what may seem to be the trivia of daily occurrence.

6.3.1 Personnel Motivation

There are, of course, some actions which can and should be accomplished to facilitate this data collection. Roadblocks should not be deliberately set up and, when discovered, should be speedily removed. While this seems obvious, it is surprising how often such obstacles are actually established without thought of their long-range consequences. If an employee accidentally drops a delicate part, he should feel free to disclose the accident so that the part can be repaired. If he feels the slightest hesitation to report the accident, from fear of reprimand, it will never be reported. It takes only one sharp word of criticism to set back 6 months of hard effort in the area of data reporting. Airline pilots were once asked to report all near misses. After a number of reprimands were issued, even though many of them may have been justified, there were considerably fewer reports of near misses—not that near misses did not occur, they just were not being reported. As a consequence of the reduction in near-miss reports, there developed a lack of data needed for airport and flight pattern improvement. Even though greater safety would eventually result, it required a repeated policy statement to the effect that no reprimand or disciplinary action would result from such reports before the reluctance of flight personnel to make these reports was overcome. There is no wonder then that when an employee has nothing to gain from reporting unfavorable events, he will have little inclination to do so; and any feeling of hesitancy will result in the withholding of information.

6.3.2 Forms Design

Just as roadblocks must be removed, the data collection system should be designed to facilitate the recording of information. Forms and reports must be tailored to fit the needs of both the particular program in which they are to be used and the personnel who are to prepare them. Sometimes these needs conflict with each other. On the one hand it is necessary that the information items satisfy the purposes for which the data are to be used, permit easy classification of information, and facilitate the data evaluation. On the other hand, when personnel are familiar with the forms and information format they are less likely to make errors or omit data, and standard forms are therefore desirable. Even when several programs are involved, standard forms are sometimes used in order to keep the data recording as simple as possible.

Wording on forms must not be ambiguous. For example, one of the simplest and yet most often misinterpreted items is the DATE. This can refer to either the date on which the report was prepared or the date on which the failure or other event occurred. These two dates are not necessarily the same, and the form should clearly specify which date is meant. There are many other information items which can be misinterpreted or which lead to confusion. On a failure report, the failure classification can result in different entries by different people. It is seldom that the test inspector or technician has in mind exactly the same definitions and criteria for the various classes of failure as does the analyst or engineer. More consistency in classification of failures results when the failure report describes what happened to the component and system and provides auxiliary information and the analyst determines the classification from his review and investigation of the failure.

Forms may be either single- or multipurpose, very detailed or relatively simple, require extensive fill-in or merely a checkoff of various information items, and be single- or multicopied. The type depends on the purposes and distribution. When an activity is repetitive or is extremely complex, single-purpose forms are most often used. Flight reports, maintenance reports, reports of operating hours, and the like are single-purpose reports used for repetitive activities which occur on a scheduled or routine basis. Failure reports are more or less single-purpose when the equipment being operated or tested is repeatedly of the same type. (Failure reports, of course, are always single-purpose in that they record failures; however, in one sense they are multipurpose when the same report form is used for a multitude of different components or systems.) Multipurpose forms find more use in development programs where there is a variety of activity with relatively little repetition.

There are advantages and disadvantages to both types. Single-purpose forms facilitate familiarization and reduce the amount of training required to obtain accurate information. However, if there are too many activities performed by the same personnel and each activity has its own form, the ease of familiarization is lost and the information is less accurate. Multipurpose forms initially necessitate more training, but they can result in better information when there is a variety of data to be recorded. A typical example is DD-787 for reporting electronic failures, Fig. 6.1. Chapter 9 of the "Reliability Handbook" shows a number of both single- and multipurpose forms, and examples of other forms are found in Calabro and ARINC Research Corporation.*

Forms which are used to record data pertinent to complex operations are often combined with the operations sheet and provide for recording the quality or reliability information directly on the sheet, as in Fig. 6.2. Such forms are quite detailed and specific. At the other extreme are logbooks in which periodic, sometimes continuous, records are kept of a system operation, test loop, or similar activity. Logbook sheets, Fig. 6.3, usually require a minimum of specific entries, but provide considerable space for

---

*ARINC Research Corporation, "Reliability Engineering," Prentice-Hall, Inc., Englewood Cliffs, N.J., 1964; S. R. Calabro, "Reliability Principles and Practices," McGraw-Hill Book Company, New York, 1962; W. G. Ireson, "Reliability Handbook," McGraw-Hill Book Company, New York, 1966.

ELECTRONIC EQUIPMENT FAILURE/REPLACEMENT REPORT DD-787 (PROPOSED) REPORT BUSHIPS 10550-1

1. DESIGNATION OF SHIP OR STATION

3. TYPE OF REPORT (CHECK ONE)
1. ☐ OPERATIONAL FAILURE
2. ☐ PREVENTIVE MAINTENANCE (POMSEE)
3. ☐ PREVENTIVE MAINTENANCE (NOT POMSEE)
4. ☐ STOCK DEFECTIVE
5. ☐ REPAIR OF REPLACEABLE UNIT OR PLUG-IN ASSEMBLY
6. ☐ OTHER

4. TIME FAIL. OCCURRED OR MAINT. BEGAN
MONTH | DAY | YEAR | TIME

5. TIME FAIL. CLEARED OR MAINT. COMPL.
MONTH | DAY | YEAR | TIME

2. REPAIRED OR REPORTED BY
NAME | DATE | AFFILIATION
1. ☐ U.S. NAVY 2. ☐ CONTRACTOR 3. ☐ CIVIL SERVICE

EQUIPMENT

6. MODEL TYPE DESIGNATION

7. EQUIP. SERIAL NO.

8. CONTRACTOR (NAVY CODE OR COMPLETE NAME)

9. FIRST INDICATION OF TROUBLE (CHECK ONE)
1. ☐ INOPERATIVE
2. ☐ OUT OF TOLERANCE, LOW
3. ☐ OUT OF TOLERANCE, HIGH
4. ☐ INTERMITTENT OPERATION
5. ☐ UNSTABLE OPERATION
6. ☐ NOISE OR VIBRATION
7. ☐ OVERHEATING
8. ☐ VISUAL DEFECT
9. ☐ OTHER, EXPLAIN

10. OPERATIONAL CONDITION (CHECK ONE)
1. ☐ OUT OF SERVICE
2. ☐ OPERATING AT REDUCED CAPACITY
3. ☐ UNAFFECTED

11. TIME METER READING
A. HIGH VOLTAGE
B. FILAMENT /ELAPSED

12. REPAIR TIME MAN-HOURS | TENTHS

REPLACEMENT DATA

13. LOWEST DESIGNATED UNIT (U) OR SUB-ASSEMBLY (SA)	14. LOWEST DES. U/SA SERIAL NO.	15. REFERENCE DESIGNATION (V-101, C-14, R111, ETC.)	16. FEDERAL STOCK NUMBER OF REMOVED ITEM	17. MFR. OF REMOVED ITEM	18. TYPE FAILURE	19. PRIMARY OR SECOND-ARY FAIL.?	20. CAUSE OF FAILURE	21. DISPOSITION OF REMOVED ITEM	22. REPL. AVAILABLE LOCALLY?
						☐P ☐S			☐Y ☐N
						☐P ☐S			☐Y ☐N
						☐P ☐S			☐Y ☐N
						☐P ☐S			☐Y ☐N
						☐P ☐S			☐Y ☐N

23. REPAIR TIME FACTORS
CODE	DAYS	HOURS	TENTHS	CODE	DAYS	HOURS	TENTHS

24. REMARKS
(CONTINUE ON REVERSE SIDE IF NECESSARY)

FIG. 6.1 Form DD-787, Naval Bureau of Ships.

recording a detailed description of whatever events occur. The sheet numbers are serialized. The amount of detail preprinted on a form, then, is related to the complexity of the operation for which the form is used and to the degree of prior knowledge of the content of the information. When the operation is

OPER	INSP	BUILDUP NO.		STAMP	
NO.	NO.	OPERATION DESCRIPTION		OPER	INSP
0220		Tighten thrust washer locknut (093073-1) to 90 to 100 in.-lb with the 095667-1 wrench and the 3/16 Allen wrench (see 0150E).			
		Note: Do not bend the tabs on the lockwasher (094079-1).			
	0225	Verify torque and record _____ in.-lb.		◯	
0230		Rotate motor housing to recirculation-end-up position.			
	0235A	(1) Measure dimension "A" of Figure 8 (see 0155B) with the shaft in the down position of maximum axial play, and record: _____ (2) Subtract 0155B measurement: _____ (3) Difference _____		◯	
	0235B	Calculate required thickness for the 093212-5 spacer: (1) Subtract 0.010 from difference recorded in operation (0235A(3)): _____ (2) Subtract this value (0235B(1)) from the spacer thickness (0145B): _____		◯	
0240A		Rotate the housing to the horizontal position.			
0240B		Remove recirculation pump spacer (093212-5) by removing the nut, washer, rpm generator rotor, impeller, and key.			
0240C		Grind spacer to required thickness: 0235B(2) ±0.001, parallel within 0.0005 TIR. S.O. _____			
0240D		Re-mark 093212-5 spacer with the buildup serial number per ASD 5215C. Record P/N (093200-9) and serial number on Parts List Requisition.			
	0245A	Measure 093200-9 spacer and record: _____ Limits: 0235B(2) ±0.001		◯	
	0245B	Measure parallelism: (0.0005 TIR): _____		◯	
0250		Tighten the 093073-1 locknut to 40 to 45 ft-lb with the 095667-1 wrench and the 095666-1 shaft support tool at the recirculation end.			
	0255A	Measure and record thrust-bearing total axial clearance (see 0195B Limits: 0.002 to 0.004).		◯	
	0255B	Verify torque and record: Preliminary _____ ft-lb Final _____ ft-lb		◯	
0260A		Bend tabs into locknut and spacer slots.			
0260B		Lockwire six 094525 bolts and three MS24673-3 screws per MS33540 with MS20995C32.			
0260C		Vacuum clean the assembly.			

Page 14

FIG. 6.2 Operation and inspection record.

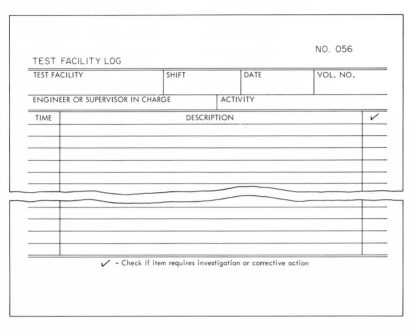

FIG. 6.3 Serialized logbook sheet.

complex and the content of the information to be recorded is known, the form can be quite detailed. Such detail helps to prevent errors and omissions. When the future events are unknown, the forms must be less specific and the data will require more review and verification.

Checklists are easy to complete and result in fewer errors and omissions than do fill-in-type forms. However, unless the operation is very repetitive, information from checklists may be incomplete and may require further investigation to clarify details. Fill-ins are more time-consuming and more subject to error but, when properly completed, often provide more information on circumstances and conditions which existed when the data were generated. The two types are frequently combined where a checklist is provided for certain items to ensure that they are recorded, and fill-in information is also required. Such combination forms are generally the most desirable because they can be designed to ensure the recording of all pertinent data with sufficient explanation to permit verification and classification of the information.

Finally, the number of copies is also dependent on the nature and distribution of the data. Failure reports are usually distributed to the Quality Assurance or Reliability organization and to cognizant design and test groups. Purchasing, Systems Engineering, Materials, and other support groups sometimes receive copies of applicable reports. Two disadvantages are inherent in large distributions of reports. First, some copies may be illegible, particularly when there are more than four. Second, those reports that may be important to a particular organization can be "lost in the shuffle" when copies of all reports (applicable or not) are received. Reproduction and distribution of additional copies can best be handled by the organization with primary responsibility for action. This organization can determine the proper distribution for reports and can ensure that copies are transmitted to other groups that have an interest. It is suggested that the original report be limited to four copies (or fewer if possible).

6.3.3 Personnel Training

Properly designed forms are necessary if accurate data are to be obtained in the most efficient manner, but they are only another step toward this end. The motivation and training of personnel who are going to complete the forms is perhaps the most important step. Motivation has been touched on briefly with respect to factors which discourage reporting of information. The principal means by which personnel are impressed with the importance of the data, and are motivated to provide them, is direct instruction from their supervisors and management. When program managers acknowledge the need for and importance of the data, the personnel responsible for the recording and transmittal of the data will have the inducement to complete these tasks. But if management displays a lack of interest, nothing short of an act of Providence will guarantee the information's accuracy and completeness. The adage that "the boss gets what he wants" is exemplified in the area of information collection and processing. The task of the reliability engineer is to provide meaningful reports which are useful to management. The preparation and dissemination of timely analyses and evaluations which can be used in formulating decisions is one of the best methods of securing management support for the information system.

In conjunction with the data collection program, line personnel can be given short training sessions. These can be limited to discussions of the data and their use, and to the means of collecting, analyzing, and distributing the information. The forms themselves should be explained to clarify the information that is wanted and to ensure that items are not misinterpreted. Sometimes films are used for training purposes, but this type of instruction is usually not suitable for training in data collection. An important aspect of the training is the conducting of brief, periodic reviews to answer any questions that may arise and to demonstrate a continued interest in the information.

In those programs where large quantities of data are being generated, or when the generation of data is a primary objective as in a research project, resident engineers or technicians are assigned to the field or test area to ensure the validity and transmittal of information. These technical personnel can be thoroughly trained in the applicable aspects of the data system and, because this is a principal function of their assignment, they are easily motivated in the desired direction. In addition to the training and motivation considerations, these personnel can make timely decisions with respect to unanticipated problems and can provide selective attention to developing those details which are most important to the evaluation. Data recorded and collected in this manner have a minimum of errors, and discrepancies can be readily detected and corrected.*

6.4 DATA SOURCES

The sources of data are many and varied. As shown in Fig. 6.4, they range all the way from the suppliers of raw materials and parts to the field support and operations activities. Sometimes even post-operations functions such as salvage can provide some useful information. The first sources, chronologically, are the suppliers of raw materials. The data on raw materials frequently prove quite helpful in later failure investigations, and failures have been traced to subtle differences in chemical composition, physical properties, and treatments such as plating. Raw materials certifications are commonly required by both commercial and military users. They should not always be taken for granted and accepted without question. A periodic check or verification by an in-house or an independent laboratory can provide assurance of the validity of such reports. Suppliers of parts and raw materials have occasionally been known to routinely make such certifications and provide chemical analyses by looking at the specification and completing the forms with acceptable entries without ever performing an actual materials analysis or requesting certification from the mill. When government source inspection is reduced or eliminated as it was in 1967 and 1968, the need for in-house verification increases.

Suppliers of standard parts constitute a second set of data sources. When parts are made in large quantities the manufacturer is best equipped to carry out the numerous performance, environmental, storage, and life tests which verify capability and longevity. Such tests are normally performed on MIL-STD and high-reliability parts. These parts are used in applications requiring this high reliability and have proved highly successful. Data relative to such parts are acceptable for use in reliability calculations, and are readily available. A partial list of data sources is provided in Appendix C. Source data relating to commercial parts are also useful when there is verification of their validity. Reputable suppliers provide verifiable information when periodic tests are routinely performed.

---

*This portion of the discussion is based in part on a more detailed discussion of the use of technical personnel and of the planning of a data system, ARINC Research Corporation, *op. cit.*, Sec. 4.2, by permission.

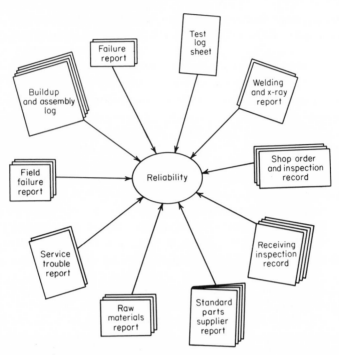

FIG. 6.4   Some typical data sources.

In-house data, from receiving inspection to final check-out, can provide a wealth of information applicable to reliability analyses. Much of the previous discussion pertains primarily to the collection and analysis of in-house data and to criteria for ensuring its applicability. The majority of forms will be used within the company, and it is seldom possible to impose desired information formats on suppliers and even less possible with respect to customers. Motivation and training activities are, of course, almost always limited to company personnel except in those instances when the company is required and funded to provide field support. The various data sources within the company include, in addition to receiving inspection and final check-out, assembly and in-process records such as operating and equipment logbooks, shop orders which contain inspection data, process records, laboratory reports of all types including non-destructive test records, and checks and inspections of subsystems and systems in all phases of the manufacturing process. It is not suggested that all data from all sources are necessary; in fact, selectivity must be exercised in choosing information which is pertinent, meaningful, and *useful* and in rejecting that which, although valid, is not needed. Prior specification of data sources saves much effort in the selection. Time is saved not only in the processing and analysis of the data but also in the training of personnel, development of forms, and monitoring of the data system. Perhaps even more important is that more useful reports result and management support is more easily obtained.

Field data include not only test and operational information but also maintenance, replacement, and failure reports. Most forms for recording field information are designed by the user for his own needs and may require considerable review and editing to determine and summarize applicable information. As indicated earlier, when field support personnel are involved they can usually be instructed in data recording procedures so that the data are most meaningful. Some programs associated with national defense have specific funding for information systems and provide large quantities of useful data.

6.5 APPLICABILITY

As indicated in the introductory paragraphs, applicability of the data, and particularly of the failures, is perhaps the most important consideration. Accuracy of the reliability calculations requires that the data be both valid and cogent.

6.5.1 Test Conditions

Conditions under which tests are performed have a profound effect on the usability of the data. The data are most applicable when generated in actual field operation and when such operation is under normally anticipated conditions. Tests run under conditions simulating field operation also yield very useful data. However, while temperature, pressure, humidity, voltage, power, and perhaps even dynamic loads of shock and vibration can be simulated, such conditions as absolute vacuum, zero gravity, and radiation which might be encountered in actual use cannot easily be reproduced in testing; and tests may not be truly representative of use conditions. Sometimes the effects of these additional parameters are known to be insignificant and the data can be used without modification. Other times, extensive modifications must be made before the data are usable. When such changes are required, the relationship of parameter values at operating conditions to values at test conditions must be known if the correction factors are to be valid.

Multiplication factors are available for converting failure data from controlled tests at ambient conditions to conditions encountered in various types of field operation. These factors for electronic parts are very definitive for such conditions as temperature, voltage, and power, but the multipliers are much less precise for dynamic loads and other types of environments. Consequently, the conditions existing when data are generated must be defined in detail to permit proper data evaluation and necessary modification.

It is desirable that the decision as to whether or not to include results of particular tests in a reliability analysis be made prior to actual test performance. When tests are exploratory in nature or are run at off-design conditions, the test data may be inapplicable with respect to some or all of the characteristics. Discussions with design and test engineers are recommended to designate which parameters will yield usable data and which will not. Operating times and failures resulting from tests at levels other than normal operation may or may not be applicable. If one were to wait until after a test was completed before deciding whether or not to include the data in the analysis, there could be a temptation to delete that data which "did not look right" even though no anomaly occurred during the testing which would invalidate the results. Of course, if some unforeseen incident occurred such as a gross operator error or major malfunction of test equipment which affected the test results, it would be perfectly legitimate to delete the results from the other data which were being analyzed.

Data from tests at all stages of component or system development as well as from field tests may be used, but the conclusions reached should be carefully derived and are frequently limited to the current status of the equipment. However, if no significant changes in design are made after the initial development, the data generated throughout the entire test program may be quite useful for reliability calculations made later in the program. Data generated prior to significant design changes, including both failure and success data, are usually not acceptable for use in final reliability calculations.

6.5.2 Criteria for Excluding Failures

Failures are generally excluded from reliability assessments when they arise from causes external to the equipment, on the basis of these criteria:

1. Failures resulting from human errors in setup or operation when such errors will not normally be experienced when the equipment is used by trained personnel. Principles of human engineering should be applied to equipment design and in the preparation of operating and maintenance manuals to minimize these errors.

2. Failures resulting from the inadequacy or malfunction of test equipment.

3. Subjection of a component or assembly to conditions or environments beyond its stated capability warrants exclusion of any resulting failure. Mishandling falls into this category. Careful evaluation and specification of all potential operating conditions should be accomplished prior to equipment design to reduce the possibility of occurrence of these failures.

4. Failures caused by unscheduled maintenance actions or adjustments when such activities are performed as a matter of convenience or expediency in place of routine, scheduled maintenance actions. However, it is usually necessary to verify that the failure was directly caused by the unscheduled

action and was not an incipient failure whose occurrence was only somewhat hastened by such action.

5. Secondary failures are likewise excluded. A secondary failure is one which results from the failure of some other item of equipment as distinguished from a primary failure which does not result from the failure of another item. For example, the shorting of a resistor (primary failure) may result in overloading a diode or transistor which then burns out (secondary failure). The resistor failure is counted while the semiconductor failure is not.

6.6 THE DATA PROGRAM

Mathematics may be compared to a mill of exquisite workmanship which grinds you stuff of any degree of fineness, but nevertheless what you get out depends on what you put in—and as the grandest mill in the world will not extract wheat flour from peapods so pages of formulae will not get a definite result out of loose data.

T. H. Huxley

The planning of an overall data program should include the elements of purpose, sources, evaluation, and criteria which have been discussed, and should also include a statement of objectives, a delineation of reports to be prepared, a list of forms and procedures, and an indication of any facilities and equipment that will be used in the recording and processing of the information. Care should be taken in defining the needs and uses of the various reports because much of the utilization is subjective and members of management must be presented with cogent arguments to ensure that they are fully aware of both the needs for and the benefits to be derived from a good data system. Such a system provides the foundation for effective utilization of reliability mathematics.

ADDITIONAL READINGS

ARINC Research Corporation: "Reliability Engineering," Prentice-Hall, Inc., Englewood Cliffs, N.J., 1964.

Calabro, S. R.: "Reliability Principles and Practices," McGraw-Hill Book Company, New York, 1962.

Ireson, W. G.: "Reliability Handbook," McGraw-Hill Book Company, New York, 1966.

Lloyd, D. K., and M. Lipow: "Reliability: Management, Methods, and Mathematics," Prentice-Hall, Inc., Englewood Cliffs, N.J., 1962.

Shewhart, W. A.: "Economic Control of Quality of Manufactured Product," D. Van Nostrand Company, Inc., Princeton, N.J., 1931.

7

LOGIC DIAGRAMS

7.1 NEED AND DESCRIPTION

Knowledge of the operational relationships of the various elements of a system is a prerequisite to the accomplishment of reliability activities. We cannot improve or even evaluate the reliability of a system unless we have a thorough understanding of how each of its components functions and how these functions affect system operation. Accurate representation of these relationships is an integral part of this understanding and is particularly needed for meaningful predictions, apportionments, and assessments. Logic diagrams provide this representation.

While the *system* diagram depicts the *physical* relationships of the system elements, the reliability block or *logic* diagram shows the *functional* relationships and indicates which elements must operate successfully for the system to accomplish its intended function. The logic diagram consists of groups of blocks or *elements* connected together in one or more series. A series may contain as few as one element or as many as several hundred. Each series depicts the means whereby a particular function or group of functions is performed by the system. In order for the function(s) to be completed successfully, no element in the series may fail. If *any* element does fail, the function cannot be performed by the particular series. The series can be thought of as a chain with each element as a link. The failure of any link constitutes failure of the chain. Figure 7.1 depicts a simple series consisting of two elements *A* and *B*. If *either A* or *B* fails, the series fails.

FIG. 7.1 Logic diagram of a simple series system of two elements.

To reduce the probability of a function failing, redundancy is often employed. That is, there is more than one series of elements which can perform the function. Every series must fail before the function fails; if only one series succeeds, the function can be performed. Figure 7.2 is a logic diagram which shows a function that can be performed in any of three ways, as indicated by three series in parallel. It should be noted that the number of elements in each series need not be the same as long as each series represents all requirements for performing the same function. Furthermore, the function which is performed may be the simple action of a switch which opens or closes a circuit or may be a very complex activity such as the guidance of a spacecraft.

The first step in preparing a logic diagram is to determine the way or ways in which the system operates or the function is performed in the intended mission. The parts, components, and assemblies which are necessary to successful completion of the mission are also determined.

The various possible modes of failure of each part or component are ascertained next. The effects of each of these failure modes on the mission or function being depicted are found. When a failure-modes-and-effects analysis (FMEA; see Secs. 10.1.1 and 10.1.2) which delineates each mode and its effects has been made, it can be used for these determinations. (The FMEA is

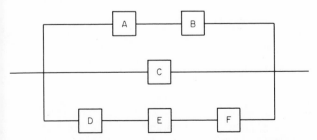

FIG. 7.2 Logic diagram of three series in a parallel configuration.

quite useful by itself and provides information that can be used to advantage in improving system reliability.)

Each mode of failure which, when occurring, will by itself result in loss or premature termination of the mission or function is shown as a series element in the related logic diagram. This depiction as a series element is independent of the actual system configuration. Frequently there are components which are in parallel in the system schematic diagram which have failure modes that cause loss of the function or mission, and these modes are therefore shown in series in the logic diagram.

Failure modes which, when occurring, do not result in function or mission loss are shown as parallel elements in the logic diagram. If there is an alternate means of performing the function or mission, then this alternate method is shown in parallel. Just as there are parallel components which have series failure modes, there may be components which are in series in the system schematic diagram which have failure modes that are shown in parallel in the logic diagram because there are alternate methods of performance.

It should be pointed out that an element may be in series with respect to a function and, at the same time, may be in parallel with respect to a mission. This occurs when the element is necessary for the function, but either the function is not vital to the mission or there are alternate means available for its performance.

7.2 EXAMPLES

The distinction between physical and functional relationships is an important one, and it can perhaps best be described by providing some examples.

Example 1. One portion of a fluid system physically consists of a pump and two check valves in series. The series check valves provide redundancy against flow in the reverse direction when the pump is not operating and the downstream pressure exceeds the upstream pressure. The system diagram depicting this portion of the system is shown in Fig. 7.3. However, the *logic* diagram (using circuit symbols) for this portion of the system would be drawn as shown in Fig. 7.4*a*.

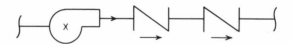

FIG. 7.3 System diagram of a pump and two check valves in series.

Actual logic diagrams use blocks as in Figs. 7.1 and 7.2 to represent the various elements or modes of failure, with notations indicating the element or mode represented by each block. Figure 7.4*b* shows the logic block diagram for the system diagram in Fig. 7.3. In this chapter, circuit symbols are used in

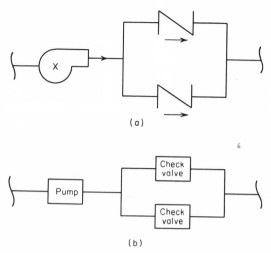

(a)

(b)

FIG. 7.4   Logic diagram for the system in Fig. 7.3 (*a*) using circuit symbols; (*b*) using blocks.

logic diagrams, when appropriate, to provide clarity; it should be remembered that the standard practice is to depict all elements as blocks and to provide adequate notations. These figures show that either check valve will perform the required function and that both valves must fail for reverse flow to occur. (We are assuming that the usual mode of failure is reverse leakage, and failing in the closed position—when forward flow is required—is not considered here.)

Example 2A. Because no single capacitor of sufficient capacitance is available, an electrical system contains a bank of 10 capacitors wired together in parallel as shown in Fig. 7.5. If we consider the primary mode of failure to be shorting, then the failure of any one capacitor will fail this portion of the system. Hence, the *logic* diagram (using circuit symbols), Fig. 7.6, depicts the 10 capacitors in series.

Example 2B. To reduce the probability of a system failure due to capacitor shorting, the designer incorporates a fuse in series with each capacitor. If a capacitor shorts out, excessive current will occur and the fuse will open. (It is assumed that sufficient capacitance will still be available if

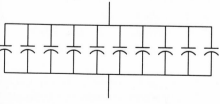

FIG. 7.5   System diagram of 10 capacitors in parallel.

not more than two capacitors or fuses open. For purposes of this example, the probability of more than two opening and thus failing the system because of insufficient total capacitance is negligible, and failure due to opening is not considered.) The system diagram of the fused capacitor bank is shown in Fig.

FIG. 7.6   Logic diagram for the system in Fig. 7.5.

7.7. However, with respect to reliability logic, each fuse provides redundancy for the associated capacitor for the shorting failure mode. The shorting of *either* a capacitor (if the fuse opens) *or* a fuse (if the capacitor does not short) can occur without causing system failure, while shorting of a capacitor

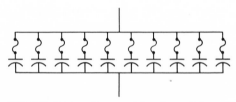

FIG. 7.7   System diagram of fused, parallel capacitors.

together with failure of the corresponding fuse to open results in shorting the entire circuit. Thus, there are 10 pairs of redundant fuses and capacitors. The 10 pairs are depicted in series in the logic diagram in Fig. 7.8.

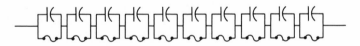

FIG. 7.8   Logic diagram for the system in Fig. 7.6.

Example 3. One particular multistage turbine* incorporates the use of three seals in its first stage. The seals prevent gas leakage around the second wheel or to the outside of the turbine case. The first (primary) seal is designed to prevent *any* leakage outside the wheel area. This seal is backed up by two additional seals. In case of failure of the first seal, one backup seal prevents leakage around the next wheel while the second backup seal prevents leakage outside the case. Figure 7.9 depicts this portion of the system design.

---

*A multistage turbine is one which contains more than one turbine wheel. Gas exiting from the first-stage wheel is directed, through a nozzle-diaphragm assembly, against the second-stage wheel, etc.

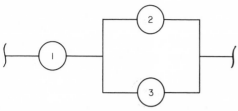

FIG. 7.9   System diagram of primary turbine seal and two backup seals.

Seal number 1 is the primary seal, and numbers 2 and 3 are the backup seals. The corresponding logic diagram, Fig. 7.10, shows two parallel paths with seals 2 and 3 in series, paralleling seal 1. The interpretation of this logic diagram is that either seal 1 alone or seals 2 and 3 together can perform the required sealing function.

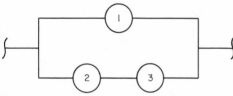

FIG. 7.10   Logic diagram for turbine seals in Fig. 7.9.

7.3   LOGIC DIAGRAMS FOR REPRESENTATIVE SYSTEMS

We shall consider next some typical systems and their associated logic diagrams and shall then review some other configurations which cannot be depicted logically with absolute accuracy but which can be mathematically defined. Approximate logic diagrams with suitable explanatory notes are used for the latter cases.

The simplest system or subsystem is one in which all components are in series and the failure of any one component results in failure of the system. Figure 7.11 depicts the logic diagram for such a system (as did Fig. 7.1). An electrical generator might be a typical example. It could consist of a turbine

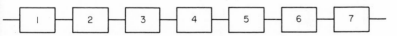

FIG. 7.11   Logic diagram of a simple series system of seven elements.

(which itself may consist of several series parts), a rotor, stator, bearing assemblies, lubricating system, and frequency and voltage controls. If any of these components fails, the generator fails.

Another simple system would be one which consisted of two or more components in parallel, all of which operate together but any one of which is sufficient to perform the function. An example would be the dereefing cutters on a spacecraft parachute. After the chute is released from its hatch on a spacecraft during the descent phase, the chute is kept in the reefed (closed) position until the proper altitude is reached. The line which prevents it from opening is then cut automatically. There are usually two or more cutters so that the failure of one cutter will not prevent the chute from opening. Each cutter consists of a cutter blade, explosive charge, housing, and mounting. The programmer and electrical system which activate the charge are not considered as part of the dereefing system. Assuming a three-cutter system, the logic diagram is depicted in Fig. 7.12.

The next step in complexity of a logic diagram is one for a simple series-parallel system as was depicted in Figs. 7.4 and 7.8. In each of these diagrams, the parallel elements are in groups of two, and each group is in series with another element (Fig. 7.4) or group of elements (Fig. 7.8). Figure 7.10 is only slightly more complex, having more than one element in one of the two parallel legs. In each of these three logic diagrams, the element(s) in any one leg is (are) sufficient to perform the required function. (Remember that, in Fig. 7.8, adequate capacitance is assumed, and the function being depicted is protection against shorting.)

Proceeding another step in complexity, we will briefly examine a combination of series and parallel elements. Suppose that a capacitor bank similar to that in Example 2A is in one part of an automatic control system and that this part of the automatic system is backed up by a manual subsystem. The logic diagram might appear as in Fig. 7.13. (We will consider only three capacitor-fuse groups for ease of illustration.)

In actual operation, the ammeter, voltmeter, and frequency meter are correctly inserted periodically into the system to provide information on current, voltage, and frequency. When the system operator has reason to question the automatic controls, he can check the instruments. If they confirm a failure in the automatic controls, the operator activates the switch

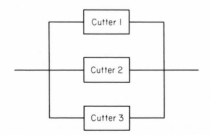

FIG. 7.12 Logic diagram of three-cutter system (one required).

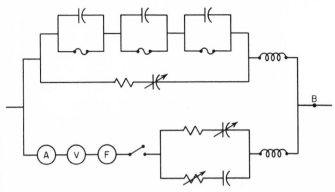

FIG. 7.13 Logic diagram of parallel automatic and manual control
system.

and manually controls resistance and capacitance to provide correct electrical outputs. Regardless of actual circuitry, in the *logic* diagram, the meters and switch are drawn in series in a leg parallel to the automatic controls, showing that they form a redundant means for controlling the system. For instance, even though the ammeter when used might be connected in series at point *B* (beyond the junction of the two parallel legs), the logic diagram shows it in the parallel leg because (1) failure of the ammeter does not fail the automatic control system, and (2) the ammeter is used in conjunction with the manual controls to set the proper operating values.

Another form of complexity involves different failure modes of a part or component. When redundant parts are used, they are generally connected to provide protection against the most prevalent mode of failure. In Example 1, the two check valves are in series in the system (Fig. 7.3) to provide more protection against reverse fluid flow. They were shown in parallel in the logic diagram (Fig. 7.4) since they are redundant for this mode of failure. However, there is a very small probability of sticking or leaking externally and this probability is doubled when there are two valves.* The complete logic diagram for one leg of the system is shown in Fig. 7.14. The notations provide clarity about the modes of failure being depicted, and they are important parts of complex logic diagrams.

Another example of the type of complexity where different logic configurations depict different modes of failure is taken from a frequently

---

*When the probability of failure in the series modes is less than half the total failure probability, this type of redundancy increases overall reliability in almost all cases. Very rarely, when the failure rate is very high and when the probabilities of failure of the two modes are about the same, the overall reliability will be adversely affected.

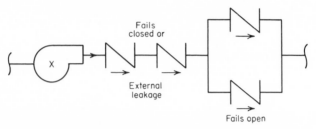

FIG. 7.14   Logic diagram for the system in Fig. 7.3 showing all modes of failure.

used electronic configuration. When high reliabilities are required, diodes are arranged in groups of four, commonly called *quads,* to provide protection against both shorting and opening. Two similar quad configurations are employed, the choice depending on the relative probabilities of occurrence of the two failure modes and the effects of a circuit failure. The only difference between the two configurations is the presence or absence of a connection across the center of the group. If there is no connection (Fig. 7.15*a*), there is more protection against a short than against an open; if a connection is made (Fig. 7.15*b*), more protection is provided against an open.

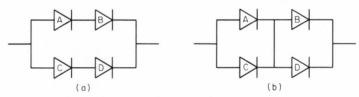

FIG. 7.15   System diagrams of diodes in "quad" configurations (*a*) without cross connection; (*b*) with cross connection.

In circuit 7.15*a*, there are two combinations where the shorting of two diodes will short out the group: *A* and *B* together, or *C* and *D* together. However, there are four ways that the opening of two diodes will open the circuit: *A* and *C*; *B* and *C*; *A* and *D*; or *B* and *D*. In circuit 7.15*b*, the reverse is true: there are four ways of shorting and only two ways of getting an open quad with two diode failures. The corresponding logic diagrams are modifications of the combination of the two possible system configurations with appropriate notations regarding the mode of failure being depicted. It is important to remember that, in a logic diagram, the failure of any element in a chain results in failure of the chain.* Also, unless otherwise noted, when

*The terms "chain," "leg," and "path" are used interchangeably.

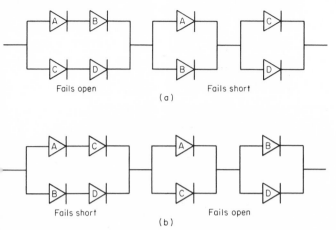

Fails open   Fails short
(a)

Fails short   Fails open
(b)

FIG. 7.16   Logic diagrams of diode quad configurations (a) for Fig. 7.15a (no cross connection); (b) for Fig. 7.15b (with connection).

there is more than one chain in parallel, the function can be performed as long as at least one of the parallel chains remains unfailed.

Figure 7.16a shows the logic diagram associated with configuration 7.15a. Note that the diodes B and C are shown in different positions on the "fails short" side of the logic diagram, indicating that the shorting of A and B together or of C and D together shorts the circuit. The "fails open" side of the logic diagram shows that the opening of either diode in a parallel leg fails that leg, and therefore there are four combinations of two open diodes that open the circuit.

The logic for configuration 7.15b is shown in Fig. 7.16b. Opening of diode pair A and C or of pair B and D opens the circuit (since all parallel legs in a group will have failed) and the right side of the logic diagram depicts this. Either A or C shorting or B or D shorting shorts out a part of the circuit and the left side depicts this, showing four possible combinations of two shorted diodes shorting the entire circuit.

Some of the complexities of logic diagrams have been indicated. Complete system logics combine many such diagrams into an overall pictorial description of the functions of the system. Separate logic diagrams are frequently prepared for each phase of a mission and define the essential elements of the phase.

7.4 SOME APPROXIMATIONS

Some functional relationships cannot be pictorially defined with absolute correctness, and the logic diagram is an approximate description of the

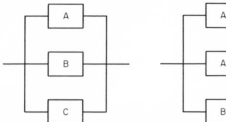

FIG. 7.17 Logic diagram of three elements in parallel.

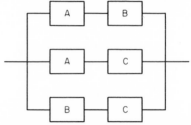

FIG. 7.18 One type of logic diagram for a two-out-of-three system.

system. For instance, the logic diagram in Fig. 7.17 shows that any one of three redundant elements *A, B,* or *C* is capable by itself of performing the function. But what if it requires any *two* of the three elements to do the job? (For example, an electrical system might contain three batteries, but two are sufficient to provide required power. All three are connected together, allowing one to fail without affecting system performance.) Two useful approximations may be employed to represent this situation.

The diagram in Fig. 7.18 shows that any pair of elements (*A* and *B, A* and *C,* or *B* and *C*) form a success path; but it can be misinterpreted to imply that there are two each of *A, B,* and *C* rather than only one.

A second possibility is given by Fig. 7.19. The notation beneath the diagram is used to clarify the actual performance requirement. Either diagram is acceptable if properly annotated.

Another actual situation which cannot be pictured with complete accuracy is illustrated by a parachute system with two reefing lines. In order to reduce the probability of premature dereefing, redundant lines are sometimes used so that accidental cutting of one line will not open the chute. The system diagram of the cutters is shown in Fig. 7.20.

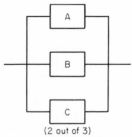

(2 out of 3)

FIG. 7.19 A second type of logic diagram for a two-out-of-three system.

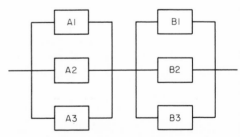

FIG. 7.20 System diagram of three cutters in parallel for each of two reefing lines.

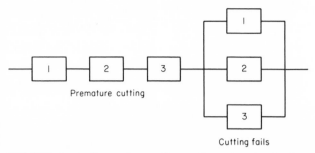

FIG. 7.21   Logic diagram for one three-cutter system showing two modes of failure.

Figure 7.21 shows the dereefing logic diagram for both failure modes for one cutter. (Figure 7.12 depicted only one mode of failure—failing to cut.)

We now wish to combine the two failure modes for two lines A and B. Figure 7.22 shows that premature cutting is now redundant and that both lines must be cut for the chute to open. However, it indicates that even if one line is cut prematurely, the line must again be cut at the proper time. Actually, if line A is accidentally cut early, then it need not be cut again, and the first group of three redundant blocks ($A1$, $A2$, and $A3$ in parallel) is extraneous. The same is true, of course, for line B and the second group of three blocks in parallel if this line is cut accidentally. The diagrams in Figs. 7.23 and 7.24 have the more serious disadvantage of stating pictorially that the proper cutting of either line satisfies the function, which is not true. In this situation, the best diagram to use is the first one (Fig. 7.22) with adequate notation.

Still another example of a modification which is necessarily employed relates to the independent depiction of each phase of a mission. Suppose that there are two batteries in a system and that both are required to supply sufficient power for all operations in a late phase of the mission but only one

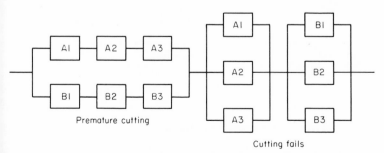

FIG. 7.22   One type of logic diagram for the redundant cutter system in Fig. 7.20.

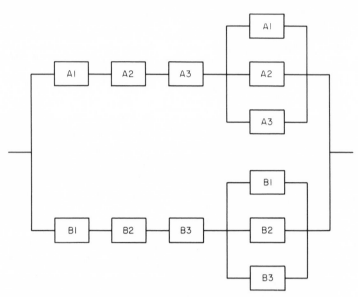

FIG. 7.23   A second type of logic diagram for the system in Fig. 7.20.

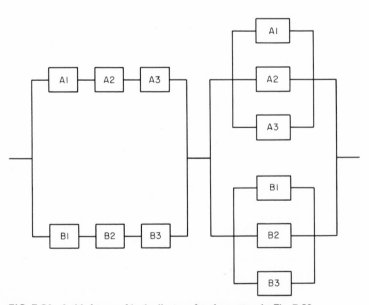

FIG. 7.24   A third type of logic diagram for the system in Fig. 7.20.

battery is needed during an earlier phase. The segment of the logic diagram for the batteries in the later phase is shown in Fig. 7.25. But what is it for the early phase? The true logic is shown in Fig. 7.26, since either battery can provide sufficient power at that time. However, if one fails, the mission

FIG. 7.25   Logic diagram for batteries in late mission phase.

ultimately fails at a later time because of insufficient battery power. Therefore, it is preferred practice to show the batteries in series in the earlier phase also, even though only one is actually required. This modification, of course, does not result from any deficiency in logic diagrams but from the desire to consider later mission requirements.

The sequential operation of redundant components will serve as a last example of approximations used in reliability logic diagrams. There is no completely correct way, without using descriptive notations, of depicting the case of two parallel legs when only one is operating and the second is on standby.* Figure 7.13 is one of the more correct illustrations of this type of logic. However, a thorough analysis will disclose particular elements which are not correctly depicted; e.g., the meters are used for checking both the automatic and manual controls but are used only occasionally and are not necessary if the automatic controls are working correctly. Again, adequate notation will greatly reduce the possibility of misinterpretation.

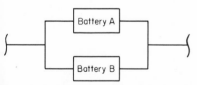

FIG. 7.26   True logic diagram for batteries in early mission phase.

The fact that 100 percent correct logic diagrams may not be practical should not deter one from constructing the best possible diagram. As stated in the opening paragraph, the overall process of preparing the logic diagram results in gaining a necessary knowledge of the system operation; and the properly annotated logic diagram provides a very useful description of the functional relationships of the system elements. Effects of proposed system

*This situation, which is quite common, is known as *sequential redundancy*. When all parallel paths operate at the same time, the term *active redundancy* is used.

modifications or changes can be readily determined if good logic diagrams are available for reference. Trade-off studies involving such factors as reliability, cost, weight, and schedules require this capability in order to arrive at rational, quantitative decisions.

ADDITIONAL READINGS

ARINC Research Corporation: "Reliability Engineering," Prentice-Hall, Inc., Englewood Cliffs, N.J., 1964.

Bazovsky, I.: "Reliability Theory and Practice," Prentice-Hall, Inc., Englewood Cliffs, N.J., 1961.

Bureau of Naval Weapons, Chief of: "Handbook: Reliability Engineering," U.S. Government Printing Office, 1964.

Hiltz, P. A., and J. L. Gaffney: "Statistical Techniques for Reliability," North American Aviation, Inc., Downey, Calif., 1965.

Lloyd, D. K., and M. Lipow: "Reliability: Management, Methods, and Mathematics," Prentice-Hall, Inc., Englewood Cliffs, N.J., 1962.

8

MATHEMATICAL MODELS

8.1 SERIES SYSTEMS

The logic diagram defines the functional relationships of the system elements pictorially; the mathematical model expresses the relationships analytically. The math model states, in terms of individual element failure rates and reliabilities, the probability of mission success. The model is usually developed from the logic diagram although, for simple systems, it is sometimes possible to prepare the model directly from knowledge of component relationships. Knowledge of quantitative reliability values is not necessary since the model provides algebraic relationships.

Although there are some cases where the logic diagram approximates a relationship, the math model almost always can define it exactly. Only in extremely complex systems does it become impossible to develop an exact mathematical model. The difficulty arises not from any deficiency in the mathematics but from the complexity of the system. The electrical subsystem of a manned spacecraft, for instance, involved more than 2,000 separate success paths for just one phase of the mission, and a complete, exact spacecraft-mission model was not possible.

This chapter shows how mathematical models are prepared and presents the models for commonly encountered relationships. Models for some of the logic diagrams discussed in Chap. 7 are then developed. Later chapters depict the utilization of math models in predictions and assessments of system reliability. The Methods of Bounds for reliability prediction, discussed in Chap. 10, explains how logic diagrams and mathematical models are used to determine upper and lower limits of predicted reliability when the system complexity precludes development of an exact model.

8.1.1 Reliability

The mathematical model for a simple series logic diagram is the product of the individual reliabilities. Thus, for a logic diagram such as that depicted in Fig. 7.11, the mathematical model is given by Eq. (8.1). In this and subsequent mathematical models, it is assumed that each element is *independent*—that the success or failure of one element does not affect the success or failure of any other element.

$$R_T = R_1 \times R_2 \times R_3 \times \cdots \times R_i \times \cdots \times R_n \qquad (8.1)$$

where R_T = the total reliability
R_1 = the reliability of the first unit
R_i = the reliability of the ith unit
R_n = the reliability of the nth (last) unit

This can be written in a more condensed form, Eq. (8.1a),

$$R_T = \prod_1^n R_i \qquad (8.1a)$$

where $\prod_1^n R_i$ means the product of all R_i from 1 through n.

As explained in Sec. 3.4, the exponential equation $R = e^{-\lambda t}$ is most often applicable to the reliability of components. We can represent the product of the constant failure rate λ_i and the particular mission operating time t_i for a component by the single uppercase letter F_i: $R_i = e^{-F_i}$. When individual element reliabilities are computed using this equation, the product of the reliabilities can be found more easily by summing the mission failure rates and raising e to the negative power of the sum. Since any $R_i = e^{-F_i}$, then specifically $R_1 = e^{-F_1}$, $R_2 = e^{-F_2}$, etc., and

$$R_1 \times R_2 \times R_3 \times \cdots \times R_i \times \cdots \times R_n$$

$$= e^{-F_1} \times e^{-F_2} \times e^{-F_3} \times \cdots \times e^{-F_i} \times \cdots \times e^{-F_n}$$

$$R_T = e^{-(F_1 + F_2 + F_3 + \cdots + F_i + \cdots + F_n)}$$

or

$$R_T = \exp\left(-\sum_{i=1}^{n} F_i\right) \tag{8.2}$$

When $t_1 = t_2 = \cdots = t_n$, then Eq. (8.2) can be written as

$$R_T = \exp\left(-t \sum_{i=1}^{n} \lambda_i\right) \tag{8.2a}$$

8.1.2 Mean Time to Failure

It was stated without proof in Sec. 3.4 that, in a constant-failure-rate series system, the mean time to failure m was the reciprocal of the system failure rate λ. We shall now show why this is true. A rather elaborate proof is required.

The reliability of a component is its probability of survival. If a large number of components are put on test, then the reliability at any time t is equal to the ratio of the number of units still operating at that time (surviving) N_S, divided by the intitial total number N_T, as given by Eq. (8.3):

$$R(t) = \frac{N_S}{N_T} \tag{8.3}$$

Since the number of units still operating N_S is equal to the total number N_T minus the number which have failed N_F, Eq. (8.3) may be rewritten as

$$R(t) = \frac{N_S}{N_T} = \frac{N_T - N_F}{N_T} = 1 - \frac{N_F}{N_T} \tag{8.4}$$

The total number of units N_T is constant while the number which have failed increases with time. We apply calculus and differentiate Eq. (8.4) as follows:

$$\frac{dR}{dt} = \frac{d(1 - N_F/N_T)}{dt} = -\frac{1}{N_T}\frac{dN_F}{dt} \tag{8.5}$$

By rearranging the terms in Eq. (8.5) we obtain

$$\frac{1}{N_T}\frac{dN_F}{dt} = -\frac{dR}{dt} \tag{8.6}$$

In Eq. (8.6), dN_F/dt is the frequency at which failures occur when the total number of units N_T remains constant (no replacement of failed units). When plotted on a graph as a function of time t, we obtain the time distribution of failures. If we divide dN_F/dt by the initial total number of units N_T, we obtain the distribution of failures, or failure frequency curve, *per component*. Such a unit failure distribution curve is called a *failure density function*, or simply $f(t)$.* Substituting this term into Eq. (8.6), we obtain

$$f(t) = -\frac{dR}{dt} \tag{8.7}$$

Equation (8.7) applies to all possible failure density functions and not just to the case of a constant failure rate.

Now, the mean time to failure m, like any mean value, is the first moment about the origin of the parameter being considered. In this case it is the average time at which failure occurs and can be found by operating all units to failure, summing the times to failure, and dividing by the number of units.

$$m = \frac{\sum\limits_1^n \text{times to failure}}{n} = \frac{\sum\limits_1^n t_i N_{F_i}}{N_T} \tag{8.8}$$

where $\Sigma N_{F_i} = N_T$. Equation (8.8) defines the average time to failure for one component. In the limit, as the number of units becomes infinitely large, the summation process becomes an integration process. Also, the distribution of failures per component N_{F_i}/N_T becomes defined by the failure density function $(dN_F/dt)/N_T$, and m becomes the integral of the product of the density function and time:

$$m = \int_0^\infty t\, f(t)\, dt \tag{8.9}$$

Equation (8.9), like Eq. (8.7), is applicable to all possible density functions. Since $f(t) = -dR/dt$ [from Eq. (8.7)], Eq. (8.9) can be rewritten

$$m = \int_0^\infty t\left(-\frac{dR}{dt}\right) dt = \int_0^\infty -t\, dR \tag{8.10}$$

*The development of this approach is credited to Igor Bazovsky, "Reliability Theory and Practice," and is presented here with the kind permission of the author and his publisher, Prentice-Hall, Inc., Englewood Cliffs, N.J.

Integrating Eq. (8.10) by parts,* we obtain

$$m = -[tR]_0^\infty + \int_0^\infty R\,dt \tag{8.11}$$

We can show that the first term on the right side of Eq. (8.11) is zero. (A detailed proof is provided in Appendix A.3.) Therefore, Eq. (8.11) reduces to

$$m = \int_0^\infty R\,dt \tag{8.12}$$

This means that, in the general case, the mean time to failure can be obtained by integrating the reliability function over the time interval 0 to ∞. This is an important point as we shall see in succeeding sections of this chapter when we are computing the mean time to failure of redundant systems.

Returning to the specific case of a constant-failure-rate component (from Appendix A.3),

$$R = \exp\left(-\int_0^t \lambda_i\,dt\right) = \exp\left(-\lambda_i \int_0^t dt\right) = \exp(-\lambda_i t)$$

which, of course, we already know. From Eq. (8.12),

$$m = \int_0^\infty R\,dt = \int_0^\infty e^{-\lambda_i t}\,dt = \int_0^\infty e^{-\lambda_i t}\frac{-\lambda_i}{-\lambda_i}\,dt$$

$$= -\frac{1}{\lambda_i}\int_0^\infty e^{-\lambda_i t}\,d(-\lambda_i t) = -\frac{1}{\lambda_i}\left[e^{-\lambda_i t}\right]_0^\infty$$

$$= -\frac{1}{\lambda_i}[e^{-\infty} - e^0] = -\frac{1}{\lambda_i}(0 - 1) = \frac{1}{\lambda_i} \tag{8.13}$$

---

*Let $-t = u$ and $dR = dv$. Then $du = -dt$ and $v = R$.

$$\int u\,dv = uv - \int v\,du \qquad \int -t\,dR = -tR - \int R(-dt) = -tR + \int R\,dt$$

In a series system, λ is merely the sum of the failure rates of the individual components, so that

$$m = \frac{1}{\lambda_1 + \lambda_2 + \cdots + \lambda_n} = \frac{1}{\lambda} \tag{8.13a}$$

8.2 ACTIVE PARALLEL SYSTEMS

Up to this point, no accounting was made for the difference between base (generic) failure rates and application failure rates. While we have not yet discussed the difference (which is explained in Chap. 9), it is well to briefly mention that the application failure rate is derived from the basic failure rate by considering environments, utilization, derating, and other factors. As long as these factors are the same for all units, *in a series system,* the system reliability may be computed in either of two ways. It makes no difference whether the component base rates are used to obtain a base failure rate for the system and the factors are then applied to the overall system failure rate, or whether the application failure rate is first computed for each unit and then the system application rate is determined. The results will be the same.

However, this is not true for a system which contains any redundancy. *When there is redundancy, the application rates for the components must be determined first.* This requirement applies to failure rates, failure probabilities, and reliabilities. If base rates were used and then the system base rate were corrected afterward, serious errors in the overall system reliability and mean time to failure would result. *Therefore, all discussions which follow assume the use of application rates.* The reader may wish to verify the difference in results by assuming any component failure rate and application factor (other than unity) and calculating the resulting system reliability using the two different methods. An example is shown in the footnote in Sec. 9.3.

8.2.1 Two-unit Parallel System

In a simple parallel system consisting of two units in parallel with both operating but only one required, both must fail for the system to fail. The probability of the first unit failing is Q_1 which is equal to $1 - R_1 = 1 - e^{-F_1}$,* and the probability of the second unit failing is $Q_2 = 1 - e^{-F_2}$. The probability of both failing is

---

*The actual failure rate when both units are operating might be slightly less than the rate after one unit failed because of somewhat lower stress levels (e.g., current) in the former situation. However, for practicality, this slight difference is almost always neglected and one failure rate is used for the operating state.

$$Q_1 \times Q_2 = \left(1 - e^{-F_1}\right) \times \left(1 - e^{-F_2}\right)$$

$$= 1 - e^{-F_1} - e^{-F_2} + \left(e^{-F_1} \times e^{-F_2}\right)$$

This is better expressed as

$$1 - e^{-F_1} - e^{-F_2} + e^{-(F_1 + F_2)} \tag{8.14}$$

The reliability is equal to one minus the probability of failure, or

$$R_T = e^{-F_1} + e^{-F_2} - e^{-(F_1 + F_2)} \tag{8.15}$$

In terms of the reliability of the individual units, the overall reliability can be expressed as

$$R_T = R_1 + R_2 - R_1 R_2 \tag{8.16}$$

An alternate form, derived directly from the fact that the system fails only when both units fail, is given by

$$R_T = 1 - (Q_1 \times Q_2) \tag{8.17}$$

When both units are the same, Eqs. (8.15), (8.16), and (8.17) simplify as follows:

$$R_T = 2e^{-F} - e^{-2F} \tag{8.15a}$$

$$R_T = 2R - R^2 \tag{8.16a}$$

$$R_T = 1 - Q^2 \tag{8.17a}$$

The mean time to failure (MTTF)* for a two-unit parallel configuration is found by integrating the appropriate reliability function. Equation (8.15) is utilized in the integration, but in the form using λt rather than F.

$$m = \int_0^\infty R_T \, dt = \int_0^\infty \left(e^{-\lambda_1 t} + e^{-\lambda_2 t} - e^{-(\lambda_1 + \lambda_2)t}\right) dt \tag{8.18}$$

We can integrate each term independently. Since the failure rate for each

*MTTF and m are used interchangeably as abbreviations for mean time to failure.

individual component is constant, we have

$$
m = \int_0^\infty e^{-\lambda_1 t} \times \frac{-\lambda_1}{-\lambda_1} \, dt + \int_0^\infty e^{-\lambda_2 t} \times \frac{-\lambda_2}{-\lambda_2} \, dt
$$

$$
- \int_0^\infty e^{-(\lambda_1 + \lambda_2)t} \times \frac{-(\lambda_1 + \lambda_2)}{-(\lambda_1 + \lambda_2)} \, dt
$$

$$
= -\frac{1}{\lambda_1} \int_0^\infty e^{-\lambda_1 t} d(-\lambda_1 t) - \frac{1}{\lambda_2} \int_0^\infty e^{-\lambda_2 t} d(-\lambda_2 t)
$$

$$
+ \frac{1}{(\lambda_1 + \lambda_2)} \int_0^\infty e^{-(\lambda_1 + \lambda_2)t} d(-\lambda_1 + \lambda_2) t
$$

$$
= -\frac{1}{\lambda_1} \left[e^{-\lambda_1 t} \right]_0^\infty - \frac{1}{\lambda_2} \left[e^{-\lambda_2 t} \right]_0^\infty + \frac{1}{\lambda_1 + \lambda_2} \left[e^{-(\lambda_1 + \lambda_2)t} \right]_0^\infty
$$

$$
= -\frac{1}{\lambda_1}(0 - 1) - \frac{1}{\lambda_2}(0 - 1) + \frac{1}{\lambda_1 + \lambda_2}(0 - 1)
$$

$$
= \frac{1}{\lambda_1} + \frac{1}{\lambda_2} - \frac{1}{\lambda_1 + \lambda_2} \tag{8.19}
$$

Now that the general procedure for evaluating an integral has been presented in detail, some intermediate steps will be omitted from future calculations of m.

In the case where $\lambda_1 = \lambda_2$, Eq. (8.19) reduces to

$$
m = \frac{1}{\lambda} + \frac{1}{\lambda} - \frac{1}{2\lambda} = \frac{3}{2\lambda} \tag{8.19a}
$$

8.2.2 Three-unit Parallel System, One Required

When there are three units in parallel and only one is required as in Fig. 7.12, all three must fail for the system to fail, and the probability of system failure is

$$
Q_1 \times Q_2 \times Q_3 = \left(1 - e^{-F_1}\right) \times \left(1 - e^{-F_2}\right) \times \left(1 - e^{-F_3}\right)
$$

This expands to

$$1 - e^{-F_1} - e^{-F_2} - e^{-F_3} + \left(e^{-F_1} \times e^{-F_2}\right) + \left(e^{-F_1} \times e^{-F_3}\right)$$

$$+ \left(e^{-F_2} \times e^{-F_3}\right) - \left(e^{-F_1} \times e^{-F_2} \times e^{-F_3}\right)$$

The reliability is, of course, one minus the failure probability and can be written

$$R_T = e^{-F_1} + e^{-F_2} + e^{-F_3} - e^{-(F_1 + F_2)} - e^{-(F_1 + F_3)}$$
$$- e^{-(F_2 + F_3)} + e^{-(F_1 + F_2 + F_3)} \tag{8.20}$$

This, in turn, can be expressed as

$$R_T = R_1 + R_2 + R_3 - (R_1 \times R_2) - (R_1 \times R_3)$$
$$- (R_2 \times R_3) + (R_1 \times R_2 \times R_3) \tag{8.21}$$

In terms of failure probabilities,

$$R_T = 1 - (Q_1 \times Q_2 \times Q_3) \tag{8.22}$$

Equations (8.20), (8.21), and (8.22) for three units in parallel with only one required reduce to Eqs. (8.20a), (8.21a), and (8.22a), when the three units are alike.

$$R_T = 3e^{-F} - 3e^{-2F} + e^{-3F} \tag{8.20a}$$

$$R_T = 3R - 3R^2 + R^3 \tag{8.21a}$$

$$R_T = 1 - Q^3 \tag{8.22a}$$

For a three-unit parallel system where only one is required to operate, the MTTF is computed as follows.

In the general case when the failure rates are different, Eq. (8.20) is used in the integration with λt used rather than F.

$$m = \frac{1}{\lambda_1} + \frac{1}{\lambda_2} + \frac{1}{\lambda_3} - \frac{1}{\lambda_1 + \lambda_2} - \frac{1}{\lambda_1 + \lambda_3}$$
$$- \frac{1}{\lambda_2 + \lambda_3} + \frac{1}{\lambda_1 + \lambda_2 + \lambda_3} \qquad (8.23)$$

When the failure rates are the same, Eq. (8.20a) is used.

$$m = \frac{3}{\lambda} - \frac{3}{2\lambda} + \frac{1}{3\lambda} = \frac{18}{6\lambda} - \frac{9}{6\lambda} + \frac{2}{6\lambda} = \frac{11}{6\lambda} \qquad (8.23a)$$

8.2.3 Three-unit Parallel System, Two Required

In the foregoing derivations for parallel systems, the reliability was found most easily by subtracting the probability of failure from unity because there was only one way in which a system could fail, i.e., if all parallel units failed, and this was the simplest approach. However, if two of three parallel units are required to operate as in Fig. 7.19, the reliability equations are derived in a different manner because there are four ways in which the system can fail instead of only one. It becomes simpler to derive the reliability by considering the ways in which the system can succeed. System success is achieved if not more than one parallel unit fails. Consequently, there are four success cases: no failures; unit A fails (and B and C are good); unit B fails (and A and C are good); and unit C fails (and A and B are good). Algebraically, these cases are expressed as: $R_1 \times R_2 \times R_3$ (no failures) $+ Q_1 \times R_2 \times R_3$ (A fails) $+ R_1 \times Q_2 \times R_3$ (B fails) $+ R_1 \times R_2 \times Q_3$ (C fails). The expression for the reliability can be simplified as follows:

$$R_T = (R_1 \times R_2 \times R_3) + (Q_1 \times R_2 \times R_3) + (R_1 \times Q_2 \times R_3)$$
$$+ (R_1 \times R_2 \times Q_3)$$
$$= (R_1 \times R_2 \times R_3) + \left(Q_1 \times R_2 \times R_3 \times \frac{R_1}{R_1} \right)$$
$$+ \left(R_1 \times Q_2 \times R_3 \times \frac{R_2}{R_2} \right) + \left(R_1 \times R_2 \times Q_3 \times \frac{R_3}{R_3} \right)$$
$$= (R_1 \times R_2 \times R_3) + \left(R_1 \times R_2 \times R_3 \times \frac{Q_1}{R_1} \right)$$
$$+ \left(R_1 \times R_2 \times R_3 \times \frac{Q_2}{R_2} \right) + \left(R_1 \times R_2 \times R_3 \times \frac{Q_3}{R_3} \right)$$

$$R_T = R_1 \times R_2 \times R_3 \times \left(1 + \frac{Q_1}{R_1} + \frac{Q_2}{R_2} + \frac{Q_3}{R_3}\right) \qquad (8.24)$$

When all units are alike, Eq. (8.24) simplifies to

$$R_T = R^3\left(1 + 3\frac{Q}{R}\right) \qquad (8.24a)$$

A more commonly used form of the equation for like units is derived directly from the first line in the development of Eq. (8.24):

$$R_T = R^3 + R^2Q + R^2Q + R^2Q = R^3 + 3R^2Q \qquad (8.24b)$$

This, too may be rewritten as

$$R_T = R^3 + 3R^2(1 - R) = R^3 + 3R^2 - 3R^3 = 3R^2 - 2R^3 \quad (8.24c)$$

Equation (8.24c) is probably the most frequently encountered form of the reliability expression for a two-out-of-three system.

The MTTF in a three-unit parallel system when only one failure is permitted can be found from the first line of the reliability expression in Eq. (8.24) by using λt in place of F.

$$
\begin{aligned}
m &= \int_0^\infty (R_1R_2R_3 + Q_1R_2R_3 + R_1Q_2R_3 + R_1R_2Q_3)\, dt \\[2mm]
&= \int_0^\infty \left[e^{-(\lambda_1 + \lambda_2 + \lambda_3)t} + \left(1 - e^{-\lambda_1 t}\right)e^{-(\lambda_2 + \lambda_3)t} \right. \\
&\qquad\quad \left. + \left(1 - e^{-\lambda_2 t}\right)e^{-(\lambda_1 + \lambda_3)t} + \left(1 - e^{-\lambda_3 t}\right)e^{-(\lambda_1 + \lambda_2)t} \right] dt \\[2mm]
&= \int_0^\infty \left(e^{-(\lambda_1 + \lambda_2 + \lambda_3)t} + e^{-(\lambda_2 + \lambda_3)t} - e^{-(\lambda_1 + \lambda_2 + \lambda_3)t} \right. \\
&\qquad\quad + e^{-(\lambda_1 + \lambda_3)t} - e^{-(\lambda_1 + \lambda_2 + \lambda_3)t} + e^{-(\lambda_1 + \lambda_2)t} \\
&\qquad\qquad\qquad\qquad\qquad \left. - e^{-(\lambda_1 + \lambda_2 + \lambda_3)t} \right) dt \\[2mm]
&= \int_0^\infty \left(e^{-(\lambda_1 + \lambda_2)t} + e^{-(\lambda_1 + \lambda_3)t} + e^{-(\lambda_2 + \lambda_3)t} - 2e^{-(\lambda_1 + \lambda_2 + \lambda_3)t} \right) dt \\[2mm]
&= \frac{1}{\lambda_1 + \lambda_2} + \frac{1}{\lambda_1 + \lambda_3} + \frac{1}{\lambda_2 + \lambda_3} - \frac{2}{\lambda_1 + \lambda_2 + \lambda_3} \qquad (8.25)
\end{aligned}
$$

When all units are alike, Eq. (8.25) reduces to

$$m = \frac{3}{2\lambda} - \frac{2}{3\lambda} = \frac{5}{6\lambda} \tag{8.25a}$$

The same result can, of course, be obtained from Eq. (8.24a) or Eq. (8.24b) or, more easily, from Eq. (8.24c).

8.2.4 Multiple-unit System, One Required

In the large majority of cases, when there are more than three units in parallel, all units are the same. Consequently, the following derivations consider only cases of identical units. The reader may, if desired, develop equations to cover multiple redundancy of non-identical units by following the procedures indicated above for three-unit dissimilar configurations. Three cases will be considered here: (1) only one unit required, (2) only one failure permitted, and (3) the general case in which any number of failures may be permitted.

The equations for the reliability and mean time to failure for additional redundancy when only one operable unit is required can be found by expanding the derivations for Eqs. (8.20) through (8.23a) to cover the additional units. In the equations, n is the total number of units and $n - 1$ failures are permitted. The three equivalent reliability expressions are

$$R_T = ne^{-F} - \frac{n(n-1)}{2} e^{-2F} + \frac{n(n-1)(n-2)}{2 \times 3} e^{-3F}$$
$$- \frac{n(n-1)(n-2)(n-3)}{2 \times 3 \times 4} e^{-4F} + \cdots (\pm) e^{-nF} \tag{8.26}$$

$$R_T = nR - \frac{n(n-1)}{2} R^2 + \frac{n(n-1)(n-2)}{3!} R^3$$
$$- \frac{n(n-1)(n-2)(n-3)}{4!} R^4 + \cdots (\pm) R^n \tag{8.27}$$

$$R_T = 1 - Q^n \tag{8.28}$$

The mean time to failure, derived from the complete expansion of Eq. (8.26), after integrating, combining terms, and simplifying is found to be

$$m = \frac{1}{\lambda} + \frac{1}{2\lambda} + \frac{1}{3\lambda} + \frac{1}{4\lambda} + \cdots + \frac{1}{n\lambda} \tag{8.29}$$

It can be seen from the preceding equations [and from Eq. (8.28) in particular] that as more units are added in parallel, the total reliability rapidly approaches unity, differing only by Q^n. However, as shown by Eq. (8.29), m does *not* increase correspondingly and each additional parallel redundant unit contributes less to the system MTTF than did its predecessor.

8.2.5 Multiple-unit System, One Failure Permitted

When only one failure is permitted, the expressions for reliability and MTTF can be found by extending Eqs. (8.24a) to (8.25a). The three equivalent reliability equations for an n-unit parallel system with no more than one failure are

$$R_T = R^n \left(1 + n\frac{Q}{R}\right) \tag{8.30}$$

$$R_T = R^n + nR^{n-1}Q \tag{8.30a}$$

$$R_T = nR^{n-1} - (n-1)R^n \tag{8.30b}$$

The mean time to failure is found by directly integrating Eq. (8.30b) with R equal to $e^{-\lambda t}$.

$$m = \frac{n}{(n-1)\lambda} - \frac{n-1}{n\lambda} \tag{8.31}$$

8.2.6 Multiple-unit System, Two or More Failures Permitted

When there are more than three units in parallel and more than one failure is permitted, the binomial expansion can be used to calculate the reliability and the MTTF. We recall (Sec. 3.2) that the terms of the expansion indicate the probabilities of exactly zero, one, two, three, . . . , r failures in an active parallel system.* If there are n parallel units, there are a total of $n + 1$ terms; and if r failures are permitted, the system reliability is calculated by summing the first $r + 1$ terms.

---

*R (reliability) should not be confused with r (number of failures *permitted*). These two letter symbols are used in this text because they are commonly employed in the literature, and care should be taken to avoid confusion. Note also that in Chap. 13 on attributes assessment methods and in Table C.3, the letter f is used instead of r to denote number of failures actually *experienced.*

$$R_T = R^n + nR^{n-1}Q + \frac{n(n-1)}{2}R^{n-2}Q^2 + \cdots$$

$$+ \frac{n!}{r!(n-r)!}R^{n-r}Q^r \tag{8.32}$$

$$
\begin{aligned}
m &= \int_0^\infty \Bigg\{ e^{-n\lambda t} + n[e^{-(n-1)\lambda t}(1 - e^{-\lambda t})] \\
&\quad + \frac{n(n-1)}{2}\left[e^{-(n-2)\lambda t}(1 - e^{-\lambda t})^2\right] + \cdots \\
&\quad + \frac{n!}{r!(n-r)!}\left[e^{-(n-r)\lambda t}(1 - e^{-\lambda t})^r\right]\Bigg\} dt \\
&= \int_0^\infty \Bigg\{ e^{-n\lambda t} + n\left[e^{-(n-1)\lambda t} - e^{-n\lambda t}\right] \\
&\quad + \frac{n(n-1)}{2}\left[e^{-(n-2)\lambda t} - 2e^{-(n-1)\lambda t} + e^{-n\lambda t}\right] \\
&\quad + \cdots + \frac{n!}{r!(n-r)!}\left[e^{-(n-r)\lambda t} - re^{-(n-r+1)\lambda t}\right. \\
&\quad + \cdots (\pm)\, e^{-n\lambda t}\Bigg]\Bigg\} dt \\
&= \frac{1}{n\lambda} + n\left[\frac{1}{(n-1)\lambda} - \frac{1}{n\lambda}\right] + \frac{n(n-1)}{2}\left[\frac{1}{(n-2)\lambda} - \frac{2}{(n-1)\lambda}\right. \\
&\quad + \frac{1}{n\lambda}\Bigg] + \cdots + \frac{n!}{r!(n-r)!}\left[\frac{1}{(n-r)\lambda} - \frac{r}{(n-r+1)\lambda} + \cdots\right. \\
&\quad\quad\quad\quad\quad\quad\quad\quad\quad\quad\quad\quad (\pm)\frac{1}{n\lambda}\Bigg]
\end{aligned}
\tag{8.33}
$$

The first term of Eq. (8.33) is, of course, $1/n\lambda$. The second term simplifies as follows:

$$n\left[\frac{1}{(n-1)\lambda} - \frac{1}{n\lambda}\right] = \frac{n}{(n-1)\lambda} - \frac{n}{n\lambda} = \frac{n^2 - (n^2 - n)}{n(n-1)\lambda}$$

$$= \frac{n}{n(n-1)\lambda} = \frac{1}{(n-1)\lambda}$$

The third term simplifies to $1/(n - 2)\lambda$; etc. Equation (8.33) thereby simplifies to Eq. (8.33a):

$$m = \frac{1}{n\lambda} + \frac{1}{(n - 1)\lambda} + \frac{1}{(n - 2)\lambda} + \cdots + \frac{1}{(n - r)\lambda} \tag{8.33a}$$

where n is the total number of units in parallel and r is the number of failures which are allowed.

Note that, when only one failure is allowed and only the first two terms are used, Eq. (8.33) reduces to Eq. (8.31):

$$\frac{1}{n\lambda} + n\left[\frac{1}{(n - 1)\lambda} - \frac{1}{n\lambda}\right] = \frac{1}{n\lambda} + \frac{n}{(n - 1)\lambda} - \frac{n}{n\lambda}$$

$$= \frac{n}{(n - 1)\lambda} - \frac{n - 1}{n\lambda}$$

Also note that when only one operating unit is needed so that $r = n - 1$, Eq. (8.33a) is identical to Eq. (8.29), but in reverse order.

8.3 SEQUENTIAL SYSTEMS

We recall that sequential systems are those in which a constant number of units are in operation and, when one unit fails, it is either repaired and restored to its original condition or replaced with another unit. The mathematical models for the reliability of these systems are more complex than other models. For the development of these models the reader is referred to Sec. 3.3 and to Bazovsky,* chaps. 9 and 12.

8.3.1 Systems with Equal Components and No Standby Failure Rate

The simplest models are for those systems in which all units are the same, where the failure detection (sensing) and switching devices are assumed to be 100 percent reliable, and where the standby failure rate is zero. The equation for such a two-unit system (one standby unit) is

$$R_T = e^{-\lambda t}(1 + \lambda t) \tag{8.34}$$

The reliability of a three-unit system (two on standby) is

$$R_T = e^{-\lambda t}\left[1 + \lambda t + \frac{(\lambda t)^2}{2}\right] \tag{8.35}$$

*Op. cit.

The expression for a system with any number of like units, n, with $n - 1$ units on standby is

$$R_T = e^{-\lambda t}\left[1 + \lambda t + \frac{(\lambda t)^2}{2} + \frac{(\lambda t)^3}{6} + \cdots + \frac{(\lambda t)^{n-1}}{(n-1)!}\right] \qquad (8.36)$$

The mean time to failure of a sequentially redundant system is computed in the same manner as we have computed the MTTF of an active parallel system: by integrating the reliability function. Integration of Eqs. (8.34) to (8.36) follows. From Eq. (8.34),

$$R_T = e^{-\lambda t}(1 + \lambda t) = e^{-\lambda t} + \lambda t\, e^{-\lambda t}$$

$$m = \int_0^\infty [e^{-\lambda t} + \lambda t e^{-\lambda t}]\, dt = \int_0^\infty e^{-\lambda t}\, dt + \int_0^\infty \lambda t e^{-\lambda t}\, dt$$

$$= \int_0^\infty e^{-\lambda t}\, dt - \int_0^\infty t e^{-\lambda t}\, d(-\lambda t)$$

The first term, as we know, is merely $1/\lambda$. The second term is integrated by parts.*

$$-\int_0^\infty t e^{-\lambda t}\, d(-\lambda t) = -\left\{ [t e^{-\lambda t}]_0^\infty - \int_0^\infty e^{-\lambda t}\, dt \right\}$$

$$= -[t e^{-\lambda t}]_0^\infty + \int_0^\infty e^{-\lambda t}\, dt$$

But $e^{-\lambda t} = R$ so that $t e^{-\lambda t} = tR$, and the value of $[tR]_0^\infty$ is, from the proof of Eq. (8.12), identically zero. Since $\int_0^\infty e^{-\lambda t}\, dt$ is again $1/\lambda$, the value of the second term is $0 + 1/\lambda = 1/\lambda$. The total MTTF of a two-unit sequential system is therefore

*Let $u = t$ and $dv = e^{-\lambda t}\, d(-\lambda t)$. Then $du = dt$ and $v = e^{-\lambda t}$.

$$\int u\, dv = uv - \int v\, du \qquad \int_0^\infty t e^{-\lambda t}\, d(-\lambda t) = [t e^{-\lambda t}]_0^\infty - \int_0^\infty e^{-\lambda t}\, dt$$

$$m = \frac{1}{\lambda} + \frac{1}{\lambda} = \frac{2}{\lambda} \tag{8.37}$$

Integrating Eq. (8.35),

$$R_T = e^{-t}\left[1 + \lambda t + \frac{(\lambda t)^2}{2}\right] = e^{-\lambda t} + \lambda t e^{-\lambda t} + \frac{(\lambda t)^2}{2} e^{-\lambda t}$$

$$m = \int_0^\infty \left[e^{-\lambda t} + \lambda t e^{-\lambda t} + \frac{(\lambda t)^2}{2} e^{-\lambda t}\right] dt$$

$$= \int_0^\infty e^{-\lambda t}\, dt + \int_0^\infty \lambda t e^{-\lambda t}\, dt + \int_0^\infty \frac{(\lambda t)^2}{2} e^{-\lambda t}\, dt$$

We already know, from Eq. (8.37), that the sum of the first two terms is $1/\lambda$ + $1/\lambda$. Integration of the third term is again performed by parts.*

$$\int_0^\infty \frac{(\lambda t)^2}{2} e^{-\lambda t}\, dt = \left[-\frac{\lambda t^2}{2} e^{-\lambda t}\right]_0^\infty - \int_0^\infty t e^{-\lambda t}\, d(-\lambda t)$$

Again using the proof of Eq. (8.12) given in Appendix A.3, the value of the first part is identically zero; the value of the second part is, from the development of Eq. (8.37), equal to $1/\lambda$. Therefore, the MTTF of a three-unit sequential system is

$$m = \frac{1}{\lambda} + \frac{1}{\lambda} + \frac{1}{\lambda} = \frac{3}{\lambda} \tag{8.38}$$

As we might well suspect, integration of Eq. (8.36) and the proof in Appendix A.3 will show that the value of each last term is $1/\lambda$. Therefore, the MTTF of an n-unit sequentially redundant system with one unit operating

---

*Let $u = -\lambda t^2/2$ and $dv = e^{-\lambda t}\, d(-\lambda t)$. Then $du = -\lambda t\, dt$ and $v = e^{-\lambda t}$.

$$\int_0^\infty \frac{(\lambda t)^2}{2} e^{-\lambda t}\, dt = \int_0^\infty \frac{-\lambda t^2}{2} e^{-\lambda t} d(-\lambda t) = \left[\frac{-\lambda t^2}{2} e^{-\lambda t}\right]_0^\infty - \int_0^\infty -\lambda t e^{-\lambda t}\, dt$$

$$= \left[\frac{-\lambda t^2}{2} e^{-\lambda t}\right]_0^\infty - \int_0^\infty t e^{-\lambda t}\, d(-\lambda t)$$

and $n - 1$ identical units on standby is given by Eq. (8.39).

$$m = \frac{1}{\lambda} + \frac{1}{\lambda} + \cdots + \frac{1}{\lambda} = \frac{n}{\lambda} \tag{8.39}$$

In retrospect, if the mean time to failure of one unit is m, and the mean time to failure of a second unit is also m, and the second unit is not operated until the first unit fails and cannot fail when not operating, then we can conclude that the combined mean time to failure for the two units operated sequentially is, indeed, $2 \times m = 2/\lambda$. Similarly, the combined MTTF for three units each with MTTF m is $3 \times m = 3/\lambda$, and for n units it is merely $n \times m$, or n/λ.

8.3.2 Systems with Unequal Components and No Standby Failure Rate

The reliability of a system with two sequentially operated units (one standby unit) with different constant operating failure rates λ_1 and λ_2 and no standby failure rate is given by Eq. (8.40). Its derivation can be found in Appendix A.4.

$$R_T = e^{-\lambda_1 t} + \frac{\lambda_1}{\lambda_1 - \lambda_2}\left(e^{-\lambda_2 t} - e^{-\lambda_1 t}\right) \tag{8.40}$$

This form of the equation shows the reliability contributed by the first unit and the additional reliability resulting from the standby unit. For ease of extending the equation to cover two or more units on standby, an alternate form of the equation is presented.

$$R_T = \left(\frac{\lambda_2}{\lambda_2 - \lambda_1} \times e^{-\lambda_1 t}\right) + \left(\frac{\lambda_1}{\lambda_1 - \lambda_2} \times e^{-\lambda_2 t}\right) \tag{8.40a}$$

The equation for a three-unit system with two on standby is given by Eq. (8.41). This can be derived just as were Eqs. (8.40) and (8.40a).

$$R_T = \left(\frac{\lambda_2}{\lambda_2 - \lambda_1} \times \frac{\lambda_3}{\lambda_3 - \lambda_1} \times e^{-\lambda_1 t}\right)$$

$$+ \left(\frac{\lambda_1}{\lambda_1 - \lambda_2} \times \frac{\lambda_3}{\lambda_3 - \lambda_2} \times e^{-\lambda_2 t}\right)$$

$$+ \left(\frac{\lambda_1}{\lambda_1 - \lambda_3} \times \frac{\lambda_2}{\lambda_2 - \lambda_3} \times e^{-\lambda_3 t}\right) \tag{8.41}$$

The extension of this equation to cover any number of standby units is

$$R_T = \left(\frac{\lambda_2}{\lambda_2 - \lambda_1} \times \frac{\lambda_3}{\lambda_3 - \lambda_1} \times \cdots \times \frac{\lambda_n}{\lambda_n - \lambda_1} \times e^{-\lambda_1 t} \right)$$

$$+ \left(\frac{\lambda_1}{\lambda_1 - \lambda_2} \times \frac{\lambda_3}{\lambda_3 - \lambda_2} \times \cdots \times \frac{\lambda_n}{\lambda_n - \lambda_2} \times e^{-\lambda_2 t} \right)$$

$$+ \left(\frac{\lambda_1}{\lambda_1 - \lambda_3} \times \frac{\lambda_2}{\lambda_2 - \lambda_3} \times \cdots \times \frac{\lambda_n}{\lambda_n - \lambda_3} \times e^{-\lambda_3 t} \right) + \cdots$$

$$+ \left(\frac{\lambda_1}{\lambda_1 - \lambda_n} \times \frac{\lambda_2}{\lambda_2 - \lambda_n} \times \frac{\lambda_3}{\lambda_3 - \lambda_n} \times \cdots \times \frac{\lambda_{n-1}}{\lambda_{n-1} - \lambda_n} \times e^{-\lambda_n t} \right)$$

$$(8.42)$$

where n is the total number of units and $n - 1$ units are on standby.

The mean time to failure for a sequentially redundant system with different failure rates is again computed by integrating the applicable reliability expression. For two-, three-, and n-unit systems the integrations of Eqs. (8.40a), (8.41), and (8.42) are given by Eqs. (8.43), (8.44), and (8.45), respectively. Integrating Eqs. (8.40a) we have

$$m = \int_0^\infty \left(\frac{\lambda_2}{\lambda_2 - \lambda_1} e^{-\lambda_1 t} + \frac{\lambda_1}{\lambda_1 - \lambda_2} e^{-\lambda_2 t} \right) dt$$

$$= \frac{\lambda_2}{(\lambda_2 - \lambda_1)\lambda_1} + \frac{\lambda_1}{(\lambda_1 - \lambda_2)\lambda_2}$$

$$= \frac{\lambda_2^2}{(\lambda_2 - \lambda_1)\lambda_1\lambda_2} + \frac{\lambda_1^2}{(\lambda_1 - \lambda_2)\lambda_1\lambda_2}$$

$$= \frac{\lambda_2^2}{(\lambda_2 - \lambda_1)\lambda_1\lambda_2} - \frac{\lambda_1^2}{(\lambda_2 - \lambda_1)\lambda_1\lambda_2}$$

$$= \frac{\lambda_2^2 - \lambda_1^2}{(\lambda_2 - \lambda_1)(\lambda_1\lambda_2)} = \frac{(\lambda_2 - \lambda_1)(\lambda_2 + \lambda_1)}{(\lambda_2 - \lambda_1)\lambda_1\lambda_2}$$

$$= \frac{\lambda_2 + \lambda_1}{\lambda_1\lambda_2} = \frac{\lambda_2}{\lambda_1\lambda_2} + \frac{\lambda_1}{\lambda_1\lambda_2} = \frac{1}{\lambda_1} + \frac{1}{\lambda_2} \qquad (8.43)$$

and integrating Eq. (8.41) we have

$$
m = \int_0^\infty \left[\frac{\lambda_2 \lambda_3}{(\lambda_2 - \lambda_1)(\lambda_3 - \lambda_1)} e^{-\lambda_1 t} + \frac{\lambda_1 \lambda_3}{(\lambda_1 - \lambda_2)(\lambda_3 - \lambda_2)} e^{-\lambda_2 t} \right.
$$
$$
\left. + \frac{\lambda_1 \lambda_2}{(\lambda_1 - \lambda_3)(\lambda_2 - \lambda_3)} e^{-\lambda_3 t} \right] dt
$$

$$
= \frac{\lambda_2 \lambda_3}{(\lambda_2 - \lambda_1)(\lambda_3 - \lambda_1)\lambda_1} + \frac{\lambda_1 \lambda_3}{(\lambda_1 - \lambda_2)(\lambda_3 - \lambda_2)\lambda_2}
$$
$$
+ \frac{\lambda_1 \lambda_2}{(\lambda_1 - \lambda_3)(\lambda_2 - \lambda_3)\lambda_3}
$$

It can be shown (Appendix A.5) that, with simplification and combination of terms, this expression reduces to

$$
m = \frac{1}{\lambda_1} + \frac{1}{\lambda_2} + \frac{1}{\lambda_3} \tag{8.44}
$$

From Eq. (8.42),

$$
m = \int_0^\infty \left[\frac{\lambda_2 \lambda_3 \cdots \lambda_n}{(\lambda_2 - \lambda_1)(\lambda_3 - \lambda_1) \cdots (\lambda_n - \lambda_1)} e^{-\lambda_1 t} \right.
$$
$$
+ \frac{\lambda_1 \lambda_3 \cdots \lambda_n}{(\lambda_1 - \lambda_2)(\lambda_3 - \lambda_2) \cdots (\lambda_n - \lambda_2)} e^{-\lambda_2 t}
$$
$$
+ \frac{\lambda_1 \lambda_2 \cdots \lambda_n}{(\lambda_1 - \lambda_3)(\lambda_2 - \lambda_3) \cdots (\lambda_n - \lambda_3)} e^{-\lambda_3 t} + \cdots
$$
$$
\left. + \frac{\lambda_1 \lambda_2 \cdots \lambda_{n-1}}{(\lambda_1 - \lambda_n)(\lambda_2 - \lambda_n) \cdots (\lambda_{n-1} - \lambda_n)} e^{-\lambda_n t} \right] dt
$$

This expression reduces to

$$
m = \frac{1}{\lambda_1} + \frac{1}{\lambda_2} + \cdots + \frac{1}{\lambda_n} \tag{8.45}
$$

As we might have deduced from the discussion in Sec. 8.3.1, the MTTF of a sequentially redundant system with no standby failure rate, whether the individual units are alike or different, is the sum of the individual mean times:

$$m = m_1 + m_2 + \cdots + m_n = \frac{1}{\lambda_1} + \frac{1}{\lambda_2} + \cdots + \frac{1}{\lambda_n}$$

This, of course, is the same result we have just obtained through a strictly mathematical proof.

8.3.3 Sequentially Redundant Parallel System

When a system has more than one unit in required operation with other units on standby to be used as individual replacements, the units are almost always of the same type. Consequently, the reliability of such a multiredundant system will be derived only for the case of similar units.

If there are l units required, then the failure rate for the system is $l\lambda_i$ and the reliability is $\exp(-l\lambda_i t)$. When one unit fails, it is replaced by a similar unit. Since it is assumed that all units have the same failure rate λ_i and are operated during the period when this failure rate is constant (i.e., after burn-in and prior to onset of wear-out), the failure rates of the unfailed units as well as that of the new unit are all still λ_i. Consequently, the system failure rate also remains constant at $l\lambda_i$. We can therefore consider a sequential redundant system with failure rate $l\lambda_i$ and use Eq. (8.36) to evaluate its reliability. If there are n *replacements* so that n failures are permitted, the system reliability is

$$R = e^{-l\lambda_i t} \left[1 + l\lambda_i t + \frac{(l\lambda_i t)^2}{2} + \frac{(l\lambda_i t)^3}{6} + \cdots + \frac{(l\lambda_i t)^n}{n!} \right] \quad (8.46)$$

The mean time to failure is again found by integration

$$m = \frac{1}{l\lambda_i} + \frac{1}{l\lambda_i} + \cdots + \frac{1}{l\lambda_i} = \frac{n+1}{l\lambda_i} \quad (8.47)$$

Note that in Eqs. (8.46) and (8.47), there are n *replacements* (and $n + l$ total units).

8.3.4 Sequential Systems with Imperfect Switching

Two additional complexities are now introduced. First, the failure detection and switching devices are not perfectly reliable and their probabilities of failure must be considered. If we assume that the system design is such that the switching and sensing functions are associated only with the standby units and do not affect the first operating unit, Eqs. (8.34) and (8.40) for two-unit systems (one standby unit) are modified to Eqs. (8.48) and (8.49), respectively.

$$R_T = e^{-\lambda t} \times (1 + R_{ds}\lambda t) \tag{8.48}$$

$$R_T = e^{-\lambda_1 t} + R_{ds} \times \frac{\lambda_1}{\lambda_1 - \lambda_2} \times \left(e^{-\lambda_2 t} - e^{-\lambda_1 t} \right) \tag{8.49}$$

where R_{ds} is the reliability of detection and switching. Similarly, Eqs. (8.35), (8.36), (8.41), and (8.42) for three-unit and multiple-unit standby redundancy can be modified to account for failure detection and switching by multiplying each term after the first by the appropriate R_{ds}. In the case of unlike units, Eqs. (8.41) and (8.42) must first be put into the form of Eq. (8.40) so that the first term is $e^{-\lambda_1 t}$. Considerable algebraic manipulation is required, but by following the steps in Appendix A.6, the necessary form can be derived. This detection and switching reliability term may be exponential or may be independent of time to reflect a single-cycle operation, or both, depending on the nature and design of the detection/switching device.

The mean time to failure of such systems can again be found by integrating the applicable reliability equation. Since the detection and switching reliability expressions may be different in that detection may have a constant failure rate while switching may be a "one-shot" device, R_{ds} will be $R_d \times R_s$ where R_d is the detection reliability and R_s is the switching reliability. The actual expressions are functions of the particular system, and the integration will not be shown here. However, it follows the general methods already discussed or presented in the Appendix.

8.3.5 Sequential Systems with Standby Failure Rates

Second, non-operating units frequently have a standby failure rate greater than zero. Mechanical components which rotate when operating can sustain bearing damage due to shock or vibration when on standby; electronic components may be subjected to high temperature or other environmental stresses which could result in failure while on standby; hydraulic and lubricating systems might fail due to increased viscosity or contamination of the fluid; and so on. Consequently, the following formulas are presented to account for the standby failure rate. Both formulas are for two-unit systems. Equation (8.50) assumes perfect failure detection and switching, while Eq. (8.51) includes the reliability of these functions. In the equations, λ_1 and λ_2 are the operating failure rates and λ_3 is the standby failure rate of the second unit.

$$R_T = e^{-\lambda_1 t} + \frac{\lambda_1}{(\lambda_1 + \lambda_3) - \lambda_2} \times \left(e^{-\lambda_2 t} - e^{-(\lambda_1 + \lambda_3)t} \right) \qquad (8.50)$$

$$R_T = e^{-\lambda_1 t} + R_{ds} \times \frac{\lambda_1}{(\lambda_1 + \lambda_3) - \lambda_2} \times \left(e^{-\lambda_2 t} - e^{-(\lambda_1 + \lambda_3)t} \right)$$

$$(8.51)$$

The equations for three-unit and multiple-unit systems with standby failure rates can also be derived from the corresponding expressions for systems without such standby rates. Again, it is best to first put the expression in the form of Eq. (8.51) showing the reliability associated with the first unit and the additional reliability contributed by the sequential units. The derivation given in Appendix A.6 can be used as a guide.

8.3.6 Comparison of Parallel and Sequential Redundancy

We have gone into some detail in the presentation of mathematical models of active and sequentially redundant configurations. A comparison of their relative reliabilities is warranted. Either type of redundancy significantly increases system reliability and, in many cases, reliability requirements cannot be met without some form of redundancy. A system which consists of a single unit with a reliability of 0.90 has a reliability of 0.990 if two similar units are in active redundancy, and it has a reliability of 0.995 if the units are used sequentially *provided that failure detection and switching are perfect and that the standby failure rate is zero.*

The words in italics are significant. In order for a two-unit sequentially redundant system to be as reliable as a two-unit active parallel system, the failure probability Q of the detection and switching functions must not exceed 50 percent of the failure probability of one redundant unit.* In the preceding example, the combined R_{ds} reliabilities must be at least 0.95 *just to break even* with an active system. Furthermore, it is frequently true that when two units are actively performing the function that one is capable of, both are operating at a lower stress level and hence have a lower failure rate and higher reliability. For these reasons, when a choice is possible, the active redundant system is often preferred. However, the form of redundancy is generally dictated by the system's performance characteristics and it may be necessary to operate the units in sequence in order to maintain required values of operating parameters. A detailed discussion of the advantages and disadvantages of both types, including maintenance considerations, is given in Chap. 16.

*The "break-even" percentage decreases rapidly when there are more than two units.

8.4   SOME MATHEMATICAL MODELS FOR CHAPTER 7

We will now consider the reliability expressions for some of the logic diagrams in Chap. 7. The simplest systems of all series or all parallel redundant elements will not be mentioned because these can be evaluated using the methods of the preceding sections. Also, mean times to failure will not be derived, but they can be easily determined from the formulas which we have developed.

When considering a series-parallel configuration such as in Fig. 7.13, the mathematical model is compiled through a series of steps, starting at the lowest level depicted. The lowest level is defined as that segment of the most complex portion of the logic diagram which cannot be further subdivided. In Fig. 7.13, there are two main parallel legs. The upper leg is divided into two smaller parallel legs connected together and in series with a single element (a coil). The upper leg of these two is further subdivided into three series groups, each group consisting of a capacitor and a fuse in parallel. These groups cannot be further subdivided, and thus they comprise the lowest level depicted.

The first step, then, is to compute the reliability of each capacitor-fuse group. We will assume that we do not know whether the exponential equation applies, and we will derive the overall reliability in terms of the reliability of each element rather than in terms of failure rates. Using Eq. (8.16) for a simple two-unit parallel group, the reliability R_p of each capacitor-fuse pair is

$$R_p = R_{c1} + R_f - (R_{c1} \times R_f)$$

where R_{c1} is the reliability of the capacitor and R_f is the reliability of the fuse. Since there are three pairs in series, the reliability of the three groups together is defined by Eq. (8.1) as $R_{p1} \times R_{p2} \times R_{p3}$, where R_{p1} is the reliability of the first pair, etc. We can assume that the three pairs are equivalent so that this reduces to $(R_p)^3$. This set of three groups is in parallel with a resistor and a variable capacitor whose reliabilities may be defined as $R_{r1} \times R_{vc1} = R_{rc}$. Thus, the next calculation involves finding the reliability R_y of a parallel group consisting of the set of three pairs in one leg and the resistor and variable capacitor in the other. Equation (8.16) is used again:

$$R_y = (R_p)^3 + R_{rc} - [(R_p)^3 \times R_{rc}]$$

The entire leg consists of this subassembly in series with a coil, and the reliability of the leg, R_{L1}, is found by again employing Eq. (8.1):

$$R_{L1} = R_y \times R_{coil1}$$

Similarly, the reliability R_{L2} of the lower leg is

$$R_{L2} = R_A \times R_V \times R_F \times R_S \times [(R_{r2} \times R_{vc2}) + (R_{pot} \times R_{c2})$$
$$- (R_{r2} \times R_{vc2})(R_{pot} \times R_{c2})] \times R_{coil\,2}$$

where R_A, R_V, R_F, R_S, R_{r2}, R_{vc2}, R_{pot}, R_{c2}, and $R_{coil\,2}$ are the reliabilities of the ammeter, voltmeter, frequency meter, switch, second resistor, second variable capacitor, potentiometer, second capacitor, and second coil, respectively.

The overall reliability is found by combining the reliabilities of the two main legs, once again using Eq. (8.16):

$$R_T = R_{L1} + R_{L2} - (R_{L1} \times R_{L2}) \tag{8.52}$$

The completely expanded form, using the reliabilities of each individual component, consists of too many terms to conveniently set down here and is left to the reader as an exercise if he so desires.

Another complication arises when different failure modes are depicted as separate elements in a logic diagram. In a series system, it is not necessary to depict individual modes, but a redundant system usually provides different degrees of protection for different modes of failure and the modes are shown separately. Since different elements represent modes of failure for the same component, the elements are no longer independent. If a component fails in one mode (one element fails), it usually cannot also fail in another mode; e.g., if a valve fails open, it cannot also fail closed at the same time.

One method of accounting for the lack of independence is to neglect it. For practical purposes no significant error results. The failure rates for the various modes of failure are not accurate enough to warrant the usually minor modification resulting from lack of independence. Certainly, the use of environmental multipliers and application factors (which will be discussed in the next chapter) does not justify modifying the failure rates for dependency. Also, in those instances when failure rates are available by mode, the rates are those actually experienced and already take into account the dependency considerations. Finally, since the affected elements will be redundant, slight changes will not affect the overall result. Consequently, even though there may be dependence between some of the elements in the diagram which we are about to discuss, we treat them as independent.

The mathematical model associated with Fig. 7.14 is again a simple one. The reliability of each logic element is determined with respect to only the failure mode(s) depicted. Thus, the reliability of the series element for one check valve considers only freedom from failure in the leakage or closed modes.

Although the starting point makes little difference in this example, to be consistent the reliability with respect to failing in the open position should be considered first since it is the most complex part of the logic diagram. Using Eq. (8.16), the reliability R_{open} is $2R_a - R_a^2$, where R_a is the reliability of one of the parallel elements. If we define R_{leak} to include both the closed and leakage failure modes, the overall reliability R_T is

$$R_T = R_{\text{pump}} \times R_{\text{leak}}^2 \times (2R_a - R_a^2) \tag{8.53}$$

We may use the mathematical models already discussed in preparing the model for the logic diagram in Fig. 7.16a. Because there are two elements in each leg, the reliability of each leg of the "open" mode is R_o^2 and the combined reliability is $R_o^2 + R_o^2 - (R_o^2 \times R_o^2)$, which reduces to $2R_o^2 - R_o^4$. The reliability with respect to shorting is $2R_s - R_s^2$ for each pair of "short" elements. For both pairs in series it is $(2R_s - R_s^2)^2$. The overall reliability R_T is given by Eq. (8.54).

$$R_T = (2R_o^2 - R_o^4) \times (2R_s - R_s^2)^2 \tag{8.54}$$

The model for Fig. 7.16b is the same with the exception that R_{open} and R_{short} are interchanged throughout.

Most of the functional relationships which cannot be correctly depicted in a logic diagram can be accurately modeled mathematically. Thus, the two-out-of-three requirement for a parallel system with similar units (Fig. 7.19) is mathematically modeled by Eqs. (8.24a), (8.24b), and (8.24c).

When it is desirable to consider dependency, appropriate mathematical models can be developed. In the case of the reefing line cutters (Fig. 7.22), as in most cases of dependency, the model is more complex. To develop this model, it is desirable to determine all mutually exclusive success paths. The total reliability is the sum of the reliabilities of the various paths. A success is achieved if:

1. Neither line is cut prematurely and both are cut when required; or if
2. Line A is cut prematurely, line B is not cut prematurely, and line B is cut when required; or if
3. Line B is cut prematurely, line A is not cut prematurely, and line A is cut when required.

The probability of *not being* cut prematurely (the reliability) is the probability (reliability) that no cutter will fail in this mode.

$$R_{\substack{\text{premature} \\ \text{cutting}}} = R_{\substack{\text{cutter 1} \\ \text{prem.}}} \times R_{\substack{\text{cutter 2} \\ \text{prem.}}} \times R_{\substack{\text{cutter 3} \\ \text{prem.}}} = (R_{\substack{\text{cutter} \\ \text{prem.}}})^3$$

This may be abbreviated to R_{cp}^3.

The probability of *being* cut prematurely (the unreliability) is one minus the reliability $= 1 - R_{cp}^{\,3}$.

The probability of cutting when required is the probability that at least one cutter works which, from Eq. (8.22a), is $1 - Q_{cutting}^{3}$. Since $Q = 1 - R$, this probability in terms of the reliability is

$$1 - (1 - R_{cutting})^3 \quad \text{or} \quad 1 - (1 - R_c)^3$$

We then combine these three probabilities into sets according to the three success cases listed above.

$$R_T = R_{cp_A}^3 \quad \times \quad R_{cp_B}^3 \quad \times 1 - \left(1 - R_{c_A}\right)^3 \times 1 - \left(1 - R_{c_B}\right)^3$$

(line A is not $\times$ (line B is not $\times$ (line A is cut $\times$ (line B is cut
cut prematurely)   cut prematurely)   when required)   when required)

$$+ \ 1 - R_{cp_A}^3 \quad \times \quad R_{cp_B}^3 \quad \times 1 - \left(1 - R_{c_B}\right)^3$$

(line A is cut $\times$ (line B is not $\times$ (line B is cut
prematurely)   cut prematurely)   when required)

$$+ \ 1 - R_{cp_B}^3 \quad \times \quad R_{cp_A}^3 \quad \times 1 - \left(1 - R_{c_A}\right)^3$$

(line B is cut $\times$ (line A is not $\times$ (line A is cut
prematurely)   cut prematurely)   when required)

Since lines A and B and their associated cutters are assumed to be identical, we can delete these subscripts and combine terms as follows:

$$R_T = (R_{cp})^6 \times \left[1 - (1 - R_c)^3\right]^2$$
$$+ 2\left\{(1 - R_{cp})^3 \times R_{cp}^{\,3} \times \left[1 - (1 - R_c)^3\right]\right\} \qquad (8.55)$$

We have examined the mathematical models for various types of systems, beginning with simple series systems and progressing through more complex active and sequential parallel configurations. We have also developed models for combination series-parallel systems, using the logic diagrams of Chap. 7 as examples. In succeeding chapters we will discuss how the models are used in the prediction, apportionment, and assessment of the reliability of complex systems.

ADDITIONAL READINGS

Amstadter, B. L.: "Calculations of Reliability Predictions for the Apollo Spacecraft," North American-Rockwell Corporation, Downey, Calif., SID 66-744, 1966, IDEP no. 347.40.00.00-F1-47.

ARINC Research Corporation: "Reliability Engineering," Prentice-Hall, Inc., Englewood Cliffs, N.J., 1964.

Bazovsky, I.: "Reliability Theory and Practice," Prentice-Hall, Inc., Englewood Cliffs, N.J., 1961.

Bureau of Naval Weapons, Chief of: "Handbook: Reliability Engineering," U.S. Government Printing Office, 1964.

Calabro, S. R.: "Reliability Principles and Practices," McGraw-Hill Book Company, New York, 1962.

Dummer, G. W., and N. B. Griffin: "Electronics Reliability: Calculation and Design," Pergamon Press, Ltd., London, 1966.

Lloyd, D. K., and M. Lipow: "Reliability: Management, Methods, and Mathematics," Prentice-Hall, Inc., Englewood Cliffs, N.J., 1962.

Shooman, M. L.: "Probabilistic Reliability: An Engineering Approach," McGraw-Hill Book Company, New York, 1968.

Thomas, G. B., Jr.: "Calculus and Analytic Geometry," 3d ed., Addison-Wesley Publishing Company, Inc., Reading, Mass., 1960.

9

PREDICTION OF
COMPONENT RELIABILITY

9.1 DEFINITION AND USE

Reliability prediction is the process of forecasting, from available failure-rate information, the realistically achievable reliability of a part, component, subsystem, or system and the probability of its meeting performance and reliability requirements for a specified application. Predictions use data from programs that are either in the operational phase or have already been completed. They represent the obtainable reliability of the equipment in field use, not the reliability at the current state of development.

Reliability predictions can provide several advantages to the overall reliability effort. Since the objective of the reliability activity is to achieve and maintain the highest practical reliability within the performance, weight, cost, schedule, and other program constraints, the predictions should be used to further this objective. To be most beneficial, they should be accomplished in a timely, useful manner at the conceptual and early design stages of the program. However, pure numerical values in themselves provide little benefit. The meaning of these values, the relationships among the reliabilities of the various system elements, and the recommendations for system improve-ment in light of the prediction results are the real contributions of the predictions. System optimization is achieved only when these objectives are accomplished.

Realistic predictions include, wherever possible, considerations of parts applications. Operating stresses (voltage, power, temperature, mechanical stress reversal, etc.) have significant effects on component failure rates and longevity. The effects on reliability of many of these stresses are given in failure-rate-data handbooks which provide either multiplying (or derating) factors or direct failure-rate values as functions of stress levels. These should be taken into account in the prediction. Parts which are being employed in applications beyond recommended stress levels or in environments for which they were not intended are disclosed, and they can be replaced by more suitable components. It is sometimes possible to reduce the stresses by adding redundancy or by similar action. We can state this function of reliability prediction more concisely as *parts application analysis.*

Predictions are extremely valuable when deciding among alternative candidate designs. Several designs for a particular system, subsystem, or function may be under consideration, and one of the factors which influences the choice is their relative reliabilities. This is equally true when various types of redundancy are being compared and when different components are being considered. An illustration of the former comparison is given in Chap. 10; an example of the latter was the reliability evaluation of rotary versus solid-state inverters for converting the dc output of a battery to 400-Hz ac power. In both instances, reliability was an important consideration.

Predictions also indicate those components and assemblies which contribute most toward the total system failure probability. By defining potential problem areas, necessary corrective action can be taken to reduce the probability of failure and improve the system. We can concentrate our design effort where it will be most beneficial to the system. Other corrective measures such as application of redundancy, changes in components, incorporation of periodic maintenance procedures, and strict control of manufacturing processes and inspection can be carried out. Also, when the prediction indicates that, with normal development effort, the reliability objective will be met easily, we can save unnecessary development costs or can transfer funds to other areas which may require more concentrated effort.

These are only some of the important uses of predictions in improving the reliability of the system. Other uses will become apparent as we proceed through this and following chapters. It is wise to recall our earlier statement that a prediction, by itself, is of little value; it is only when we put it to use that we derive its benefits.

9.2 COMPONENT FAILURE RATES

Several factors considered in predicting reliability include the basic failure rate under laboratory conditions (usually designated as the "generic" rate), the effects of environments on this failure rate (the environmental multiplier), the effects of the equipment's operating level (derating factor), the number of ON-OFF cycles as well as the total number of operating hours, and the nature of the failure distribution. The system configuration (type and degree of redundancy, nature and frequency of maintenance actions, the modes of failure and the effects of each component failure mode on the system) must also be considered.

The prediction of system reliability is usually (although not always) based on the prediction of the reliabilities of its constituent parts and assemblies. After the reliabilities of all components are determined, they are suitably combined to arrive at the system prediction. Three methods of combining these individual predictions are discussed in the next chapter. This chapter is concerned primarily with the individual reliability predictions.

Most reliability predictions utilize the exponential distribution, discussed in Chap. 3. The vast majority of available failure-rate data are derived or at least predicated on the assumption of a constant failure rate. This assumption is realistic because components are usually pretested before being put into service and they are then operated within their normal useful lives. The assumption also facilitates calculation of reliability predictions at the system level by making possible the appropriate combinations of reliabilities of the various components.

The initial step is the determination of a basic failure rate. Failure-rate information is available from both industry and government sources. Unfortunately, much of it is not exact or is somewhat inconsistent, and judicious selection of appropriate numerics is necessary if the ultimate prediction is to be meaningful. On the other hand, since predictions are usually used either to obtain a general idea of the reliability of a component or system or to compare alternative candidate designs, this limitation is not serious and useful prediction numerics can be obtained. In the case where only one component is being evaluated, we may be interested in the order of magnitude or degree of reliability, that is, whether the reliability is 0.90 or 0.99, and the data are sufficiently accurate to provide this distinction. Although they may not be accurate enough to discern whether the reliability stated as 0.99 is really 0.991 or 0.989 or even 0.98, it is seldom that we are concerned, for an individual prediction, with this degree of accuracy.

When we are trying to decide among two or more candidate configurations, we are interested not so much in absolute prediction values as in relative values. As long as we are consistent in our choice and utilization of failure-rate data, the relative merits of the different candidate configurations can be determined even though we may not be sure of the absolute prediction values. Thus, although failure-rate data may not be all that we desire, it is still extremely useful.

9.2.1 Applied Failure Rates

Two approaches to obtaining applied failure rates for parts are available. The first utilizes the generic failure rates and appropriate factors (multipliers) for adjusting these rates for different environmental applications. Most frequently the factors are for a number of environments in combination, as is experienced in a missile, for example. Some multipliers are available for electronic parts for temperature and radiation, but separate factors for such dynamic environments as shock, vibration, and acoustic noise are not as readily found in the literature as are multipliers for combined environments.

The second approach uses failure-rate data from actual applications. However, the exact environment may not be known because information on such things as the use of isolation mounts for an item of equipment installed in an aircraft is unavailable. The application failure rate is sometimes modified directly for the new application, or it may be divided by the appropriate environmental factors to obtain a generic failure rate which is later converted to a failure rate for the new application.

Either approach involves an element of uncertainty. The determination of environmental multipliers is the one aspect of reliability prediction which is most subject to judgment. When both types of failure-rate data—generic and applied—are available, they should both be used to confirm the final rate and reduce the possibility of a gross error in the prediction. The examples in this chapter will be limited to the use of generic failure rates and environmental multipliers, recognizing that applied failure rates provide equally valid results.

9.2.2 Environmental Effects

We have mentioned that exact environmental multipliers are not available and, in the case of applied data, we have raised the question of the knowledge of the applied environment. Shall we then neglect the environment of the new application? Certainly not. The fact that we may have reason to be concerned about some of the failure-rate data which we use in predictions does not relieve us of the responsibility for ascertaining the new environment to which the component will be exposed. Because our original failure rate may be 50 percent in error, we should not compound the problem by neglecting the new environment and perhaps obtaining a 500 percent error. Rather, we should attempt to be accurate to *avoid* having an even greater error than we may already have.

Also, a good knowledge of the new environment can be used to provide worthwhile trade-off information. Perhaps we can suggest a slight additional weight to provide cooling or vibration isolation and thereby greatly reduce the failure rate. These considerations are independent of the original failure rate. Information on the new environment is essential if we are going to make recommendations for reliability improvement. Of what value is a prediction if it is not used for just such purposes?

This brief discussion on the need for knowledge of the new environment should not be interpreted to mean that all elements of the environment must be precisely known. Knowledge of the environment is necessary, but the difference in a vibration level between 6 and 7 g, for example, is normally not sufficient to warrant the excessive expenditure of funds to determine the exact level, even if this were possible. In this case, a good estimate of the general level is sufficient. On the other hand, the difference between 190 and 210°C temperature is very significant for some types of electronic parts (e.g., semiconductor diodes) and can mean the difference between success and failure. Each environment should be considered separately, and consultation with a specialist in environments is suggested. His recommendations could result in considerable savings in test expenditures or in significant improvements in reliability. In general, small changes (less than 25 percent) in dynamic environments do not lead to large changes in failure rates, while changes in temperature, particularly near the recommended maximum, can have significant effects on component life.

Once the environments have been determined to the degree practicable, the effects of these environments on the failure rates must be considered.

As stated earlier, the selection of appropriate environmental multipliers—k factors—is the most subjective and therefore perhaps the most difficult aspect of reliability prediction. While small errors may occur in the determination of appropriate generic failure rates, the choice of an incorrect k factor could result in an error of up to one hundred times the actual failure rate. Consequently, the effects of the environmental exposure should be established as precisely as practicable when the prediction is performed. In a complex mission such as the launch and orbit of a satellite which involves more than one set of environments (e.g., ground, launch, separation, and orbit) the mission is usually divided into phases and an appropriate k factor is used for each phase.

Several sources of k factors are available. Some of these are included with the failure-rate data. Others may be found in the reliability manuals of various companies (many of these manuals are obtainable on request). Some data sources such as the "Navy Failure Rate Data Handbook" (FARADA) provide total observed failure rates experienced in service. These may first be converted to generic (laboratory) rates and then appropriately multiplied for the new application or, when they are applicable, they can be used directly. The FARADA data include information on the application (e.g., aircraft or vehicle type), but such data as mounting method, vibration isolation, or actual operating temperature are not readily available. A list of failure-rate sources generally available is provided in Appendix C. The list is not complete since new tabulations are continually being published.

9.2.3 Derating

In addition to the use of appropriate environmental multipliers, derating factors must be considered. Generic failure rates, particularly those for electronic parts, are often based on operation at rated value (temperature, voltage, power, etc.). However, most parts are operated below these full ratings and this derating has the effect of lowering the failure rate and increasing the reliability. The amount of decrease in the failure rate is not directly proportional to the decrease in operating level, and there is a level below which further derating has little additional benefit. The effects of derating must be determined individually. Several sets of derating curves and tables are available for electronic equipment which either define the percent of the failure rate existing at various derated operating levels and temperatures or provide actual failure rates at the various percents of full rated operation. The FARADA tables provide failure-rate data for derated operation of many commonly used parts and are a good source of failure-rate data for parts made to applicable government specifications.

9.3 MISSION FAILURE RATES

The generic failure rate for a part or component is usually given either in percent per thousand hours or in failures per million hours. The first step in calculating a part's failure rate for a mission is to convert this into the failure rate per hour. The failure rate for the part or component is then found by taking its failure rate per hour and multiplying this by the appropriate derating factor. (When the derated failure rate is given, this figure is used directly.) The resulting failure rate is then multiplied by the environmental k factor to arrive at a basic part hourly failure rate for the mission conditions.

The next step is the calculation of the part failure rate for the mission. The basic part failure rate per hour as determined above is multiplied by the number of *operating* hours per mission. As a first simple example illustrating this step-by-step procedure, assume the following conditions for a part:

Generic failure rate . . . . 0.05 percent per 1,000 hr at full rate operation

Derating factor . . . . 40 percent—operating conditions reduce rate *to* 40 percent of generic rate

Environmental k factor . . . . . 50—includes such conditions as shock, vibration, etc.

Mission time . . . . 100 hr

Operating time . . . .   75 percent of mission
Non-operating failure rate   0

1. The generic failure rate is first converted into a failure rate per hour:
   0.05 percent per 1,000 hr = 0.0005 failure per 1,000 hr = 0.0000005
   failure per hour (= 0.5×10^{-6} failure per hour)
2. This is multiplied by the derating factor: (40 percent = 0.40)
   $0.5 \times 10^{-6} \times 0.40 = 0.2 \times 10^{-6}$ failure per hour
3. The derated failure rate is then multiplied by the k factor:
   $0.2 \times 10^{-6} \times 50 = 10 \times 10^{-6} = 1.0 \times 10^{-5}$ failure per hour
4. The number of operating hours per mission is found and the hourly
   failure rate is multiplied by this value:
   100-hr mission $\times$ 75 percent operating time = 75.0 hr of operation per
   mission
   $1.0 \times 10^{-5} \times 75.0 = 7.5 \times 10^{-4} = 0.00075$ (failure rate per mission)

The mission reliability is $e^{-0.00075} = 0.99925$.

The failure probabilities for the mission for all the various parts are later
appropriately combined according to the logic diagram and mathematical
model to arrive at a mission probability of failure which is converted to a
reliability value.

Sometimes shortcuts are available that will reduce the amount of
arithmetic. Two possibilities which are sometimes encountered relate to the k
factor and to the operating time, but they can be used only for *series*
elements. If the same k factor is applicable to all parts, then it is easier and
quicker to find the combined failure rate for all *series* parts before
multiplying each part's failure rate by k. The combined rate for the series
parts is then multiplied by k, thereby saving one arithmetic step in the
calculation of each part's failure rate.*

Second, if all series parts of a component or subsystem have the same
operating time, then the multiplier for operating time can be applied once to

---

*It is easily shown that the shortcut leads to erroneous results if it is applied to groups
of parallel parts. Assume a simple logic diagram consisting of two elements:

The derated failure rate for each element is assumed to be 0.001 per hour and the k
factor is 100. When we multiply the derated failure rate by the k factor, the resulting
failure rate per element is 0.1 per hour, and the system failure rate is 0.01 since both A
and B must fail for the system to fail ($0.1 \times 0.1 = 0.01$). If the system failure rate were
determined before multiplying by k and were then multiplied by 100, the resulting
system failure rate would be ($0.001 \times 0.001) \times 100 = 0.0001$, which is incorrect.

the combined rate for the series parts rather than to each individual part. This saves a second step in the calculation of failure rates for series parts.*

9.3.1 Non-operating Failure Rates

We will progress from this simple example of a single part in one environment through a series of more and more complex systems and missions to illustrate the various considerations and pitfalls that may be encountered in failure-rate determination.

First of all, it was assumed in the previous example that the component when not operating had a zero failure rate. However, in many instances there is a failure rate associated with a component when it is not in operation such that, when operation is attempted later in the mission, the component is found to have failed.† For example, a pyrotechnic device may detonate prematurely due to susceptibility to radio frequencies; or a solenoid valve required to be in the closed position may open due to shock or vibration; or a pump may sustain damage to its bearings during transportation which renders it inoperable during the mission.

A next possible step, then, in increasing complexity is application of a failure rate to the non-operating state. Let us assume, in the previous example, that the non-operating failure rate is 10 percent of the generic operating failure rate, and we shall recalculate the failure rate for the mission on this basis.

1. The generic hourly failure rate is first multiplied by 10 percent (0.1):
   $0.1 \times (0.5 \times 10^{-6}) = 0.05 \times 10^{-6}$ failure per hour
2. The derating factor is not applicable and this step is omitted.
3. The generic non-operating failure rate is multiplied by the k factor:
   $0.05 \times 10^{-6} \times 50 = 2.5 \times 10^{-6} = 0.25 \times 10^{-5}$
4. The number of non-operating hours is found and the non-operating failure rate is multiplied by this value:
   100-hr mission $-$ 75.0 hr of operation = 25.0 hr non-operating
   $0.25 \times 10^{-5} \times 25.0 = 0.625 \times 10^{-4} = 0.00006$ (failure rate per mission)
5. The operating and non-operating failure rates constitute a series relationship and, therefore, are added to obtain the total failure rate for the mission:
   $0.00075 + 0.00006 = 0.00081$ (failure rate per mission)

The mission reliability, considering both the operating and non-operating failure rates, is $e^{-0.00081} = 0.99919$.

---

*If this shortcut were applied to groups of parallel parts, the same type of error just discussed would result.

†The non-operating state and the storage state are frequently treated as identical, except for possible environmental multipliers. In this chapter they are assumed to be the same when the environmental conditions are similar.

In this example, the non-operating failure rate for the mission is so much smaller than the operating failure rate that it could be omitted without significantly affecting the overall prediction.* It should be noted that non-operating failure rates are difficult to obtain and usually must be estimated. Some failure rates at high temperatures and in radiation environments are available for electronic parts and components.

Non-operating failure rates for a mission are not always much smaller than operating rates, and care should be exercised before arbitrarily omitting non-operating rates from the calculations. In the operation of a satellite, for example, much of the equipment may not operate during the greater portion of the mission. Even though the non-operating hourly failure rate is much lower than the operating hourly failure rate, when the relative times are considered the failure rate for the non-operating portion may be higher than that for the operating part of the mission.

As another example, the environment encountered during a launch is much more severe than that encountered in orbit. For a short-duration mission the overall failure rate for the launch portion for a component which is not operated can be much greater than for the orbit portion when it is in operation. This situation is usually covered in another manner: by dividing the mission into phases. Each phase is then considered separately. This is the next step in the order of complexity.

9.3.2 Mission Phases

We have been discussing the launch and orbit of a spacecraft or satellite. Such a mission involves at least three phases: ground operation, launch, and orbit. A manned spacecraft would also include deorbit, reentry, landing, and post-landing phases. Also, if the launch vehicle has more than one stage, the launch phase may be subdivided, and an orbit injection and separation function may be added. If the mission includes maneuvers in space, the orbit phase may be broken into several phases. For our purposes it will be sufficient to consider only the first three phases mentioned. Appropriate environmental factors and times are determined for the three mission phases and are listed in Table 9.1.

*Accuracy of ± 50 percent is often considered acceptable due to the uncertainty of the data being used—generic rates, k factors, etc. Accuracy of ± 20 percent meets the requirements of most programs.

TABLE 9.1 Phases, times, and environmental multipliers

Phase	k factor	Phase time
Ground . . .	5	24 hr
Launch . . .	400	0.10 hr
Orbit	1	100 days

We will use the same generic failure rate of 0.5×10^{-6} failure per hour and the same derating factor of 0.40, and we will use operating times of 25 percent on the ground (during check-out and launch preparation), 0 percent during launch, and 50 percent during orbit. The non-operating failure rate is assumed to be 5 percent of the operating rate.

Mission phase	Mode	Failure rate	Derating factor	k factor	Phase time hr	Fraction of time	Mission failure rate
1. Ground							
	Operating	$(0.5 \times 10^{-6}) \times 0.40$		$\times$ 5	$\times$ 24	$\times$ 0.25	= 0.000006
	Non-operating	$(0.5 \times 10^{-6}) \times 0.05$		$\times$ 5	$\times$ 24	$\times$ 0.75	= 0.000002
2. Launch		$(0.5 \times 10^{-6}) \times 0.05$		$\times$ 400	$\times$ 0.1	$\times$ 1.00	= 0.000001
3. Orbit							
	Operating	$(0.5 \times 10^{-6}) \times 0.40$		$\times$ 1	$\times$ 2,400	$\times$ 0.50	= 0.00024
	Non-operating	$(0.5 \times 10^{-6}) \times 0.05$		$\times$ 1	$\times$ 2,400	$\times$ 0.50	= 0.00003

Total = 0.000279
= 0.00028
(failure rate per mission)*

*Note that only two-place accuracy is retained.

Usually, it is more convenient to determine the rate of failure for each phase for all components and then to determine the overall rate for the mission. When the logic diagram for any system or subsystem changes from phase to phase, it is necessary to consider one phase at a time. Then the total mission failure rate for the component is used only for information but is not used in calculating the overall mission prediction. This will become clearer in a more complex case which we will examine next.

9.4 DETAILED EXAMPLE

We will again consider a mission of only three phases (ground, launch, and orbit), recognizing that these are made up of several subphases of various durations and environments. We will assume that the launch must take place within a predetermined time period and that therefore a failure during ground check-out can fail the mission. Additional simplifying assumptions, for purposes of illustration, concern the number and configuration of the various systems and subsystems. For example, other systems and subsystems such as a programmer and sequence controller and an attitude control system are not shown in order to avoid too lengthy a discussion. The following logic diagrams will apply to the three phases. Notes provide assumptions and reasons for depicting the logic diagrams as shown.

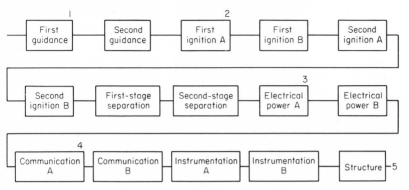

(1) Launch will not be initiated unless all systems are satisfactory. (2) There are two sets of igniters for each engine; premature ignition of any engine system fails the mission. (3) Although either electric power system is adequate, launch will not be attempted unless both are satisfactory. (4) Comment 3 on the electric power system also applies to the communication and instrumentation systems. (5) Propulsion cannot be checked on ground and is assumed to be unsusceptible to ground failure.

FIG. 9.1   Logic diagram for ground check-out phase.

An actual mission logic diagram is usually even more complex. For example, one electrical system might be associated with only one-half of the ignition system, e.g., electrical system A might activate ignition A, and electrical system B might activate ignition B, as shown in Fig. 9.4. There is no way of depicting the true logic relationships. The electric power systems are shown twice (once dotted so that they are not numerically counted twice). This type of logic diagram results in the smallest chance of misinterpretation. The diagram in Fig. 9.5 would be less correct because the failure of first ignition A does not preclude the use of second ignition A, and the possibility of misinterpretation would be greater. Problems of logic diagrams are discussed more fully in Chap. 7.

In general, we will use the same k factors and mission phase times as in the previous example for the failure-rate calculations. Some k factors, however, will vary because of the nature of the components. The failure rates and percent operating times are listed in Table 9.2.

The failure-rate calculations for each of the subsystems are then computed on a phase-by-phase basis (Table 9.3). Failure probabilities for each phase for both the operating and non-operating states are listed in the right-hand column of the table. The probabilities for all subsystems and phases can be suitably combined to obtain an overall failure probability (and reliability) for the complete system. Three methods for combining component and subsystem reliabilities are discussed in the next chapter.

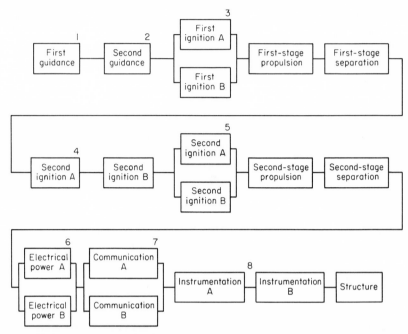

(1) First guidance is assumed to be launch guidance. Full launch time applies.
(2) Second guidance is assumed to be orbit guidance. It is assumed to be non-operating during launch and updated by ground control at start of orbit. (3) Either first ignition system will ignite first-state propulsion. (4) Second ignition systems must not function prematurely; therefore, they are shown in series during first-stage propulsion. (5) Either second ignition system will ignite second-stage propulsion. (6) Either electric power system is sufficient. However, if one fails it is not available during orbit. (7) Comment 6 also applies to the communication system. (8) If either instrumentation system fails, essential data are lost. Instrumentation system A applies only to propulsion and separation characteristics.

FIG. 9.2   Logic diagram for launch phase.

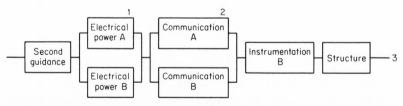

(1) If one electric power system failed during the launch phase, it will not be available for orbit. Overall mission-failure-rate calculations must reflect this. (2) Comment 1 applies to the communication system. (3) Systems and subsystems not shown are not required in this phase.

FIG. 9.3   Logic diagram for orbit phase.

TABLE 9.2 Failure Rates and Percent Operating Times for the Various Subsystems

| Subsystem | Percent of phase time operating[a] | | | Generic failure rate per 1,000 hr | |
	Ground	Launch	Orbit	Operating	Non-operating
First guidance	25	100	Not applicable	0.01	0.0001
Second guidance	25	0	100	0.01	0.0001
First ignition	0	S.S.D.[b]	Not applicable	0.1[g]	0.001
Second ignition	0	(c)	Not applicable	0.1[g]	0.001
First-stage separation . .	0	(c)	Not applicable	0.01[g]	0.0001
Second-stage separation . .	0	(d)	Not applicable	0.01[g]	0.0001
Electrical	75	100	100	0.02	0.0002
Communications	40	50	50	0.05	0.0005
Instrumentation A	40	100	Not applicable	0.01	0.0005
Instrumentation B	60	0	100	0.01	0.0005
First-stage propulsion . . .	Not applicable	50[e]	Not applicable	0.05[g]	Not applicable
Second-stage propulsion . . .	Not applicable	50	Not applicable	0.05[g]	0.001
Structure	100[f]	100	100	0.001	Not applicable

[a] Non-operating percent = 100 − operating percent unless otherwise noted.
[b] Single-shot device.
[c] Inactive for 50 percent of launch phase, then S.S.D.
[d] Inactive for 100 percent of launch phase, then S.S.D.
[e] No non-operating period during launch phase for first-stage propulsion.
[f] Structure k factors are: ground, 1; launch, 2,000; orbit, 0.1.
[g] Failure per 1,000 operations; launch phase k factor is 100 because dynamics are self-induced

161

TABLE 9.3 Failure Rates for Each Subsystem in Each Phase

Subsystem	Mission phase	Operating/non-operating	Generic rate per 1,000 hr	Time*	k factor†	Phase total
First guidance	Ground	Operating	0.01	6	5	0.0003
	Ground	Non-operating	0.0001	18	5	0.000009
	Launch	Operating	0.01	0.1	400	0.0004
Second guidance	Ground	Operating	0.01	6	5	0.0003
	Ground	Non-operating	0.0001	18	5	0.000009
	Launch	Non-operating	0.0001	0.1	400	0.000004
	Orbit	Operating	0.01	2,400	1	0.024
First ignition	Ground	Non-operating	0.001	24	5	0.00012
	Launch	Operating	0.1	1 oper.	100	0.01
Second ignition	Ground	Non-operating	0.001	24	5	0.00012
	Launch	Operating	0.1	1 oper.	100	0.01
	Launch	Non-operating	0.001	0.05	400	0.00002
First-stage separation	Ground	Non-operating	0.0001	24	5	0.000012
	Launch	Operating	0.01	1 oper.	100	0.001
	Launch	Non-operating	0.0001	0.05	400	0.000002
Second-stage separation	Ground	Non-operating	0.0001	24	5	0.000012
	Launch	Operating	0.01	1 oper.	100	0.001
	Launch	Non-operating	0.0001	0.1	400	0.000004
Electrical	Ground	Operating	0.02	18	5	0.0018
	Ground	Non-operating	0.0002	6	5	0.000006
	Launch	Operating	0.02	0.1	400	0.0008
	Orbit	Operating	0.02	2,400	1	0.048

Component	Phase	State				
Communications	Ground	Operating	0.05	9.6	5	0.0024
	Ground	Non-operating	0.0005	14.4	5	0.000036
	Launch	Operating	0.05	0.05	400	0.001
	Launch	Non-operating	0.0005	0.05	400	0.00001
	Orbit	Operating	0.05	1,200	1	0.06
	Orbit	Non-operating	0.0005	1,200	1	0.0006
Instrumentation A	Ground	Operating	0.01	9.6	5	0.00048
	Ground	Non-operating	0.0005	14.4	5	0.000036
	Launch	Operating	0.01	0.1	400	0.0004
Instrumentation B	Ground	Operating	0.01	14.4	5	0.00072
	Ground	Non-operating	0.0005	9.6	5	0.000024
	Launch	Non-operating	0.0005	0.1	400	0.00002
	Orbit	Operating	0.01	2,400	1	0.024
First-stage propulsion	Launch	Operating	0.05	1 oper.	100	0.005
Second-stage propulsion	Launch	Operating	0.05	1 oper.	100	0.005
	Launch	Non-operating	0.001	0.05	400	0.00002
Structure	Ground	Operating	0.001	24	1	0.000024
	Launch	Operating	0.001	0.1	2,000	0.0002
	Orbit	Operating	0.001	2,400	0.1	0.00024

*Time = phase time × percent applicable. Phase times are: ground, 24 hr; launch, 0.1 hr; orbit, 2,400 hr. Percentages are given in Table 9.2.
†See footnotes f and g in Table 9.2.

163

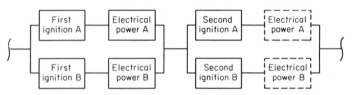

FIG. 9.4   Logic diagram when electrical systems are associated with particular ignition systems.

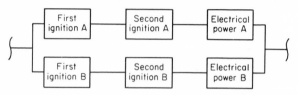

FIG. 9.5   Alternate logic diagram for system in Fig. 9.4.

9.5   EVALUATION OF THE DATA

The contribution of each system to the mission failure rate can be reviewed
and evaluated. Those systems which contribute a large percentage of the total
failure rate can be analyzed in detail and suitable action can be taken to
reduce their failure rates. The communications system, at first glance, appears
to contribute heavily toward the mission failure rate in the orbit phase.
However, the logic diagram shows that this system is redundant and its overall
failure probability during orbit is really about 0.0036. Similarly, the electrical
power system is also redundant and its failure probability during the orbit
phase is approximately 0.0023. Operation of the two ignition stages is also
redundant and the associated failure probabilities are low. This leaves the
failure rates of the second guidance system (0.024 in orbit) and the second
instrumentation system (0.024 in orbit) as the principal contributors to the
overall mission failure rate. Effort can be concentrated on these two systems
where it will be most beneficial, and trade-offs between weight, mission
purposes, and reliability can be made on an objective, quantitative basis.

If desired, the data can be rearranged by mission phase to facilitate
calculation of the failure probability during each phase of the mission. The
operating and non-operating failure probabilities can also be computed
independently. In the last example, as in the earlier one, the total of the

failure rates in the non-operating state is relatively small and may be omitted from further consideration. However, should the operating failure rates be considerably reduced by appropriate design changes, then the non-operating rates might become a more significant part of the total. As a rule of thumb, if the non-operating rate is less than 10 percent of the total and if there are no significant time or environmental differences which increase the total non-operating failure probability relative to the operating failure probability, the non-operating rate may be omitted.

ADDITIONAL READINGS

ARINC Research Corporation: "Reliability Engineering," Prentice-Hall, Inc., Englewood Cliffs, N.J., 1964.

Bazovsky, I.: "Reliability Theory and Practice," Prentice-Hall, Inc., Englewood Cliffs, N.J., 1961.

Calabro, S. R.: "Reliability Principles and Practices," McGraw-Hill Book Company, New York, 1962.

Dummer, G. W., and N. B. Griffin: "Electronics Reliability: Calculation and Design," Pergamon Press, Ltd., London, 1966.

Earles, D. R.: Reliability Growth Prediction during the Initial Design Analysis, *Proc. Nat. Symp. Reliability Quality Control, 7th, Philadelphia,* Institute of Electrical and Electronics Engineers, Inc., New York, 1961, pp. 380-393.

Shooman, M. L.: "Probabilistic Reliability: An Engineering Approach," McGraw-Hill Book Company, New York, 1968.

10

SYSTEM PREDICTION

10.1 INTRODUCTION

The prediction of system reliability utilizes many of the methods which have been presented. The procedures discussed in the preceding three chapters relating to logic diagrams, mathematical models, and component predictions are of particular importance. Knowledge of the functional relationships of the various components and estimates of their individual reliabilities are necessary for a meaningful system reliability prediction.

There are three different approaches to obtaining a system prediction: exact or semiexact mathematical models and associated calculations; simulation techniques usually termed *Monte Carlo*; and the determination of limiting values as exemplified by the *Method of Bounds*. The first method—mathematical models—is used most frequently because the majority of systems and subsystems can be described mathematically. We can usually subdivide a system into its constituent subsystems and evaluate these independently, and then appropriately combine the subsystem reliability predictions to obtain a prediction for the overall system. However, sometimes the system, its subsystems, or their interrelationships are so complex that a mathematical model is not practical or, in some cases, not possible. In these instances, simulation procedures have found application. These consist of performing large numbers of iterative calculations on a success/failure basis using a computer. Random number techniques are employed.

The Method of Bounds is a means of determining limiting values, within which falls the true reliability prediction. Limiting value techniques reduce the time and expense associated with simulation procedures and provide equivalent accuracy. They can also be used for less complex systems to provide more rapid determination of the system prediction numeric than is possible using exact mathematical models.

10.1.1 Preliminary Steps

Regardless of which method is used, for any except simple series systems, it is desirable and usually necessary to prepare a logic diagram following the guidelines discussed in Chap. 7. Also required when the system is complex is a failure-modes-and-effects analysis (FMEA). The FMEA delineates for every component each possible mode of failure, its effects on the component and on the system, and related information. A complete FMEA includes failure causes, relative probabilities of occurrence, and possible means of prevention or correction. It is used in systems evaluations, design reviews, failure analyses, and other facets of the overall reliability effort. When used with the logic diagram for preparing a system reliability prediction, the portions of the FMEA which are most needed are the modes of failure, their effects on the system, and any relevant comments. To simplify the presentation, only those portions of the FMEA needed for the prediction are shown in the example which follows.

Also required are failure rates by mode for those components which appear as parallel elements in the logic diagram. Normally these rates are not available in standard tables, although the total component failure rate may be well known for the particular application. The rates per mode can be estimated with sufficient accuracy for prediction purposes by using knowledgeable judgments of those engineers who have broad experience with the particular component. Determination of failure rates by mode will be discussed in more detail in the example. To illustrate the calculation of a system reliability prediction, the example will be worked out in detail using both a semiexact mathematical model and a limiting value method. The application of a simulation technique will also be discussed but a complete computer program will not be shown.

10.1.2 Illustrative Example

In the development history of a power conversion system to be used to convert heat energy from a nuclear reactor to electrical energy, three possible redundant configurations were considered for the primary heat-transfer loop (subsystem). These are shown schematically in Fig. 10.1. The heat-transfer fluid is a sodium-potassium eutectic mixture (NaK), which is circulated through the loop by a centrifugal pump-motor assembly (PMA). After being heated by the reactor, the NaK flows through a boiler where heat is transferred to a flow of liquid mercury (Hg) which is changed to a gaseous state and turns a turbine. In addition to the pump, reactor, and boiler, the primary heat-transfer loop contains a heat exchanger (used during system start-up) and valving. The loop is closed, recirculating the NaK continuously. (The reactor portion of the loop is not shown in the figure.)

The three redundant configurations were being considered in order to provide continuous system operation in case of failure of one of the components. After the primary heat-transfer loop, which was to have two

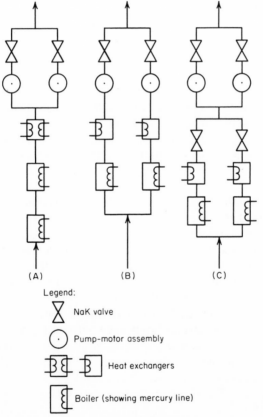

(A)   (B)   (C)

Legend:

NaK valve

Pump-motor assembly

Heat exchangers

Boiler (showing mercury line)

FIG. 10.1 System diagrams of three candidate configurations for the primary heat-transfer loop.

boilers and two pumps, the remaining subsystems contained identical redundancy regardless of the first loop's configuration. Consequently, the study was limited to a comparison of the three configurations of the first loop.

Ground rules and criteria were developed to facilitate calculations and enable comparison of the three configurations. It is important that these applicable considerations be explicitly stated to provide firm bases for both present and future evaluations. These considerations were

1. Any of the three configurations will provide adequate system performance.

2. Only one PMA may operate at one time in order to maintain constant system operating conditions. Therefore, redundancy is sequential.

3. If the valves in both PMA legs are open, internal circulation (between the two parallel legs) develops and the system fails.

4. Moderate internal leakage through a valve does not constitute a failure.

5. The standby (non-operating) failure rates with respect to functional operation are zero. However, the standby failure rates with respect to external leakage and NaK plugging of the boiler or pump are the same as the operating failure rates. (These are reflected in the logic diagram by depicting these failure modes in series.)

6. Sufficient redundancy is available for failure detection and switching controls, and these functions are assumed to be 100 percent reliable. (Internal boiler leaks cannot be detected; the valve operations themselves are not controls.)

In conjunction with the stated ground rules, the FMEA was prepared. An abbreviated version of the FMEA, showing only those aspects necessary to the present discussion, is shown in Table 10.1.

Logic diagrams were prepared along with the FMEA and ground rules. Three diagrams were required, one for each configuration. Figure 10.2 shows the logic diagram for configuration A of Fig. 10.1. Two of the three methods of predicting system reliability, discussed in this chapter, will be used to evaluate this configuration.

After delineation of ground rules and preparation of the FMEA and logic diagrams, the next step was the determination of failure rates for all modes of failure. While this determination involves an element of engineering judgment, it should not be construed as introducing inaccuracies. The percentage of total failures of each component in each mode is estimated by those engineers who are most familiar with the component, its operation, and its past failure history. These include system, development, design, test, and reliability engineers with applicable experience. Their pooled opinions are used, and the resulting percentages are multiplied by the overall component failure rate to obtain a failure rate for each mode. In those instances where this procedure has been used, there has been excellent agreement among the estimates even

TABLE 10.1 Failure Modes and Effects Analysis—Abbreviated System Effects (Ref. Fig. 10.1)

Failure mode	Configuration A	Configuration B	Configuration C	Comments†
(1) PMA fails to operate or degrades	Switch to 2d PMA (both boilers still available)	Switch to 2d PMA and 2d turbine* (1st boiler not available)	Switch to 2d PMA (both boilers still available)	Configuration A could be provided with two heat exchangers. Heat exchanger failure rate is relatively low
(2) PMA leaks externally	System fails	System fails	System fails	
(3) Heat exchanger fails internally	System fails (20%) No effect (80%)	Switch to 2d PMA and 2d turbine (20%) (1st boiler and PMA not available). No effect (80%)	Switch to 2d boiler and 2d turbine (20%) (both PMAs still available). No effect (80%)	
(4) Heat exchanger fails externally	System fails	System fails	System fails	
(5) Boiler leaks externally	System fails	System fails	System fails	
(6) Boiler leaks internally	System fails	System fails	System fails	Assumes that Hg leak is not detectable with present capabilities
(7) Boiler plugging (NaK side)	System fails	Switch to 2d PMA and 2d turbine (1st PMA not available)	Switch to 2d turbine (both PMAs still available)	
(8) Boiler plugging (Hg side)	Switch to 2d turbine (both PMAs still available)	Switch to 2d PMA and 2d turbine (1st PMA not available)	Switch to 2d turbine (both PMAs still available)	
(9) Boiler performance degradation	Switch to 2d turbine (both PMAs still available)	Switch to 2d PMA and 2d turbine (1st PMA not available)	Switch to 2d turbine (both PMAs still available)	Assumes no obstruction to NaK flow
(10) Erosion/corrosion of boiler	Switch to 2d turbine (both PMAs still available)	Switch to 2d PMA and 2d turbine (1st PMA not available)	Switch to 2d turbine (both PMAs still available)	Assumes no effect on 2d boiler or on NaK loop

Failure mode				Comments
(11) Structural failure of boiler	System fails	System fails	System fails	Assumes gross loss or contamination of NaK
(12) Valve leaks externally	System fails	System fails	System fails	
(13) Valve of PMA fails to close when required – when PMA fails	System fails	System fails	System fails	Internal NaK circulating loop is developed
(13a) Valve at PMA fails to close when required – when PMA does not fail	Not applicable	System fails	Not applicable	Internal NaK loop is developed when applicable
(14) Valve at 2d PMA fails to open when required	System fails	System fails	System fails	
(15) Valve at PMA closes inadvertently	Switch to 2d PMA (both boilers still available)	Switch to 2d PMA and 2d turbine (1st boiler not available)	Switch to 2d PMA (both boilers still available)	
(16) Valve at 2d PMA opens inadvertently	Switch to 2d PMA (both boilers still available)	Switch to 2d PMA and 2d turbine (1st boiler not available)	Switch to 2d PMA (both boilers still available)	
(17) Valve at heat exchanger fails to close when required	Not applicable	Not applicable	System fails	Except if failure is NaK boiler plug
(18) Valve at 2d heat exchanger fails to open when required	Not applicable	Not applicable	System fails	
(19) Valve at heat exchanger closes inadvertently	Not applicable	Not applicable	Switch to 2d turbine (both PMAs still available)	
(20) Valve at 2d heat exchanger opens inadvertently	Not applicable	Not applicable	Switch to 2d turbine (both PMAs still available)	

*Second turbine includes second set of all associated subsystems. (Turbines, not shown in system schematics of primary loop, are rotated by mercury vapor from boiler.)

†Switching includes all necessary valve position changes; this comment refers to all applicable failure modes.

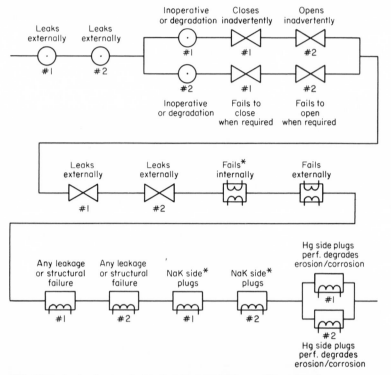

FIG. 10.2 Logic diagram (using circuit symbols) for configuration A.

*This mode is shown separately because logic diagrams of other configurations are different for these failure modes.

though they were made independently. The estimates which were made for the four components of the primary heat-transfer loop displayed this consistency. Table 10.2 shows the estimates made by seven cognizant engineers of the percentage of failures of the boiler in each mode of failure. The consensus, to the nearest 5 percent, is also shown in the table.

Failure rates per mode were then found by multiplying the operating failure rate (generic rate × environmental multiplier × application factor × time) by the percent of failures occurring in each mode. The failure rates per mode are given in Table 10.3. Note that these rates are for a 1,000-hr mission.* If the mission time were changed, the rates for each mode of failure would be changed before calculating the system reliability. As stated in Chap. 8, except

*The failure rates given here were determined early in the development program. Final reliabilities were significantly improved.

TABLE 10.2   Summary of estimates of percentage of failures in each mode for the boiler

Failure modes	Estimates							Consensus (to nearest 5%)
	1*	2*	3	4	5	6	7	
External leak (NaK to vacuum) .	10	10	10	15	30	10	5	10
Internal leak	25	25	10	20	10	10	10	20
Plugging (NaK side)	5	5	0	15	0	5	5	5
Plugging (Hg side)	20	20	5	2½	0	20	10	15
Performance degradation (not caused by plugging)	15	15	50	15	30	20	5	20
Erosion/corrosion	20	20	20	25	30	25	60	25
Structural failure (other than leaks)	5	5	2½	5	0	10	0	5
Other modes	0	0	2½	2½	0	0	5	0

*Weighted double because of depth of experience: 1 Design engineer A; 2 Design engineer B; 3 Reliability engineer; 4 Test engineer A; 5 Test engineer B; 6 Systems engineer; 7 Test engineer C.
Note: Overall boiler failure rate = 0.1 per 1,000 hr; percent x overall rate = rate per mode.

TABLE 10.3   Failure rates for a 1,000-hr mission

Item*	Component	Mode of failure	Failure rate
(1)	PMA	Fails to operate	0.196
(2)	PMA	Leaks externally	0.004
(3)	Heat exchanger	Fails internally	0.001†
(4)	Heat exchanger	Fails externally	0.005
(5)	Boiler	Leaks externally	0.010
(6)	Boiler	Leaks internally	0.020
(7)	Boiler	Plugs (NaK side)	0.005
(8)	Boiler	Plugs (Hg side)	0.015
(9)	Boiler	Performance degradation	0.020
10)	Boiler	Erosion/corrosion	0.025
11)	Boiler	Structural failure	0.005
12)	Valve	Leaks externally	0.003
13)(17)‡	Valve	Fails to close	0.064
14)(18)	Valve	Fails to open	0.064
15)(19)	Valve	Closes inadvertently	0.006
16)(20)	Valve	Opens inadvertently	0.006

*Item numbers correspond to those in Table 10.1.
†The failure rate is 0.005; only 20 percent are failures of system.
‡Items (17) through (20) are not applicable to configuration *A*.

for simple series systems, it is necessary that the failure rate for the mission be determined for each element in the logic diagram before system calculations are performed.

10.2 USE OF THE MATHEMATICAL MODEL

The first method of computing the loop reliability utilizes a mathematical model. The redundant portions of the loop are operated sequentially; that is, the second component is not used until the first one fails. Also, in configuration A, which we will be evaluating, the first redundant portion (see Fig. 10.2) has two different failure rates because the two legs contain dissimilar failure modes of the valves and these modes have different failure rates. Since failure detection and switching controls are 100 percent reliable (ground rule 6), the applicable equations are Eqs. (8.1), (8.2), (8.34), and (8.40). Equation (8.40) is used to find the reliability of the first redundant portion. Equation (8.34) is used for the second redundant portion, and Eqs. (8.1) and (8.2) are employed for finding the overall loop reliability. The complete mathematical model is given by Eq. (10.1). Figure 10.2 is redrawn to label the various elements for use in the model. Identical elements with equal failure rates are given the same identification. (See Fig. 10.3.)

$$R_T = R_A^2 \times R_X \times R_G^2 \times R_H \times R_I \times R_J^2 \times R_K^2 \times R_Y$$

$$R_X = e^{-\lambda_1 t} + \frac{\lambda_1}{\lambda_1 - \lambda_2} \left(e^{-\lambda_2 t} - e^{-\lambda_1 t} \right)$$

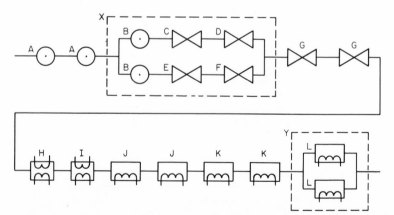

FIG. 10.3 Identification of elements of logic diagram (using circuit symbols) for use in mathematical model.

$$R_X = e^{-(F_B + F_C + F_D)} + \left[\frac{(F_B + F_C + F_D)}{(F_B + F_C + F_D) - (F_B + F_E + F_F)} \right.$$

$$\left. \times\ e^{-(F_B + F_E + F_F)} - e^{-(F_B + F_C + F_D)} \right) \Big]$$

$$R_Y = e^{-\lambda t}(1 + \lambda t) = e^{-F_L}(1 + F_L)$$

$$R_T = R_A^2 \times R_G^2 \times R_H \times R_I \times R_J^2 \times R_K^2 \times R_X \times R_Y$$

$$R_T = e^{-(2F_A + 2F_G + F_H + F_I + 2F_J + 2F_K)} \times R_X \times R_Y \qquad (10.1)$$

$$R_X = e^{-(0.196 + 0.006 + 0.006)}$$

$$+ \left[\frac{(0.196 + 0.006 + 0.006)}{(0.196 + 0.006 + 0.006) - (0.196 + 0.064 + 0.064)} \right.$$

$$\left. \times\ (e^{-(0.196 + 0.064 + 0.064)} - e^{-(0.196 + 0.006 + 0.006)}) \right]$$

$$= e^{-0.208} + \frac{0.208}{0.208 - 0.324}(e^{-0.324} - e^{-0.208})$$

$$= 0.812 + \frac{0.208}{-0.116}(0.723 - 0.812) = 0.972$$

$$R_Y = e^{-0.060}(1 + 0.060) = 0.9418(1.060) = 0.998$$

$$R_T = e^{-(2 \times 0.004 + 2 \times 0.003 + 0.001 + 0.005 + 2 \times 0.035 + 2 \times 0.005)}$$

$$\times\ 0.972 \times 0.998 = 0.905 \times 0.972 \times 0.998 = 0.878$$

10.3 THE MONTE CARLO METHOD

The simulation, or *Monte Carlo,* method is used for reliability prediction when an exact mathematical model cannot be developed economically or when it becomes too complex to permit timely evaluation. This method involves the determination of the distributions of the parameters of the various elements in a system, selection of a random sample of each element and its parameters, and combining of these samples to obtain a measure of the system performance or reliability. This implies, of course, that the effects on the system of different values of the element parameters (e.g., the relationship between component failures and system reliability) are known. (These relationships must be known regardless of which method is used to predict system reliability.) The process of random selection and determination of system effects is repeated a large number of times, perhaps as many as 10,000, each repetition resulting in another independent estimate of the system characteristic that is being measured (e.g., reliability). The law of large numbers states that, as the sample size (number of estimates) increases, the difference between the sample mean and the population mean becomes smaller and smaller, and the sample mean becomes an increasingly better estimate of μ.* Also, the distribution of the samples becomes a better and better representation of the distribution of the population.

The Monte Carlo process has been used for determining distributions of system performance characteristics, for assessing system reliability based on actual test data of components and assemblies, and for other related types of system evaluations, as well as for predicting system reliability from predicted values of element reliabilities. However, we are concerned here only with its use in prediction.

The predicted reliability of each element of the logic diagram can be represented by a set of random numbers. For example, if the reliability of an element were 0.8000, then success (satisfactory operation) of that element could be represented by all numbers from 0.0000 through 0.7999 and failure (unsatisfactory operation) by the numbers from 0.8000 through 0.9999. There are tables of random numbers which list tens of thousands of numbers from zero to one in random sequence.† The table is entered at any convenient point and a number is selected. Depending on its value, it represents either a success or a failure of the particular element. The selection of numbers is continued *in sequence,* each succeeding number representing another success or failure. In actual practice, the procedure is performed by a computer which has within its data bank a program for generating random numbers. This selection and interpretation is performed for each applicable element in the system, and by appropriately combining results, simulates the system performance (reliability). The steps in the procedure may be as follows.

The random number set for the first element in the logic diagram is entered and a number is selected, thereby simulating success or failure of the element.

---

*We learned in Sec. 3.5 that the variation of sample averages is inversely proportional to the square root of the sample size.

†Two readily available sources are W. J. Dixon and F. J. Massey, Jr., "Introduction to Statistical Analysis," 3d ed., pp. 446-450, McGraw-Hill Book Company, New York, 1969, and D. B. Owen, "Handbook of Statistical Tables," pp. 519-538, Addison-Wesley Publishing Company, Inc., Reading, Mass., 1962. A third source, which provides a million random numbers, is The Rand Corporation, "A Million Random Digits with 100,000 Normal Deviates," The Free Press, New York, 1955.

Based on the logic diagram and description of the system's operation and functional relationships of the elements, the computer program is designed so that a particular course of action is taken depending on whether a success or a failure is represented by the first random number. If the number represents a success, the set of random numbers for the next element in the same logic path is entered and the new random number value determines whether that element is a success or failure. The process is continued until a failure is encountered, which thereby fails that path in the logic diagram.

The computer is programmed to automatically backtrack to the closest junction of another parallel path when a failure occurs. It then uses the appropriate random number table for the first element in the parallel path. If this selection simulates a success for the first element, the random number set for the second element in the parallel path is used to determine (simulate) that element's success or failure. The process is continued until a complete success path has been found (denoting a system success) or a failure has been simulated in all possible paths (denoting a system failure). The process is repeated hundreds or perhaps thousands of times to obtain the simulated reliability prediction numeric for the system. The numeric is the simple ratio of successes to total trials.

Since a computer is used, the actual simulation process is relatively rapid and many simulations can be made in a comparatively short time. The length of time will depend, of course, on the complexity and number of elements in the system. It will also depend on the reliabilities of the elements. If the reliabilities are high, many trials may simulate system success using only the first possible path. If the element reliabilities are low, each trial may require trying many or all possible paths before a system success or failure is simulated.

The preparation of actual computer programs is considerably more difficult than it may appear from the preceding brief discussion. If it were as simple as indicated, it would not be necessary to use the Monte Carlo process and a system mathematical model could be developed. Several factors affect the difficulty in program development. Three of these are given here to indicate the nature and extent of the complexity.

First, the redundancy of many systems is sequential rather than active parallel. This means that the time of failure of one element influences the reliability of its sequentially redundant alternate. Therefore, the random number set representing success or failure of the redundant element is not constant but is rather a function of the particular random number which represents the time of failure of the element in the first path. Each number of the first set which represents a failure of the first element also determines which random number set is to be used for the redundant element. An alternate method is that the second random number set remains the same but the interpretation of each value as a success or a failure varies as a function of the random number in the set for the first element.

Second, complex missions frequently involve several mission phases Individual logic diagrams apply to each phase. An element which is redundan' in one phase may be in series in another or may be in a different redundan' configuration. If it fails in one phase, it is not available in succeeding phases The computer program must be designed so that the state of each componen' is retained from one phase to the next until the complete mission has beer simulated. Also, when more than one phase is involved, the determination o: one success path per phase is not sufficient. The state of *every* element mus' be determined in each phase to permit correct simulation of succeeding phases.

Third, a mission may have several objectives and the probability of meetin; each objective individually may be required. In many systems, probabilities o personnel safety and of mission success are both determined and ar' interrelated. Decisions relating to mission continuation and thus to botl mission success and personnel safety are influenced by the number o redundant paths which remain available for performing critical functions. A' the same time, the probabilities of successfully accomplishing other non-critical functions may require computation (simulation). The progran' must be designed to simultaneously determine all these interdependen' probabilities and the programming will, of necessity, be quite complex.

Since the subject of computer programming is beyond the scope of thi: text, we will not discuss the simulation of system reliability predictions t(any greater extent. It is sufficient to indicate the nature of the procedure an(its application to reliability to provide the reader with an introduction to thi' method. Should it be necessary to utilize the Monte Carlo method, th(activity should be carried out as a joint effort of the Computer Programmin; and Reliability departments. Some of the additional readings provide furthe: information on the use of computers for simulation. The Method of Bounds discussed next, is a powerful tool for determining the reliability prediction: for complex systems, and in most cases it obviates the need for using th(Monte Carlo procedure.

10.4   THE METHOD OF BOUNDS

The third method for predicting reliability—the Method of Bounds—is : limiting value procedure and will be discussed in detail.* Its distinc' advantages warrant its consideration whenever the reliability is to b(computed for any system other than series or the most simple redundan' configuration. It is an excellent time-saving substitute for the more lengthy

---

*The method was first presented publicly by this author in *Proc. Ann. Symp* *Reliability, Boston, 1968,* and is presented here by permission of the Institute o Electrical and Electronics Engineers, Inc.

mathematical model procedure, as well as a very convenient technique for highly complex systems where exact math models cannot be developed and simulation procedures have heretofore been required. It has been employed for evaluating such complex systems as the Apollo spacecraft* where its accuracy was confirmed by the use of a simulation program and on other systems for comparing alternate candidate designs.

The method involves the calculation of upper and lower limits for the prediction and requires only simple success and failure probability computations and an equally simple means of combining these values to obtain a realistically precise estimate of the true value. Probabilities of failure cases (for the system) are subtracted from unity to obtain an upper reliability prediction bound, and probabilities of success cases (for the system) are added to obtain a lower bound. The more cases that are considered, the closer together the two bounds become. As a rule of thumb, *the process is continued until the difference between the two bounds is about the same magnitude as the difference between the upper bound and unity.* It is generally possible but not very practical or necessary to have the two bounds very close for two reasons:

1. In order to close in the bounds, more and more cases must be considered. These additional cases are more complex, and the complexity grows geometrically.

2. There is no significant difference between the estimate found by combining the limit values empirically and the estimate found by further converging the bounds. The method for combining the limit values is explained following the description of bound calculations.

10.4.1 Calculating the Upper Bound

The first calculation of the upper bound considers only those element failures which individually would cause mission failure. Hence, only those blocks in the logic diagram which are in *series* are considered. Generally this is sufficient to provide a satisfactory estimate. However, if the reliabilities associated with individual blocks are not very high (for a very complex system requiring high reliability, 0.99 is a representative lower limiting block value), it may be necessary to consider system failures resulting from multiple component failures, i.e., to consider parallel blocks. We shall develop the method using a simple logic diagram, bearing in mind that its greatest value is its use for more complex systems. Assuming the logic diagram in Fig. 10.4, the algebraic equation for the simple upper bound will be derived. When only the series elements are considered, the system reliability upper bound R_{upper}

*B. L. Amstadter, Calculations of Reliability Predictions for the Apollo Spacecraft, North American-Rockwell Corporation, Downey, Calif., SID 66-744, 1966, IDEP no. 347.40.00.00-F1-47.

is the product of the reliabilities of elements A and B. (These elements may represent either non-redundant components or catastrophic failure modes of redundant ones.)

$$R_{upper} = R_A \times R_B{}^*$$

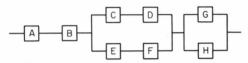

FIG. 10.4   Logic diagram for Method of Bounds.

An exponential model (constant-failure-rate system) is usually assumed. This approach is reasonable because each component is pretested and is then operated within its normal useful life. Use of the exponential model also facilitates calculations by making possible the suitable combination of reliabilities of the various components and modes.

Since

$$R_A = e^{-F_A} \quad \text{and} \quad R_B = e^{-F_B}$$

where F_A and F_B are the mission failure rates of elements A and B respectively,

$$R_{upper} = e^{-F_A} \times e^{-F_B} = e^{-(F_A + F_B)}$$

In general, the first calculation of the reliability upper bound is found by raising e to the negative of the sum of the failure rates of the *series* elements, Eq. (10.2).

*In the logic diagram in Fig. 10.4, there are only four possible combinations of success and failure of series elements:
  1. Both A and B good.
  2. A good and B fails.
  3. A fails and B good.
  4. Both fail.
Combination 1 represents mission success; combinations 2, 3, and 4 represent mission failure. Since the sum of the probabilities of the four combinations must equal unity, the probability of the first combination ($R_A \times R_B$) is equal to unity minus the sum of the probabilities of the other three combinations. Hence, $R_A \times R_B$ represents the subtraction of the probability of series failure cases from unity. This is in keeping with our earlier description of the upper bound.

$$R_{upper} = \exp\left(-\sum_{i=1}^{m} F_i\right) \tag{10.2}$$

where F_i is the mission failure rate of series element i, and m is the number of series elements.

It is seldom necessary to consider parallel elements. However, when the reliabilities of some elements are relatively low (e.g., less than 0.99), and these are functionally in parallel, the system reliability upper bound, if only series elements are considered, may be overly optimistic. It can be lowered by considering multiple failures. In the preceding logic diagram (Fig. 10.4), failure of any of the following pairs of elements results in mission failure: C and E, C and F; D and E, D and F, or G and H. If these components are not highly reliable, there may be more than a negligible chance of both elements in a pair failing, and two elements at a time should be considered. (Except in very rare instances, calculation of the probability of three failures at once is not warranted.) The bound is still the upper limit because only mutually exclusive cases are considered but not all failure probabilities have been subtracted. Care should be taken, of course, to ensure that the cases are unique, and that significant combinations are not inadvertently considered twice. The following reasoning ensures that mutual exclusion is generally maintained.

The probability of failure of a series element was calculated without regard for the non-series elements in the system. That is, regardless of whether the parallel elements were operable or had failed, failure of a series element failed the system. Therefore, no terms for parallel elements were considered in Eq. (10.2).

When system failure results from failure of two parallel elements, all series elements must be good (otherwise the system would have failed due to the failure of a series element). The series elements must therefore be included in these calculations. Hence, system failure probability resulting from failure, for example, of elements C and E in the preceding logic diagram is defined as $R_A \times R_B \times Q_C \times Q_E$, where $Q = 1 - R$ (not to be confused with F which is $- \log_e R$). Similar expressions apply to the other pairs of parallel elements enumerated above.

$$R_A \times R_B \times Q_C \times Q_F \qquad R_A \times R_B \times Q_D \times Q_E$$

$$R_A \times R_B \times Q_D \times Q_F^* \qquad R_A \times R_B \times Q_G \times Q_H$$

---

*Strictly speaking, for elements D and F alone to cause failure, C and E must not have both failed (to maintain mutual exclusion) and the expression for system failure resulting from failures of D and F is $R_A \times R_B \times [1 - (Q_C \times Q_E)] \times Q_D \times Q_F$. However, omission of the term $[1 - (Q_C \times Q_E)]$ has negligible effect on the final results and greatly simplifies the calculations. Similar simplifications apply to the other failed pairs.

The probability of system failure resulting from the failure of two non-series elements is found by adding these terms.

$$P = (R_A \times R_B \times Q_C \times Q_E) + (R_A \times R_B \times Q_C \times Q_F) + \cdots$$
$$+ (R_A \times R_B \times Q_G \times Q_H)$$

This can be simplified to

$$P = R_A \times R_B \times [(Q_C \times Q_E) + (Q_C \times Q_F) + \cdots + (Q_G \times Q_H)]$$

Since $R_A \times R_B = e^{-F_i}$, from Eq. (10.2), we can write this as

$$\left[\exp\left(-\sum_{i-1}^{m} F_i \right) \right] \left[\sum_{(k, \, k')=1}^{x} (Q_k \times Q_{k'}) \right] \tag{10.3}$$

where m is the number of series elements, Q_k and $Q_{k'}$ are the failure *probabilities* of pairs of parallel elements which together cause system failure, and x is the number of such pairs. This value is subtracted from Eq. (10.2) to obtain the system reliability upper bound when two failures are considered:

$$R_{\text{upper}} = \left[\exp\left(-\sum_{i=1}^{m} F_i \right) \right] - \left[\exp\left(-\sum_{i=1}^{m} F_i \right) \right] \left[\sum_{(k, \, k')=1}^{x} (Q_k \times Q_{k'}) \right]$$

$$R_{\text{upper}} = \left[\exp\left(-\sum_{i=1}^{m} F_i \right) \right] \left[1 - \sum_{(k, \, k')=1}^{x} (Q_k \times Q_{k'}) \right] \tag{10.4}$$

10.4.2 Calculating the Lower Bound

The lower bound is found by adding probabilities of success cases. In redundant systems, there are usually many cases where one or more element failures can occur without causing system failure. Therefore, it is always necessary when calculating the lower reliability bound of a redundant system to consider all cases with one failure, and frequently cases with two or even three failures.

The first calculation considers the case of no failures, that is, where all elements are good. This is simply the product of the reliabilities of *all* elements,

$$R_{\text{lower}} = \prod_{j=1}^{n} R_j$$

where R_j is the reliability of any element, *series or parallel,* and n is the total number of elements. In the previous logic diagram of Fig. 10.4, the first calculation of the lower bound is equal to $R_A \times R_B \times R_C \times \cdots \times R_H$,

$$= e^{-F_A} \times e^{-F_B} \times e^{-F_C} \times \cdots \times e^{-F_H} = \exp\left(-\sum_{A}^{H} F\right)$$

The general case for zero failures is expressed as follows:

$$R_{\text{lower}} = \exp\left(-\sum_{j=1}^{n} F_j\right) \tag{10.5}$$

A success is also achieved if one non-series element fails and all other elements are good. Thus, a failure of any one element $C, D, E, F, G,$ or H is still a system success. These cases are listed here for clarity, with R denoting success and Q denoting failure.

$$(R_A \times R_B \times Q_C \times R_D \times R_E \times R_F \times R_G \times R_H)$$
$$+ (R_A \times R_B \times R_C \times Q_D \times R_E \times R_F \times R_G \times R_H)$$
$$+ \quad \cdot \quad \cdot \quad \cdot \quad \cdot \quad \cdot \quad \cdot \quad \cdot \quad \cdot$$
$$\cdot \quad \cdot \quad \cdot \quad \cdot \quad \cdot \quad \cdot \quad \cdot \quad \cdot \quad \cdot$$
$$\cdot \quad \cdot \quad \cdot \quad \cdot \quad \cdot \quad \cdot \quad \cdot \quad \cdot \quad \cdot$$
$$+ (R_A \times R_B \times R_C \times R_D \times R_E \times R_F \times R_G \times Q_H)$$

These expressions can be simplified by multiplying by appropriate terms equal to unity and then simplifying, as follows:

$$= \left(R_A \times R_B \times Q_C \times \frac{R_C}{R_C} \times R_D \times R_E \times R_F \times R_G \times R_H \right)$$

$$+ \left(R_A \times R_B \times R_C \times Q_D \times \frac{R_D}{R_D} \times R_E \times R_F \times R_G \times R_H \right)$$

$$+ \quad \cdot \quad \cdot \quad \cdot \quad \cdot \quad \cdot \quad \cdot \quad \cdot \quad \cdot \quad \cdot$$
$$\cdot \quad \cdot \quad \cdot \quad \cdot \quad \cdot \quad \cdot \quad \cdot \quad \cdot \quad \cdot \quad \cdot$$
$$\cdot \quad \cdot \quad \cdot \quad \cdot \quad \cdot \quad \cdot \quad \cdot \quad \cdot \quad \cdot \quad \cdot$$

$$+ \left(R_A \times R_B \times R_C \times R_D \times R_E \times R_F \times R_G \times Q_H \times \frac{R_H}{R_H} \right)$$

$$= \left(R_A \times R_B \times R_C \times R_D \times R_E \times R_F \times R_G \times R_H \times \frac{Q_C}{R_C} \right)$$

$$+ \left(R_A \times R_B \times R_C \times R_D \times R_E \times R_F \times R_G \times R_H \times \frac{Q_D}{R_D} \right)$$

$$+ \quad \cdot \quad \cdot \quad \cdot \quad \cdot \quad \cdot \quad \cdot \quad \cdot \quad \cdot \quad \cdot$$
$$\cdot \quad \cdot \quad \cdot \quad \cdot \quad \cdot \quad \cdot \quad \cdot \quad \cdot \quad \cdot \quad \cdot$$
$$\cdot \quad \cdot \quad \cdot \quad \cdot \quad \cdot \quad \cdot \quad \cdot \quad \cdot \quad \cdot \quad \cdot$$

$$+ \left(R_A \times R_B \times R_C \times R_D \times R_E \times R_F \times R_G \times R_H \times \frac{Q_H}{R_H} \right)$$

$$= (R_A \times R_B \times R_C \times R_D \times R_E \times R_F \times R_G \times R_H)$$

$$\times \left(\frac{Q_C}{R_C} + \frac{Q_D}{R_D} + \cdots + \frac{Q_H}{R_H} \right)$$

The first term is simply the probability of all elements being good, a expressed in Eq. (10.5). The second term is the sum of the ratios of failure probability to success probability of all non-series elements. The product of these two terms can be written as

$$= \left[\exp\left(-\sum_{j=1}^{n} F_j \right) \right] \left[\sum_{k=1}^{p} \left(\frac{Q_k}{R_k} \right) \right] \tag{10.6}$$

where Q_k is the probability of failure of a non-series element, R_k is the reliability of the non-series element, and p is the number of non-series elements of the system. The equation for cases of two non-series element failures which together do not cause system failure is similarly derived:

$$= \left[\exp\left(-\sum_{j=1}^{n} F_j \right) \right] \left[\sum_{(k,l)=1}^{p'} \left(\frac{Q_k}{R_k} \times \frac{Q_l}{R_l} \right) \right] \tag{10.7}$$

where k and l are pairs of non-series elements which, failing together, do *not* cause system failure and p' is the number of such pairs. Equations (10.5), (10.6), and (10.7) are then added to give the lower-bound probability of success considering no element failures, one element failure, and two element failures.

$$R_{\text{lower}} = \exp\left(-\sum_{j=1}^{n} F_j \right) + \left[\exp\left(-\sum_{j=1}^{n} F_j \right) \right] \left[\sum_{k=1}^{p} \left(\frac{Q_k}{R_k} \right) \right]$$

$$+ \left[\exp\left(-\sum_{j=1}^{n} F_j \right) \right] \left[\sum_{(k,l)=1}^{p'} \left(\frac{Q_k}{R_k} \times \frac{Q_l}{R_l} \right) \right]$$

$$= \left[\exp\left(-\sum_{j=1}^{n} F_j \right) \right] \left[1 + \sum_{k=1}^{p} \left(\frac{Q_k}{R_k} \right) + \sum_{(k,l)=1}^{p'} \left(\frac{Q_k}{R_k} \times \frac{Q_l}{R_l} \right) \right] \tag{10.8}$$

Figure 10.5 depicts graphically the step-by-step results of applying the Method of Bounds.

10.4.3 Final Combined Prediction

The last step is the combining of the two bounds to obtain a single system reliability prediction value. The easiest method would be to take the simple

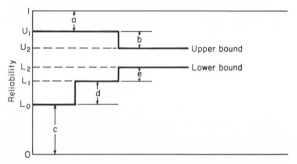

FIG. 10.5 Pictorial representation of the Method of Bounds.
(*a*) Probability of system failure resulting from the failure of
a single element. (U_1 is upper bound considering only series
element failures.) (*b*) Probability of system failure resulting
from the failure of two non-series elements. (U_2 is upper
bound considering series and two parallel element failures.)
(*c*) Probability of system success with no failures. (L_0 is
lower bound with no failures.) (*d*) Probability of system
success with one non-series element failure. (L_1 is corresponding lower bound.) (*e*) Probability of system success
with two non-series element failures. (L_2 is corresponding
lower bound.)

arithmetic average of the two bounds. However, experience has shown that
this results in an overly pessimistic value. It has been found empirically that
the true value of the system failure probability can be more closely
approximated by taking the square root of the product of the unreliabilities
associated with the upper and lower reliability bounds. This is then
subtracted from unity to obtain the single prediction value. (This is still
usually somewhat conservative but not quite so low as the prediction which
would be obtained by taking a simple average.)

$$R_{system} = 1 - \sqrt{(1 - R_{upper})(1 - R_{lower})} \tag{10.9}$$

It is important that the calculations of the two bounds be stopped at the
same point. That is, if only cases of one failure are considered for the upper
bound, then cases of only zero and one failure should be considered for the
lower bound; if two failures are considered for the upper bound, then
consider cases of two failures for the lower bound; etc. Inaccuracies will
result in Eq. (10.9) ($R = 1 - \sqrt{Q_{upper} \times Q_{lower}}$) if the calculations are not
carried to the same point. Consequently, Eq. (10.9) is calculated using *either*
the values from Eq. (10.2) and the sum of Eqs. (10.5) and (10.6) which
together cover the non-failure and single-failure cases, *or* the values from Eqs.
(10.4) and (10.8) which together cover non-failure, one failure, and two
failures.

We shall use the Method of Bounds to determine the reliability of the first configuration of Fig. 10.1, utilizing the logic diagram as shown in Fig. 10.3. The upper bound, considering only one failure and using Eq. (10.2), is

$$R_{upper} = R_A^2 \times R_G^2 \times R_H \times R_I \times R_J^2 \times R_K^2$$

$$= e^{-(2F_A + 2F_G + F_H + F_I + 2F_J + 2F_K)}$$

$$= e^{-0.100} = 0.905$$

The lower bound, considering cases of zero and one failure and using Eqs. (10.5) and (10.6), is

$$R_{lower} = \left[\exp\left(-\sum_i^n F_j \right) \right] + \left[\exp\left(-\sum_1^n F_j \right) \right] \left[\sum_1^p \left(\frac{Q_k}{R_k} \right) \right]$$

$$= \left[\exp\left(-\sum_1^n F_j \right) \right] \left[1 + \sum_1^p \left(\frac{Q_k}{R_k} \right) \right]$$

$$= e^{-(2F_A + 2F_B + F_C + F_D + F_E + F_F + 2F_G + F_H + F_I + 2F_J + 2F_K + 2F_L)}$$

$$\times \left(1 + \frac{2Q_B}{R_B} + \frac{Q_C}{R_C} + \frac{Q_D}{R_D} + \frac{Q_E}{R_E} + \frac{Q_F}{R_F} + \frac{2Q_L}{R_L} \right)$$

$$= e^{-0.752} \left[1 + \frac{2(0.178)}{0.822} + \frac{0.006}{0.994} + \frac{0.006}{0.994} + \frac{0.062}{0.938} + \frac{0.062}{0.938} + \frac{2(0.058)}{0.942} \right]$$

$$= 0.4715(1 + 0.433 + 0.006 + 0.006 + 0.066 + 0.066 + 0.123)$$

$$= 0.802$$

Using Eq. (10.9) to obtain the single prediction value, we have

$$R_{system} = 1 - \sqrt{(1 - 0.905)(1 - 0.802)}$$

$$= 1 - \sqrt{(0.095)(0.198)} = 0.863$$

The two prediction values (0.878 and 0.863), derived from the mathematical model and from the Method of Bounds, are reasonably comparable. (When the redundancy is active parallel, the Method of Bounds and the mathematical models yield almost identical prediction values.) It was necessary to use only the first step in the Method of Bounds procedure to obtain this degree of accuracy—the difference between the two bounds after the first step (0.103) was about the same as the difference between the upper bound and unity (0.095). Because of the relative simplicity of this particular logic diagram, the advantages of the Method of Bounds are not fully realized. However, when the logic diagrams are more complex and there is multiple redundancy, the usefulness of this technique becomes apparent.

10.4.4 Advantages of the Method

Three important points should be discussed. First, the Method of Bounds does not require that the elements of the logic diagram be independent. Hence, there is no hesitation to depict individual modes of failure. Note that, if the mathematical model were to be absolutely correct, it would be necessary to use conditional probabilities and thereby add considerably to the complexity of the model. We recall, from Sec. 8.4, that the mathematical model usually neglects dependency but that the errors introduced are relatively negligible. When individual failure modes are depicted, a much clearer picture of actual system operation is obtained. In the example shown, in configuration A of Fig. 10.1, the boilers are in series but they have failure modes which are actually in parallel as shown in Fig. 10.2. Similarly, the pumps and valves are in parallel in the system diagram but have failure modes which are in series in the logic diagram. The Method of Bounds encourages the depiction of individual modes of failure. In more complex systems this is a distinct advantage.

Second, the Method of Bounds can be used for either active parallel or sequentially redundant systems. When the redundancy is sequential, it is

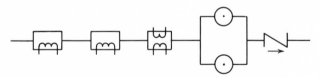

FIG. 10.6 System diagram for active parallel redundancy.

necessary to show the modes of failure of the failure detection and switching and to include their probabilities in the calculations. The detection and switching controls in the example were considered to be 100 percent reliable, but the actual switching devices (the valves) were not, and their modes of failure were depicted accordingly. If the system were capable of operating with both pumps running, a simple swing check valve might have been used and the system diagram for configuration A would be as shown in Fig. 10.6.

The swing check valve would be designed as shown in Fig. 10.7. When both pumps were operating, the gate would be as shown; when one pump failed, the gate would automatically assume the appropriate position indicated by one of the dashed lines, depending on which pump was still running. The valve would have a very small failure rate and the only failure mode shown in the logic diagram would be external leakage.

FIG. 10.7 Cross section schematic of swing check valve.

Figure 10.8 shows the logic diagram for the active parallel system using the same notations as in Fig. 10.3. External leakage of the check valve is shown by M.

Assuming that the failure rates for the elements remain the same and that the failure rate for external leakage of the check valve is the same as for the other valves (0.003), we shall calculate the reliability for an active parallel system, using the two methods. The mathematical model value is

$$R_T = R_M \times R_A^2 \times (1 - Q_B^2) \times R_H \times R_I \times R_J^2 \times R_K^2$$
$$\times (1 - Q_L^2)$$

$$1 - Q_B^2 = 1 - (1 - R_B)^2 = 1 - \left(1 - e^{-F_B}\right)^2 = 1 - (1 - 0.822)^2$$
$$= 0.968$$

$$1 - Q_L^2 = 1 - (1 - R_L)^2 = 1 - \left(1 - e^{-F_L}\right)^2 = 1 - (1 - 0.942)^2$$
$$= 0.997$$

$$R_M = e^{-0.003} = 0.997$$

$$R_A^2 \times R_H \times R_I \times R_J^2 \times R_K^2 = 0.911$$

$$R_T = 0.997 \times 0.911 \times 0.968 \times 0.997 = 0.877$$

The upper limit for this system, using the Method of Bounds, is

$$R_{\text{upper}} = e^{-0.097} = 0.908$$

The lower limit is

$$R_{\text{lower}} = e^{-0.609} \left[1 + \frac{2\,(0.178)}{0.822} + \frac{2\,(0.058)}{0.962} \right] = 0.845$$

$$R_{\text{system}} = 1 - \sqrt{(1 - 0.908)(1 - 0.845)} = 0.881$$

Note that for the active parallel redundant system, the Method of Bounds gives almost exactly the same answer as the mathematical model. This will always be true—the Method of Bounds will yield a slightly lower value than the mathematical model when the redundancy is sequential, and almost the same value for an active parallel system. The small error in the prediction of the sequential system is unimportant for most purposes and has no effect in a comparison of alternative configurations.

Finally, this method can be used for complex missions involving more than one phase and more than one objective. We shall indicate the calculations of the bounds for cases of zero and one failure to illustrate that, while there are more details when there are several subsystems and phases, the complexity is not greatly increased. As an example, we shall consider a system consisting of three major subsystems and a mission with three phases. The subsystems are denoted by the letters A, B, and C and the phases by the numbers 1, 2, and 3.

Upper bound. We recall from the derivation of Eq. (10.2) that the probability of all series elements being successful (which is the equivalent of subtracting series failure cases from unity) represents the one failure case for the upper bound. For a system with several subsystems and several phases, we merely sum all such series failure cases by taking the product of the corresponding terms. Denoting the single series failure case for subsystem A

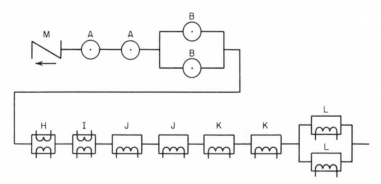

FIG. 10.8 Identification of elements of logic diagram for active parallel configuration.

in phase 1 by $R_{A,1}$, the product of all $R_{i,j}$ will yield the upper bound for the complete system for the entire mission.

$$R_{upper} = R_{A,1} \times R_{B,1} \times R_{C,1} \times R_{A,2} \times R_{B,2} \times R_{C,2}$$
$$\times R_{A,3} \times R_{B,3} \times R_{C,3} \tag{10.10}$$

Equation (10.10) can be rewritten for the general case as

$$R_{upper} = \prod_{\substack{i=1 \\ j=1}}^{\substack{j=m \\ i=n}} R_{i,j}(upper) \tag{10.11}$$

where there are n subsystems and m mission phases.

Lower bound. The lower bound is found in a similar manner. We simply add all possible cases of no element failure and only one non-series element failure in each subsystem. This is done for each subsystem individually and the results for all subsystems are multiplied together. We shall enumerate the cases for subsystem A for the example.

1. No failures in any mission phase
2. A non-series failure in phase 1 and no other failures
3. A non-series failure in phase 2 and no other failures
4. A non-series failure in phase 3 and no other failures

(*Note*: To simplify the calculations, the original logic diagram for each phase may be retained although it is recognized that a failure of a non-series element in one phase modifies the logic diagrams for succeeding phases. This approach is conservative because there will actually be fewer elements to be considered in phases subsequent to the failure.)

The sum of 1, 2, 3, and 4 above yields the lower-bound reliability for subsystem A for the entire mission. Similar sums are obtained for subsystems B and C. The products of these terms represent the lower bound for the overall system for the complete mission. In general, this may be written as

$$R_{lower} = \prod_{i=1}^{n} R_i = \prod_{i=1}^{n} \left(P_0 + \sum_{j=1}^{m} P_1 \right)_i \tag{10.12}$$

where there are again n subsystems and m phases, and where P_0 is the probability of no failures in any phase for one subsystem and P_1 is the probability of one non-series failure in one phase and no failures in any other phase for the same subsystem. As indicated by the notation, it is necessary that each subsystem probability $(P_0 + \Sigma P_1)$ be determined *first* and *then* that the product be taken for all the subsystems.

One or two actual applications of the Method of Bounds are all that are needed to become adept with this technique, and application will be enhanced with usage. Taken together, the three methods provide all necessary tools for performing the large majority of system reliability predictions.

ADDITIONAL READINGS

Amstadter, B. L.: Calculations of Reliability Predictions for the Apollo Spacecraft, North American-Rockwell Corporation, Downey, Calif., SID 66-744, 1966, IDEP no. 347.40.00-F1-47.

———: Prediction of System Reliability by Method of Bounds, *Proc. Ann. Symp. Reliability, Boston, 1968,* Institute of Electrical and Electronics Engineers, Inc., New York, 1968, pp. 423-430.

ARINC Research Corporation: "Reliability Engineering," Prentice-Hall, Inc., Englewood Cliffs, N.J., 1964.

Barlow, R. E., and F. Proschan: "Statistical Theory of Reliability," John Wiley & Sons, Inc., New York, 1965.

Bazovsky, I.: "Reliability Theory and Practice," Prentice-Hall, Inc., Englewood Cliffs, N.J., 1961.

Bosinoff, I., et al.: Mathematical Simulation for Reliability Prediction, *Sylvania Electronics Systems, Rept. F-491-1,* Waltham, Mass., 1961; ASTIA no. 271367, Dayton, Ohio.

Bureau of Naval Weapons, Chief of: "Handbook: Reliability Engineering," U.S. Government Printing Office, 1964.

Calabro, S. R., and S. Pearlman: Simulation Techniques Verify Reliability, *Proc. Nat. Symp. Reliability Quality Control in Electron., 3d, Washington,* Institute of Electrical and Electronics Engineers, Inc., New York, 1957, pp. 213-219.

Chorafas, D. N.: "Statistical Processes and Reliability Engineering," D. Van Nostrand Company, Inc., Princeton, N.J., 1960.

Dixon, W. J., and F. J. Massey, Jr.: "Introduction to Statistical Analysis," 3d ed., McGraw-Hill Book Company, New York, 1969.

Owen, D. B.: "Handbook of Statistical Tables," Addison-Wesley Publishing Company, Inc., Reading, Mass., 1962.

Rand Corporation: "A Million Random Digits with 100,000 Normal Deviates," The Free Press, New York, 1955.

Shooman, M. L.: "Probabilistic Reliability: An Engineering Approach," McGraw-Hill Book Company, New York, 1968.

11

APPORTIONMENT

11.1 PURPOSE AND REQUIREMENTS

Reliability apportionment is the process in which the failure allowance specified for a system is subdivided and equitably allocated among the system's components. The primary purpose of apportionment is to establish a reliability goal or objective for each component so that the personnel responsible for the component's development are made aware of its required reliability. This objective influences cost, weight, schedule, and other factors associated with the development. It is particularly important when the component is developed by an outside contractor because the reliability requirement will directly influence the development cost. Since awards of subcontracts are usually made on a competitive basis where the proposal submitted by each prospective supplier is evaluated for all applicable requirements, the reliability objective becomes an integral part of the overall component requirement.

Equitable allocation of allowances requires, as a beginning, that the relative probabilities of failure of the components be known. This requires knowledge of all the information necessary to make realistic predictions: generic failure rates, environments, derating, duty cycles, and so forth. That is, the relative reliability predictions constitute a major part of the total consideration.

It has sometimes been advocated that, when no information about the components is available other than duty cycle and mission phase times and associated environmental stresses, the basic failure rates should be assumed to be equal. Such an assumption negates the value of the apportionment. It can lead to unrealistic demands for improvement of "bad" components whose actual initial reliabilities are low, while at the same time "good" components whose initial reliabilities already meet or exceed their goals are tacitly accepted. Resultant product-improvement programs may impose exorbitant expenditures on "bad" components whereas significant portions of the required overall system improvement might be easily achieved by improving the "good" components for a much smaller expenditure.

Another undesirable result of apportionments made in this manner is that they contribute to the sometimes justified charge that reliability numerics are not meaningful. If reliability numerics are to be of value and the procedures are to become part of a recognized engineering discipline, then the numerics must be as accurate as possible. The assumptions which are made must be realistic, and the utilization of the results should take into account whatever real limitations exist in the data. We should not arbitrarily impose artificial limitations such as the assumption of equal failure rates.

If no information is known initially, then some effort to obtain or develop this relevant data must be expended. The cost involved in obtaining realistic data on failure rates and other factors mentioned above will be more than justified by later savings in development costs and by increased confidence in the numerics.

Thus, a principal by-product of the apportionment—one which in many cases is more important than the apportionment itself—is the determination of predicted reliabilities of the components. Even though these predictions may not be extremely accurate with respect to their absolute values, the *relative* reliabilities provide important information for trade-offs of weight, schedule, mission objectives, and cost, and these trade-offs can then be evaluated on a more quantitative basis. These considerations were discussed in the preceding chapters on reliability prediction and will be examined in more detail in Chap. 16.

The predicted reliability of a component is calculated in the same way when an apportionment is made as when a prediction is made. The generic (or application) failure rate is found from suitable failure-rate tables, the environments are determined and the appropriate environmental multiplier (k factor) is established, the operating time of the component in the particular mission is estimated, and any other pertinent factors are considered. The operating failure rate for the mission is then calculated by multiplying the base failure rate after any derating by the environmental multiplier and by the number of hours of operation. When the component has a standby or non-operating failure rate, this failure rate (including any k factor) is multiplied by the number of hours that the component will be on standby or non-operating status. This product is added to the operating failure rate to obtain the component's overall failure rate for the mission. Assuming an exponential failure distribution, the reliability for the component is found in the usual manner:

$$R \; = \; e^{-\lambda t} \; = \; e^{-(\lambda_o t_o + \lambda_s t_s)} \tag{11.1}$$

where $\lambda_o t_o$ is the product of the operating failure rate and operating time, and $\lambda_s t_s$ is the product of the standby failure rate and standby time.

11.2 SERIES SYSTEMS

A detailed example where the mission consists of a single phase and the components are in a series configuration will illustrate the basic procedure. Other factors being equal, in a simple series system each component's allowance is directly proportional to the predicted probability of failure, where the probability of failure Q is one minus the probability of success R, ($Q = 1 - R$). That is, if the predicted probability of failure of component A is twice as high as the predicted probability of failure of component B, then component A is allocated twice as large a percentage of the total failure allowance as is component B. The apportioned reliabilities are then approximately one minus the apportioned failure allowance.* These apportioned numerics become reliability goals, objectives, or requirements for the individual components. In a simple series system the sum of the individual failure allowances is approximately equal to the total system allowance,* and the product of the individual reliability goals yields the true system goal just as the product of the predicted reliabilities is equal to the predicted system reliability. The following example illustrates the apportionment procedure for a simple series system.

Assume that a system has a reliability goal of 0.90, corresponding to a failure allowance of 0.10. This allowance is to be allocated among 20 series components whose predicted reliabilities are given in the second column of Table 11.1. The corresponding predicted failure probabilities are given in the third column. The sum of these failure probabilities is 0.200. The first

---

*Since the exponential model is assumed, the actual allowance should be very slightly higher because the exact apportioned reliability is $e^{-\lambda t}$ rather than $1 - \lambda t$. However, in the great majority of cases, individual allowances are 0.01 or less and the use of $R = 1 - \lambda t$ results in an error of less than 0.00005 per component, so that it is a perfectly satisfactory approximation.

TABLE 11.1   Determination of reliability goals for a 20-component series system

(System reliability goal = 0.90; system failure allowance = 0.10)

Component number	Predicted reliability	Predicted unreliability	Failure allowance	Reliability goal
1	0.998	0.002	0.001	0.999
2	0.997	0.003	0.0015	0.9985
3	0.997	0.003	0.0015	0.9985
4	0.995	0.005	0.0025	0.9975
5	0.995	0.005	0.0025	0.9975
6	0.994	0.006	0.003	0.997
7	0.992	0.008	0.004	0.996
8	0.991	0.009	0.0045	0.9955
9	0.991	0.009	0.0045	0.9955
10	0.990	0.010	0.005	0.995
11	0.990	0.010	0.005	0.995
12	0.990	0.010	0.005	0.995
13	0.988	0.012	0.006	0.994
14	0.988	0.012	0.006	0.994
15	0.987	0.013	0.0065	0.9935
16	0.986	0.014	0.007	0.993
17	0.985	0.015	0.0075	0.9925
18	0.984	0.016	0.008	0.992
19	0.982	0.018	0.009	0.991
20	0.980	0.020	0.010	0.990
		Sum = 0.200	Sum = 0.100	Product 0.9046 $\approx$ 0.90 (difference = 0.0046)

component, which has a predicted failure probability of 0.002, is then allocated 0.002/0.200 or 2/200 of the total 0.10 allowance or 0.001; the second component is allocated 3/200 of 0.10 or 0.0015; and so on. (An equivalent way to apportion allowances to the series components is to multiply each component's failure probability by the ratio of the total failure allowance to the total failure probability. Thus the allowance for the first component is $0.100/0.200 \times 0.002$, for the second component, $0.100/0.200 \times 0.003$, etc. This is somewhat easier because all multipliers are the same.). These allocations are shown in column 4, and the corresponding goals are shown in the last column. The product of the goals obtained in this manner very closely approximates the specified system goal, for reasons explained in the previous footnote.

If it is desired to have the product of the component goals exactly equal to the system goal of 0.90, then the difference of 0.0046 can be allocated proportionally to the components, thereby slightly increasing each component's allowance. However, this is not recommended for several reasons:

1. No allowance has been given for assembling the components. The 0.0046 excess can be allocated to this.

2. We would be deluding ourselves by adding more decimal places to the goal than we can justify from the original prediction data. Even three-place accuracy is sometimes considered to be unrealistic.

3. The amount of time necessary to provide additional "accuracy" is not warranted in view of the principal utilization of the goals.

4. The designer or supplier would not change his activity, cost, or schedule in any way to reflect a difference, for instance, between a goal of 0.995 and one of 0.9952 or 0.9948. This difference could not be detected in the great majority of demonstrations of achievement of reliability goals.

5. It is often desirable to retain a small percentage of the total allowance so that those components which fail to meet their goals due to the state of the art, unanticipated development difficulties, or other justifiable causes can be granted some additional allowance.

An alternate method for apportioning the reliabilities utilizes the mission failure rates rather than the unreliabilities, that is, λt rather than Q. This method results in the product of the apportioned component goals exactly equaling the system reliability goal, with no remaining unused allowance to be apportioned. For a simple series system where the components have equal criticalities, equal capabilities for improvement, and other equivalent factors, this alternate method might be preferred if the five reasons just stated for not apportioning a remainder did not apply. The previous example with 20 series components will be used to illustrate the alternate procedure, although other considerations which will be discussed shortly will indicate why it does not simplify the calculations in most cases.

The same predicted reliabilities are assumed. Failure rates for the mission corresponding to these reliabilities are given in the third column of Table 11.2 using the relationship $R = e^{-\lambda t}$, and each component's failure allowance is based on the ratio of its failure rate to the total for the system. The component reliability goals are found by again utilizing the exponential equation. Operating times are assumed to be equal for all components.

11.3 REDUNDANT SYSTEMS

Allocation of system goals to components in more complex systems is not only more time-consuming but also does not lend itself to a simple direct solution. An iterative (repetitive) process is required. The first phase of the apportionment process is straightforward. Each group of parallel elements is suitably combined into a single element which is then considered to be one

TABLE 11.2   Determination of reliability goals using failure rates

(System reliability goal = 0.90; system failure rate = 0.10536)

Component number	Predicted reliability	Predicted failure rate*	Failure-rate allowance	Reliability goal
1	0.998	0.00200	0.00105	0.9989
2	0.997	0.00300	0.00157	0.9984
3	0.997	0.00300	0.00157	0.9984
4	0.995	0.00501	0.00262	0.9974
5	0.995	0.00501	0.00262	0.9974
6	0.994	0.00602	0.00315	0.9968
7	0.992	0.00803	0.00420	0.9958
8	0.991	0.00904	0.00473	0.9953
9	0.991	0.00904	0.00473	0.9953
10	0.990	0.01005	0.00526	0.9948
11	0.990	0.01005	0.00526	0.9948
12	0.990	0.01005	0.00526	0.9948
13	0.988	0.01207	0.00632	0.9937
14	0.988	0.01207	0.00632	0.9937
15	0.987	0.01309	0.00685	0.9932
16	0.986	0.01410	0.00738	0.9926
17	0.985	0.01511	0.00791	0.9921
18	0.984	0.01613	0.00845	0.9916
19	0.982	0.01816	0.00951	0.9905
20	0.980	0.02020	0.01058	0.9895
		Sum = 0.20123	0.10534	0.900

*For time period equal to the mission time.

series component block. For example, the group

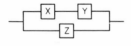

consists of two parallel legs or paths. The first leg contains elements X and Y and its reliability, $R_1, = R_X \times R_Y$. The second leg consists only of element Z and its reliability, R_2, is R_Z. The combined reliability R of the two parallel legs $= 1 - [(1 - R_1) \times (1 - R_2)]$.

After the parallel elements have been combined and the system is depicted as a series system, the allowances are allocated to the series component blocks in the manner discussed in the previous section. The allowance for each block is proportional to its probability of failure. Allowances are then determined for elements of the parallel legs.

Suppose that the elements X, Y, and Z in the above group have predicted reliabilities of 0.97, 0.95, and 0.88 and corresponding failure probabilities of 0.03, 0.05, and 0.12, respectively. The reliability of the first leg is $R_X \times R_Y = 0.97 \times 0.95 =$ approximately 0.92, corresponding to a probability of failure of approximately 0.08. The probability of failure of the second leg is 0.12. The overall reliability is $1.0 - (0.08 \times 0.12)$ or approximately 0.99, and the overall probability of failure is 0.01.

Suppose further that the series block representing these three elements has been apportioned a total allowance of 0.005. The ratio of the allowance to the predicted failure probability for the block is 0.005/0.01 or 0.50. The predicted failure probability for each element is multiplied by this ratio to determine the element's allowance. Thus the allowance for element X is 0.50 $\times$ 0.03 = 0.015; the allowance for element Y is $0.50 \times 0.05 = 0.025$; and the allowance for element Z is $0.50 \times 0.12 = 0.060$.

If there is more than one level of redundancy, the procedure is continued *one level at a time* until each individual element has been given an allowance *in proportion to its failure probability*. The procedure should be followed level by level. For instance, a complex logic diagram might be

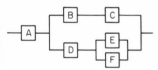

The first combination at the lowest level would yield the following block diagram:

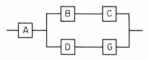

where block G represents parallel elements E and F. The next step would yield the diagram

where block H represents series blocks B and C and block I represents blocks D and G. The final series diagram would be where J represents the parallel blocks H and I. After the allowance for block J has been determined, the allowances for blocks B, C, D, E, and F are found by employing the ratio and proportion method just discussed.

After the allowance for each individual element has been determined, the entire process may be repeated, except that calculations in the iteration use allowances rather than predicted failure probabilities. This repetition is performed because it will be found that the overall system allowance has not yet been completely allocated.* A simple example will illustrate why. In complex systems there is unfortunately no simple way to allocate the entire system allowance in one trial without yielding discrepancies between the relative allowances and the relative failure probabilities.

Assume the following logic diagram:

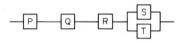

Assume also that a system allowance of 0.02 is to be allocated according to the relative probabilities of failure, and that the reliabilities of elements P, Q, and R are each 0.99 and that those of S and T are each 0.90. The corresponding failure probabilities are 0.01 and 0.10, respectively. The series block diagram is

where U represents the parallel combination of elements S and T and has a resulting reliability of $1.0 - (0.10 \times 0.10) = 0.99$. Since the four blocks P, Q, R, and U all have the same reliability, each is given one-fourth of the total allowance: $1/4 \times 0.02 = 0.005$, or an apportioned reliability of 0.995.

The ratio of the allowance of each series block including block U to its predicted failure probability is $0.005/0.01 = 0.50$. The allowances for elements S and T are then each $0.50 \times 0.10 = 0.05$. When we recalculate the apportioned reliability for block U using the allowances for S and T, we obtain $1.0 - (0.05 \times 0.05) = 0.9975$, or an allowance of only 0.0025, rather than the 0.005 allowance that we started with. The total allocation is $0.005 + 0.005 + 0.005 + 0.0025$, a total of 0.0175 instead of 0.02. The difference $(0.02 - 0.0175)$ is 0.0025.

This difference can be apportioned to the various elements in approximately the same ratio as the original allowances. An iterative procedure is

---

*When the system allowance is greater than the predicted system unreliability, the opposite result will occur. It will be found that the initial overall allocation will exceed the overall system allowance.

needed. Since 0.0025 is one-seventh (1/7) of 0.0175, the additional allowance allocated to each element is one-seventh of the element's first allowance. Another check is then made, and a final correction is accomplished. Table 11.3 shows the final allowances for elements P, Q, R, S, and T as established by this iterative procedure. Note that while 0.0025 is 1/7 of 0.0175, 0.0004 is only 1/51 of 0.0204.

TABLE 11.3   Final allowances for elements P, Q, R, S, and T of logic diagram

Element	First allocation	Additional allocation	Subtotal	Final correction	Total allocation
P	0.005	0.0007	0.0057	−0.0001	0.0056
Q	0.005	0.0007	0.0057	−0.0001	0.0056
R	0.005	0.0007	0.0057	−0.0001	0.0056
S	0.050	0.0071	0.0571	−0.0011	0.0560
T	0.050	0.0071	0.0571	−0.0011	0.0560
System (calculated)	0.0175		0.0204		0.01994 ≈ 0.02
Balance	0.0025		−0.0004		0.00
Ratio of balance to allocation	1/7		−1/51		0

11.4 MULTIPHASE MISSIONS

We have seen how the failure allowance can be apportioned to the various system components for a single-phase mission. Many missions, however, have more than one phase, and complex space missions may have as many as 25 or 30 separate phases. Even those systems which are stationary have many components which have two phases: operating and non-operating. When there are two or more phases, there may be separate reliability requirements contractually specified for each phase. Even when separate requirements are not specified, it may be desirable to divide the mission into a number of separate phases which have distinct environmental conditions. For example, a Mars probe might be divided into phases as follows (subphases are indicated in parentheses):

Launch (first stage, separation, second stage, separation, third stage, and Earth-orbit injection)
Earth orbit
Earth-Mars transfer (trans-Mars injection, coast, mid-course correction, coast)
Mars-orbit injection

Mars orbit
Mars-Earth transfer (trans-Earth injection, coast, mid-course correction, coast)
Earth-orbit injection
Powered (retro) reentry
Descent and landing
Flotation and recovery

Associated with each of these phases and subphases is a set of environments and equipment operating times. Also associated will be a reliability goal or objective.

The primary goal would be the collection, storage, and recovery of information relating to the nature of the Mars atmosphere, radiation levels, and so on. If most of this information were transmitted to earth via electronic communication equipment while a smaller portion were stored on magnetic tape or another storage medium for later earth recovery, then a high goal might be specified for the first half of the mission which includes Mars orbiting with a lower goal specified for the complete mission.

To illustrate the division of reliability apportionment and allocation by phase, we will consider a simpler mission model. An electrical power system used to provide emergency power in case of primary power failure has three principal phases. The non-operating phase may extend for a period of 1 year during which time the power system must retain operational readiness without maintenance. When called on, the system must start using battery power and gradually build up to full system power without human action. After achieving full power, it must operate without maintenance for a period of 30 days. Overall reliability should be at least 90 percent, and the 10 percent failure allowance is to be allocated among the three phases.

Since, in this particular utilization, the environment is controlled and non-operating conditions are very favorable, very little of the allowance will be allocated to the non-operating phase. Furthermore, it is important that the system maintain operational readiness during this period. Therefore, only 5 percent of the allowance is allocated to this phase. The resulting reliability requirement is approximately 0.995 (5 percent of 0.10 = 0.005).

Start-up requires the use of some components whose operation is unique to this phase. Also, start-up imposes additional stresses on the other components so that, although the time period is short (approximately 1 hr), 20 percent of the allowance is allocated to this function. This amounts to 0.02, and the corresponding reliability is about 0.98. The remaining allowance of 0.075 is allocated to the operating phase, and its reliability requirement is thus about 0.925. Reliability requirements determined in this way result in an overall reliability of 90.2 percent. Although the 0.2 percent remaining could be reallocated as explained previously, it is suggested that this small percentage be retained as a safety margin or additional allowance, to be allocated later to elements which fail to meet their goals, or for switching devices, assembly, wiring, or other elements which are not included in the logic diagram. (Component allowances are allocated within each phase by means of the methods already discussed, using the allowance established for the particular phase.)

If it is desired to allocate the entire allowance so that the product of the phase reliabilities exactly equals the overall requirement, the exponential failure distribution may be assumed. The failure rate for the entire mission is 0.10536. This is divided into 5 percent (non-operating), 20 percent (start-up), and 75 percent (operating) portions, which are 0.00527, 0.02107, and 0.07902, respectively. The corresponding phase reliability requirements are then 0.9947, 0.9792, and 0.9240.

11.5 OTHER IMPORTANT CONSIDERATIONS

The discussion to this point has considered only one set of factors influencing the apportionment: the predicted failure rates for the mission or its phases. However, there are three other important considerations: criticality of each component and its relationship to the system and mission; the state of the art of each component, that is, the capability for achieving improvement; and the performance and effects of maintenance and repair. Complexity is sometimes included as a fourth consideration. However, complexity is automatically considered because the component and subsystem failure rates are determined from the failure rates of all constituent parts and assemblies. Complexity is therefore included in the overall failure rates and need not be considered separately.

11.5.1 Criticality

The first of these three considerations—criticality—can usually be readily determined. If an aircraft radio fails, there is probably a backup communications system which can be used and which permits the aircraft to complete its flight. However, if an engine fails, even on a multiengine airplane, the flight may be terminated early or rerouted to an alternate airfield with better ground safety equipment. A manned space flight is usually designed and programmed to carry out a number of experiments. Some of these may be complete failures and still permit the mission to continue for its planned duration. However, if there are two attitude control systems and one fails, the mission will be aborted and the spacecraft will return to earth to the first opportune landing area. In the case of an experiment, there probably is no alternate means of performance, and the function is lost. Because it is not necessary to crew safety, the flight is continued. On the other hand, the attitude control function can be carried out with the backup control system, but if the backup fails, both the mission and the crew are lost. Hence, the mission is aborted when only one means of performing a vital function remains.

As a more familiar example of criticality, an automobile usually has a radio, heater, and other equipment, as well as a spare tire. If the radio fails during a trip, the driver will most likely continue his journey without it even though the car has no other communication equipment. However, if a flat tire occurs, the spare tire will be used and the driver may have the flat repaired at the first opportunity before completing his trip.

There are no hard-and-fast rules concerning relative weights assigned to components which perform functions of different criticality. One possibility for components is

Critical (a failure results in a high probability of personnel loss) . . . 100
Major (a failure results in mission loss) 10
Minor (a failure does not affect the mission or personnel safety
although the function may be lost) 1

A different weighting might be 10:5:1 or 10:3:1. The particular weighting to be used can usually be determined from the requirements of the contract, either stated numerically or described in terms of redundancy, abort criteria, and other requirements. Failure allowances are allocated in an inverse ratio to the criticalities. In this case, a critical element would have an allowance of 1/100 of the allowance for a minor element if other factors were equal. Similarly, a major element would have an allowance of ten times that allotted to a critical element but only 1/10 of that allocated to a minor one.

It frequently happens that the different modes of failure of a component have different criticalities. The explosion of a gasoline truck, although it might be extremely improbable, would still be a critical failure, while poor performance might be a major or perhaps only a minor failure, depending on the degree of deficiency. A valve failure could likewise be critical in one mode and major in another. The advantage of the logic diagram is that it can depict functions or modes of failure rather than just components, and appropriate criticalities can be assigned to each element in the diagram. This can be of great help in determining which components can or should be made redundant and which component redundancy would not benefit the system. Redundancy considerations are discussed in more detail in Chap. 16.

11.5.2 Product Improvement

The second consideration—capability for achieving improvement—is also important. If a component has been available for a long time and has experienced an extensive development program including failure analysis and corrective action of deficiencies, it may be quite difficult to further improve its reliability even if the reliability is considerably lower than is desired.

Failure rates of some valves, for example, have already been reduced to levels where further development would be extremely costly and relatively little increase in reliability would be obtained. Other components can have initially high reliabilities but further improvement may still be economically possible. This is particularly true of static components, that is, those which have no moving parts. A new type of semiconductor device, for example, may have a reliability which is higher than that of a valve but still much lower than that of other semiconductors. Experience indicates that funds expended on improvement programs for semiconductors are productive, and consequently a product-improvement program for the new type would be indicated. As a result, this capability for improvement as well as the predicted rates and criticalities would be considered when failure allowances are allocated and reliability goals are apportioned.

Caution should be exercised to avoid overweighting improvement capability. When predicted failure rates are used, they often are based on available reliabilities after extensive development and product-improvement programs. Thus, the reliability-growth factor is already considered, and it should not be considered again independently. However, when predicted failure rates are taken from actual applications, there may not have been time for extensive improvement programs, and some allowance should be made in the apportionment. Careful evaluation of the failure-rate data and their sources is required along with the exercise of good engineering judgment to estimate improvement potential.

A suggested scale of relative weights is 3:2:1. A component which may readily be improved is placed in category 3 and its predicted failure rate would be multiplied by 1/3. Another component which may be less readily improved is placed in category 2 and a multiplier of 1/2 is used. Failure allowances of components whose reliability improvement may be relatively slight are not modified.

11.5.3 Maintenance Considerations

Maintenance is the third consideration when apportioning failure allowances. A component which is periodically maintained or one which is regularly monitored or checked and repaired as necessary will have, on an average, a higher reliability than one which is not. However, the effects of maintenance and repair actions on the component and on the system must be carefully analyzed in each case to determine the nature and extent of any possible modification in the failure allocations. Some maintenance/repair actions will warrant little or no change in allowances, while others may be the bases for significant reductions. Three important factors which influence changes in allowances are (1) the nature of the failure distribution, (2) the extent and type of redundancy, and (3) the effects of failure or outage of the component on the system's performance.

1. The nature of the failure distribution determines the effectiveness of maintenance/repair actions. If the equipment is still in its infant-mortality or debugging period, then the repair of early failures effectively improves the equipment's reliability. It is assumed that the repair action is such that the mode of failure is eliminated or at least its probability of recurrence is substantially reduced. (If the repair action merely restores the equipment to its original condition without correcting the basic deficiency, the reliability will not be improved.) Similarly, if the useful life of an equipment item is shorter than the mission time so that wear-out may occur before the end of the mission, then maintenance actions including parts replacement may substantially increase the life and correspondingly increase the reliability.

However, if the equipment is operating during its constant-failure-rate period, maintenance actions are not effective. By definition, the probability of failure during this period is constant, and the probability of failure at time t_1 where $t_1 = t_0 + \Delta t$ is the same as the probability of failure at time t_0. A maintenance action which "restores" the equipment to its t_0 condition does not affect its reliability or probability of failure. An exception to this is the replacement of any failed redundant part. After a redundant part has failed, the reliability of the equipment item is reduced because a second (or additional) part failure can cause the equipment to fail. Replacement of the failed part does restore the equipment to its original, more reliable state. (Note that, in the correct mathematical sense, the equipment's failure rate is not constant during this period, but increases after the failure of a redundant part. An increasing failure rate is actually better described by the wear-out period than by the constant-failure-rate period.)

Maintenance actions during the constant-failure-rate period and repairs during debugging which do not "debug" do not affect the equipment's reliability and are not justification for changing the apportionment. Maintenance/repair actions which increase the reliability—correction of deficiencies, replacement of worn parts, and replacement of failed redundant parts—do affect the apportionment and make changes desirable. When an apportionment is changed, the allowance will be reduced. The amount of reduction will depend on the frequency of the maintenance/repair activity.

2. Redundancy significantly affects the apportionment of maintained systems. The replacement of failed redundant parts can result in a substantial upgrading of component reliability. If failed elements are repaired or replaced immediately, the reliability of the redundant segment can be essentially unity. Those failure modes which do not fail all elements simultaneously, i.e., those that are truly redundant, would be considered in reducing the allowance.

An active parallel redundant system does not usually have its own inherent failure-detecting mechanism, and failed parts must be detected during scheduled checks or maintenance. The equipment must therefore be reliable for the period between scheduled maintenance/repair actions. On the other hand, a sequentially redundant system has failure detection and switching. If the detection system's operation discloses a failure, a repair or replacement action may sometimes be initiated almost immediately.

There are two associated aspects which should be considered with either type of redundancy. First, it must be possible to make the repair or replacement. Not only must there be a replacement part, but the necessary tools and equipment must be available and a competent technician or other trained person must be on hand to effect the repair. Even though we are considering systems which are maintained, it may still take time to procure a new part or assembly. The very fact that the equipment is originally redundant sometimes indicates that considerable time may be necessary to effect a repair or to replace a failed item.

Second, the repair action itself may not be 100 percent reliable so that the equipment is not restored to its original state. Repairs made in the field are often temporary, and the equipment may be sent to a more fully equipped maintenance/repair station at a more convenient time. Even these repairs do not necessarily return the equipment to an "as new" condition. We have all had experiences wherein a repair was less than 100 percent satisfactory.

The effect of redundancy, as would be expected, is to further increase the reliability of maintained equipments which have redundant parts and to reduce their failure allowances. The reduced allowances are based on the probabilities of failure modes which do not cause simultaneous failure of redundant parts, on the length of time the equipment must operate between maintenance/repair actions, on the length of time required to make the repairs, and on the repair's effectiveness.

3. Finally, the effects on the system of an outage of the equipment must be considered. If an outage of any duration makes the system inoperable so that it cannot be restored to an operating condition even if the failed item itself is repaired, then repair or maintenance of a component whether in series or parallel is of no avail when it requires system shutdown. On the other hand, many maintenance/repair actions on parallel components can be effected without system shutdown. In Chap. 16 some system design considerations are discussed in which this capability is an important factor in the design of redundancy. Also, many systems are not required to operate during the entire mission so that system shutdown may be very feasible and maintenance/repair activities can be accomplished; indeed, maintenance actions are scheduled accordingly. If the failed component is redundant, then the system may still be available in an emergency. Of course, if it is not redundant, or if the system cannot be operated while maintenance/repair is taking place, then the system would not be available for emergency use. It is apparent that systems effects have many facets which must be carefully evaluated when an apportionment is made as well as when the system is designed.

Assuming that the system design and operating characteristics are such that maintenance and repair actions are possible and are effective in reducing the mission failure rate, the failure allowance can and should be appropriately changed. Since the change in the allowance of one component or assembly affects the allowances available for every other system element, it is necessary that maintenance/repair be considered before the first apportionment is made. The effect of the maintenance/repair action on a component is to increase its reliability and consequently to correspondingly reduce the failure allowance.

Exact mathematical expressions are not possible because the reliability increase depends on the nature of the failure distribution and on the values of the parameters which describe it. Recalling that if the failure distribution is exponential (a constant-failure-rate condition), maintenance/repair is not effective, and we are left with other distributions which are much more complex and essentially undefined. For example, if the maintenance/repair action prevents failure of a component whose principal failure mode is wear-out and even if the wear-out is normally distributed, the exact time of maintenance/repair significantly affects the failure distribution, and the new distribution cannot readily be defined mathematically. Figure 11.1a to c shows how relatively small changes in time to maintenance can have major effects on the reliability. Failure probabilities are indicated by the shaded portions of the figure.

In Fig. 11.1a maintenance/repair is performed just after the start of the wear-out period. The shaded portion of the figure indicates the probability of mission failure prior to the time that maintenance/repair is performed. The

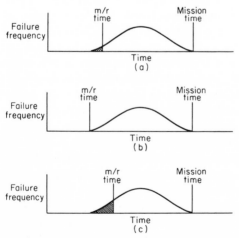

FIG. 11.1 Effect of time to maintenance/repair on failure probability.

mission time can be anywhere to the right of the beginning of the wear-out period without affecting the results of the maintenance/repair action. Maintenance/repair is required more than once if the onset of wear-out occurs before the mission is half over. In Fig. 11.1*b*, the maintenance/repair action is performed before the wear-out period is entered, the mission time is to the right of the onset of wear-out, and maintenance/repair is required more than once. In this case, the mission failure probability due to wear-out is essentially zero. In Fig. 11.1*c*, a relatively small increase in time before maintenance results in a significant increase in the probability of failure. It becomes apparent from this simple example that when repairs or replacements are made, the resulting overall system failure distribution becomes quite complex and mathematical description of the distribution becomes extremely difficult. Consequently, each set of conditions should be individually evaluated and the corresponding maintenance factors established which engineering judgments indicate are reasonable. The resulting failure rates are then used in making the system apportionment. However, the *individual* reliability goal or requirement for a *component* itself should not be changed unless the maintenance considerations are called out in the component specification. The apportionment is based on the *original* component failure rate modified to account for the maintenance.

11.6 EXAMPLE OF A DETAILED APPORTIONMENT

All these factors, as well as the sequence in which they are considered, affect the apportionment. Normally, the steps are performed in the following order. The mission phases are determined first so that the appropriate environments and failure-rate multipliers can be established. The particular systems operating in each phase are delineated and reliability requirements are established. As indicated earlier, the initial allowance or requirement for each phase is determined by contractual requirements, by the relative importance of the functions performed in the phase, and by the relative phase reliabilities. However, except for the contractual requirements, these considerations involve an element of engineering judgment, and the phase allowances may require modification if it becomes apparent that significant anomalies in the apportionment result.

The next step, which can be carried out concurrently with phase allocations, is the determination of failure rates of the various elements operating within the phase and of the non-operating elements if they have standby failure rates. This step includes the factors of basic failure rates, environmental multipliers, operating times (duty cycles), derating, and so on.

This is followed by modification for criticality. Failure rates for critical and major elements are suitably reduced or those for minor and major elements are increased. In either case, the resulting failure allowances reflect the relative criticalities of the various elements. Of course, it is necessary that the same procedure be used in all phases. If the failure rates for critical and major elements are reduced in one phase, they should be reduced in all phases.

Immediately following this, or at the same time that criticality multipliers are being established, multiplying factors for state-of-the-art (product-improvement) and maintainability considerations are determined. These multipliers are then combined with the criticality multiplier and with the failure rates which now include environment, duty cycle, and other factors. The final failure rates are now used for the apportionment, following the procedures which have been described.

We shall complete the detailed apportionment of a complex system for a somewhat complex mission. The apportionment shall be accomplished to the subsystem level and will consider many of the factors which have been discussed. The particular mission consists of an Earth-orbiting satellite used to detect radiation levels and to transmit related data. The initial phase is launch including orbit injection, followed by orbital coast during which attitude control maneuvers are performed on a daily basis. Telemetry operations are performed for 9 min during each 90-min orbit. The reliability requirement is 0.90 for 30 days and 0.50 for 6 months (180 days). This requirement reflects the need for a successful launch and accumulation of initial radiation measurements, while the accrual of additional measurements for verification and continuity is less important.

A reliability of 0.90 for 30 days would yield a reliability of 0.53 for 180 days if the failure rates were constant. Also, since the first 30-day period includes the launch phase, the requirement for the 30 days of orbit will be greater than 0.90. A 6-month reliability considerably greater than 0.53 would be required for a constant-failure-rate system. Consequently, the 30-day requirement is the more stringent and will be apportioned first. The apportionment for the remaining 5 months can then be made separately.

The mission is divided into launch, coast, and attitude control phases. Mission systems include first- and second-stage engines and separation subsystems, guidance system, propulsion electric power subsystem, radiation-detection and measuring equipment (redundant), data recording and storage equipment (redundant), telemetry and communication subsystem (redundant), radiation electric power subsystem (redundant), attitude sensing and electrical controls, attitude control engines (redundant), attitude control electric power, temperature controls, and the structure. Figure 11.2 is a logic diagram showing the redundancy relationships. Table 11.4 is a matrix depicting the subsystems which operate in each phase of the mission.

Note the nature of the redundancy in the radiation-detection portion of Fig. 11.2. There is no cross linkage so that failure of one subsystem in one leg fails the entire leg. Also note that the electric power subsystems are subdivided into portions applicable to each of the other subsystem functions

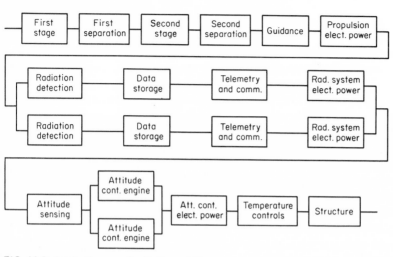

FIG. 11.2   Logic diagram of radiation satellite.

being supplied with electric energy. These portions are independent of each other so that, in effect, there are several electric power subsystems.

Actual operating times for the first 30 days are determined for those subsystems which are time or cycle dependent. This includes all subsystems except those associated with the launch, which is considered to be a single-shot operation. However, it may be necessary to subdivide the

TABLE 11.4   Subsystems operating in each mission phase

Subsystem	Phase		
	Launch	Coast	Attitude control
First-stage engine and controls	×		
First-stage separation	×		
Second-stage engine and controls	×		
Second-stage separation	×		
Guidance	×		
Propulsion and guidance power	×		
Radiation detection and measurement . .		×	×
Data recording and storage		×	×
Telemetry and communication		×	
Radiation system electric power		×	×
Attitude sensing		×	×
Attitude control engines			×
Attitude control electric power		×	×
Temperature control		×	×
Structure	×	×	×

operating times into times during coast and times during attitude control operations because the environmental multipliers for these time periods are very different. The non-operating failure rate may also have to be considered for the electronic components because the period of non-operation is during launch, when the environmental multiplier is quite high.

Assume that the launch, including orbit injection, takes a total time of 6 min (0.1 hr) and that each attitude correction takes 36 sec (0.01 hr). Since attitude correction takes place on a daily basis, the total time for these maneuvers for the 30-day period is 0.3 hr. If the environmental multipliers for launch and attitude maneuvers are 1,000 and 25, respectively, while the k factor for orbital coast is 1, and if the non-operating failure rate is 0.02 times the operating rate, the relative failure probabilities are as follows:

Launch: $1,000 \times 0.1 \times 0.02 \times \lambda = 2.0\lambda$

Maneuver: $25 \times 0.01 \times 30 \times \lambda = 7.5\lambda$

Coast: $1 \times 719.7 \times \lambda \approx 720\lambda$

where λ is the generic operating failure rate per hour. Since the launch and maneuver failure probabilities are only a small percentage (1.3 percent) of the total, for practical purposes they can be omitted for the orbiting subsystems in this instance without noticeably affecting the apportionment. Operating times and predicted failure probabilities, then, can be based on the orbital phases. The times for the various subsystems are given in Table 11.5.*

Next, hourly failure rates are obtained from standard-failure-rate handbooks. Piece part failure rates are suitably lowered to account for derated operation (determined from circuit analyses). Since, in general, we are not considering increased environmental stresses for the orbiting subsystems, the environmental multiplier for parts in these subsystems is unity. The piece part failure rates are then combined to obtain the failure rates for the various subsystems. We shall assume that the failure rates given in Table 11.6 apply, and that the non-operating failure rates are zero.

Criticality, state-of-the-art, and maintainability considerations are now evaluated and applied as appropriate. Because the system is unmanned, it might be assumed that there are no differences in criticality. However, this is not the case. Failure of some subsystems can have more serious consequences than failure of others. This is easily seen by reviewing the logic diagram. If one of the redundant subsystems related to radiation fails, most of the information will still be obtained, recorded, and transmitted by the second.

---

*Since actual operating times are less than 720 hr, the percentage of failures during launch and attitude maneuvers increases from 1.3 percent to as high as 13.0 percent. Omission of these failures, however, still does not appreciably affect the apportionment.

TABLE 11.5   Subsystem operating times for 30-day mission (not including launch)

Radiation detection and measurement . .	10 min/hr × 720 hr	= 120 hr
Data recording and storage	10 min/hr × 720 hr	= 120 hr
Telemetry and communication	9 min/orbit × 480 orbits	= 72 hr
Radiation system electric power	16 min/hr × 720 hr	= 192 hr
Attitude sensing	Full time	= 720 hr
Attitude control engines	Daily (failure rate is cycle oriented)	= 30 cycles
Attitude control electric power	Full time	= 720 hr
Temperature control	Full time	= 720 hr
Structure	(Stressed only during attitude maneuvers)	= 0.3 hr
	(Structure is also stressed during launch)	= 0.1 hr

TABLE 11.6   Subsystem failure rates considering time and environmental factor

Subsystem	Failure rate, time, and k factor	Product
First-stage engine	0.01 per cycle × 1 cycle × 1*	0.010
First-stage separation	6×10^{-6} per cycle† × 1 cycle × 500‡	0.003
Second-stage engine	0.01 per cycle × 1 cycle × 1*	0.010
Second-stage separation	6×10^{-6} per cycle × 1 cycle × 1,000 §	0.006
Guidance	0.0002 per hr × 0.1 hr¶ × 1,000	0.020
Propulsion electric power	5×10^{-5} per hr × 0.1 hr × 1,000	0.005
Structure (launch phase)	1×10^{-5} per hr × 0.1 hr × 1,000	0.001
Radiation detection and measurement . .	5.4×10^{-4} per hr × 120 hr	0.065
Data recording and storage	3.1×10^{-4} per hr × 120 hr	0.037
Telemetry and communications	8.9×10^{-4} per hr × 72 hr	0.064
Radiation system electric power	2.0×10^{-4} per hr × 192 hr	0.038
Attitude sensing	6.6×10^{-5} per hr × 720 hr	0.048
Attitude control engines	0.003 per cycle × 30 cycles	0.090
Attitude electric power	4×10^{-5} per hr × 720 hr	0.029
Temperature control	8×10^{-5} per hr × 720 hr	0.058
Structure (orbit phases)	1×10^{-5} per hr × 0.3 hr × 25	<0.0001

*A higher environmental multiplier is not used because the failure rate for rocket engines applies to a launch environment.

†This factor includes internal redundancy within the detonator; the k factor applies to the assembly.

‡The basic failure rate for explosive squibs is for a static environment; the k factor is for one stage.

§This k factor is for two stages.

¶The guidance system is used only for launch and orbit injection in this mission.

Similarly, if one attitude control engine subsystem fails, the second subsystem can perform this function. On the other hand, if a series element fails, the function performed by that element is lost. If the function is necessary for mission success, it should have a more critical (i.e., higher) classification than is given to the redundant element. Table 11.7 gives criticality ratings and weightings to the various subsystems in the example. Applicable comments are provided. A rating system of 10:3:1 is used because of the nature of the system and the consequences of a failure.

State of the art is the next consideration. The following assumptions are made. Launch systems have attained a relatively stable level of reliability and no major reliability improvements are anticipated in the near future. On the other hand, considerable work is being done in the field of power supplies and detection systems, and improvements in their reliabilities are anticipated. Data recording and telemetry equipment of the types being employed in this system are middle of the road with respect to state of the art and some improvement can be expected within the program life span. Reliabilities of attitude control, temperature control, and structural components may be

TABLE 11.7   Subsystem criticalities

Subsystem	Criticality	Rating	Comments
First-stage engine	Critical	10	Some failures could affect the safety of ground personnel
First-stage separation	Major	3	All elements whose failure would fail the mission but would not affect safety are classed as major
Second-stage engine	Major	3	
Second-stage separation	Major	3	
Guidance	Critical	10	Some failures could affect ground personnel safety
Propulsion electric power	Major	3	
Structure	Major	3	
Radiation detection and measurement .	Minor	1	All redundant subsystems are classed as minor
Data recording and storage	Minor	1	
Telemetry and communications	Minor	1	
Radiation system electric power	Minor	1	
Attitude sensing	Minor	1	Loss of attitude control could result in the loss of some data but most data would still be available
Attitude control engines	Minor	1	
Attitude electric power	Minor	1	
Temperature control	Minor	1	Temperature control is a precautionary measure to increase component life. Loss of control would not necessarily result in loss of data.

expected to remain at the current level. A factor of 3 will be applied to those subsystems which can be improved significantly; 2 will be used for those where some improvement is expected; and unity will be used for those subsystems whose reliabilities will remain fairly constant. The factors are applied in the same manner as the criticality factors: failure rates are divided by the appropriate constant.

No maintenance is possible for this unmanned system, so that maintenance considerations are not applicable to any of the subsystems. However, in order to show its effect, Eq. (11.2) includes a maintenance factor. When applicable, this factor will be less than 1, as explained in the previous section.

$$\lambda_{used} = \lambda \times \frac{1}{M_1} \times \frac{1}{M_2} \times M_3 \tag{11.2}$$

where M_1 is the criticality factor, M_2 is the state-of-the-art factor, and M_3 is the maintenance factor. Table 11.8 shows the final subsystem failure rates used in the apportionment for the first 30 days of the mission.

The last step before making the apportionment is to combine the failure rates of the redundant elements in order to obtain a series system, as was explained earlier in the chapter. All failure rates used are the final rates listed in Table 11.8. The total failure rate for one radiation-detection system (four series elements in one leg in Fig. 11.2) is $0.0217 + 0.0185 + 0.0320 + 0.0127 = 0.0849$. The corresponding reliability for one leg is 0.9186, and the reliability for the two parallel legs is $1 - (1 - R_i)^2 = 1 - (0.0814)^2 = 0.9934$. The probability of failure is 0.0066. The failure rate for one set of attitude control engines is 0.0900 and the corresponding reliability is 0.9139. The reliability for two sets in parallel is $1 - (0.0861)^2 = 0.9926$ and the probability of failure is 0.0074. The revised failure probabilities are shown in Table 11.9. Failure probabilities of actual series elements are approximated. The approximations are very close to the true failure probabilities that would be obtained by using the exponential formula $Q = 1 - e^{-\lambda t}$, and they are quite satisfactory for apportionment purposes. The ratio method described earlier is used to apportion initial allowances to the system elements. The allowances for the redundant systems are shown in parentheses to indicate that they are based on this ratio rather than having been calculated from the allowances for the components in the parallel legs, and that they therefore require revision.

It was shown that when apportionments are made to series-parallel systems an iterative process may be required. This was because, after the first apportionments are made and the system reliability is recalculated using the apportioned failure allowances, it is found that not all the allowance has been allocated. This follows from the fact that when the allowances for the redundant elements are recombined to find the allowance for the redundant

TABLE 11.8   Final failure rates used in 30-day apportionment

Subsystem	Failure rates considering time and environments	Inverse of criticality factor, $1/M_1$	Inverse of state-of-the-art factor, $1/M_2$	Maintenance factor (M_3)	Final failure rate
First-stage engine	0.010	1/10	1	1	0.0010
First-stage separation	0.003	1/3	1	1	0.0010
Second-stage engine	0.010	1/3	1	1	0.0033
Second-stage separation	0.006	1/3	1	1	0.0020
Guidance	0.020	1/10	1	1	0.0020
Propulsion electric power ...	0.005	1/3	1/3	1	0.0006
Structure	0.001	1/3	1	1	0.0003
Radiation detection and measurement	0.065	1	1/3	1	0.0217
Data recording and storage	0.037	1	1/2	1	0.0185
Telemetry and communications	0.064	1	1/2	1	0.0320
Radiation system electric power ...	0.038	1	1/3	1	0.0127
Attitude sensing	0.048	1	1	1	0.0480
Attitude control engines	0.090	1	1	1	0.0900
Attitude electric power	0.029	1	1/3	1	0.0097
Temperature control	0.058	1	1	1	0.0580

TABLE 11.9   Initial apportionment of failure allowances
to radiation measurement mission

(30-day reliability goal = 0.90; system failure allowance = 0.10)

Subsystem	Failure probability		Allowance
First-stage engine	0.0010		0.0007
First-stage separation	0.0010		0.0007
Second-stage engine	0.0033		0.0023
Second-stage separation	0.0020		0.0014
Guidance	0.0020		0.0014
Propulsion electric power	0.0006		0.0004
Structure	0.0003		0.0002
Redundant radiation detection system	0.0066		(0.0049)
Radiation detection and measurement		0.0217	0.0155
Data recording and storage		0.0185	0.0132
Telemetry and communications		0.0320	0.0228
Radiation system electric power		0.0127	0.0091
Attitude sensing	0.0480		0.0343
Redundant attitude control engines	0.0074		(0.0053)
Single attitude control engine		0.0900	0.0643
Attitude control electric power	0.0097		0.0069
Temperature control	0.0580		0.0414
Total system	0.1399		
Ratio: Failure allowance to failure probability . .	5/7		

combination, a new allowance results. (For example, 0.0155 + 0.0132 + 0.0228 + 0.0091 = 0.0606. $1 - 0.0606^2 = 0.0037$, not 0.0049 which was the initial allowance.) The results of these calculations for the radiation measurement mission are shown in Table 11.10, which also delineates associated reliability goals.

The remaining failure allowance of 0.0025 may be allocated proportionally to the various system elements if desired. However, in addition to the five reasons stated earlier for not doing so, two additional reasons are now applicable. First, the choice of criticality, state-of-the-art, and maintenance factors, while not arbitrary, is a matter of good engineering judgment. In the example, one would find it difficult to argue that the criticality ratio of 10:3:1 is much better than a ratio of 9:3:1 or 10:5:1. Since the ratio has a very significant effect on the allocation, changes resulting from allocating the remaining allowance are relatively unimportant and are not warranted.

Second, several approximations were made during the allocation process. The most frequent was the use of $R = 1 - \lambda t$ rather than $R = e^{-\lambda t}$. The use of these approximations does not affect the usefulness of the apportionment but does negate any gain in "accuracy" that would be achieved by apportioning the remaining allowance.

TABLE 11.10 **Final apportionment of failure allowances for 30-day mission**

(System reliability goal = 0.90; system failure allowance = 0.10)

Subsystem	Allocated failure probability		Reliability goal
First-stage engine	0.0007		0.9993
First-stage separation	0.0007		0.9993
Second-stage engine	0.0023		0.9977
Second-stage separation	0.0014		0.9986
Guidance	0.0014		0.9986
Propulsion electric power	0.0004		0.9996
Structure	0.0002		0.9998
Redundant radiation detection system	0.0037		
Radiation detection and measurement		0.0155	0.9845
Data recording and storage		0.0132	0.9868
Telemetry and communications		0.0228	0.9772
Radiation system electric power		0.0091	0.9909
Attitude sensing	0.0343		0.9657
Redundant attitude control engines	0.0041		
Single attitude control engine		0.0643	0.9357
Attitude control electric power	0.0069		0.9931
Temperature control	0.0414		0.9586
Total system (calculated)	0.0975		
Difference (system allowance – allocations) ...	0.0025		

After the apportionment is completed for the 30-day mission, the 6-month apportionment can be made. The procedure is to determine the allowance for the remaining 5 months and apportion it independently following the same methods that were used for the first time period. Note that the subsystems associated with the launch are no longer applicable. Also, while not necessary in this example, the relative criticalities of the remaining subsystems sometimes require reevaluation because of changing functions of the mission.

The product of individual phase reliability goals is equal to the overall goal. Therefore, the reliability goal for the second time period can be found by dividing the overall reliability goal by the reliability goal for the first period. $R_{2nd} = R_{total}/R_{1st} = 0.50/0.90 = 0.5556$. The failure allowance for the second period is then $1.000 - 0.5556 = 0.4444$. This allowance is allocated among the remaining subsystems, and reliability goals for the 5-month period are determined. Overall reliability goals can then be found by multiplying together the goals for the two time periods. However, since the allowances are not directly related to the lengths of two periods, this is not recommended.

All principal factors affecting the apportionment of failure allowances and establishment of reliability goals have been evaluated in detail and their effects have been examined. These factors of basic failure rates, non-operating failure rates, mission phases and requirements, environments, derating, duty cycle, criticality, state of the art, and maintenance/repair have significant effects on the allowances and are duly considered. The approximations which were made have a negligible overall effect and do not affect the validity or usefulness of the apportionment. When apportionments are made in the manner indicated, they can be powerful aids in determining potential problem areas, in indicating where redundancy should be added to upgrade the system reliability, and in providing realistic goals for all elements of the system.

ADDITIONAL READINGS

ARINC Research Corporation: "Reliability Engineering," Prentice-Hall, Inc., Englewood Cliffs, N.J., 1964.

Bureau of Naval Weapons, Chief of: "Handbook: Reliability Engineering," U.S. Government Printing Office, 1964.

Hiltz, P. A., and J. L. Gaffney: "Statistical Techniques for Reliability," North American-Rockwell Corporation, Downey, Calif., 1965.

Lloyd, D. K., and M. Lipow: "Reliability: Management, Methods, and Mathematics," Prentice-Hall, Inc., Englewood Cliffs, N.J., 1962.

12

RELIABILITY GROWTH

12.1 INTRODUCTION

Analysis of reliability growth and the preparation of appropriate growth curves can be considered either as an apportionment or as an assessment process. If a growth curve is prepared to depict expected levels of reliability achievement and to provide a continuum of time-oriented goals, it can be classed as an apportionment procedure. When used in this manner, it will also serve as an aid for measuring accomplishment toward program reliability objectives. Emphasis and attention can then be given to those components and assemblies whose reliability growth has not met expectations or requirements.

When reliability-growth equations and curves are prepared after a program is well into the development phase and thereby utilize actual test data to arrive at appropriate numerics, growth analysis becomes primarily an assessment process. The growth equation then reflects actual reliability status and may be used, with appropriate constraints, to forecast future reliability achievement if no significant program changes are contemplated. However, achieved reliability may have little relationship to system requirements.

The selection of appropriate reliability-growth equations and curves is more an art than a science. There are relatively few data available which can be used to define particular equations applicable to the reliability growth of specific types of components or systems. For example, the reliability of some parts, particularly structural members, may be extremely close to unity, especially for short-duration missions, and some growth curve equations assume an upper reliability limit of one. Other components, e.g., complex electronic assemblies, may have an ultimate reliability less than unity. In a similar vein, some growth curves start with an initial reliability of zero, others assume a starting reliability greater than zero, and still others do not rise above zero until after a period of time has elapsed. In fact, some growth equations can cover any of the three situations, depending on the particular values of the constraints which are chosen. Finally, reliability-growth equations can provide curves which, on standard graph paper, are concave downward throughout their length—they represent a reliability-growth *rate* which decreases as reliability increases. Other equations result in an S-shaped curve with the reliability-growth rate initially increasing, reaching a maximum rate, and then decreasing as reliability improves.

With such a variety of possible choices and so few historical data to aid in the selection, we must look elsewhere for realistic criteria for the selection. The criterion which has been emphasized throughout this book is that of practicality. A growth function which provides time- or test-related reliability goals which are needed to meet program schedule requirements and which can be readily plotted and technically justified will provide the most benefits. The most exact curve (based on available program data) is not necessarily the most practical. No one would advocate using a 9-degree equation if there were 10 data points just because a curve based on this equation would pass through all the points. Since an approximation is being made, a second-degree equation is usually selected, and we choose one which both fits the available data and the goal *and* which is readily calculated and plotted. The sacrifice of a little "accuracy" (by possibly not choosing the optimum second-degree equation from the standpoint of curve fitting) is more than made up for by the benefits which can result from choosing a simple relationship.

We have stated that a component's growth equation should reflect both the current reliability status and the reliability goal or requirement. By adopting this postulate we then determine both what the growth rate has been and what it must be in the future, and we can compare the two rates. If the current rate of growth is not sufficient to attain final reliability goals within the schedule constraints, then appropriate action can be taken. If the present growth rate equals or exceeds requirements, it may be possible to divert some effort to other components whose reliability growth rate is not satisfactory.

12.2 THREE USEFUL GROWTH EQUATIONS

We recognize that reliability growth is, in reality, a discrete function which occurs, hopefully, each time a design change is made. However, since the exact time of these changes is unknown in advance and since a curve can be fitted reasonably well to the discrete values, almost all reliability-growth curves are depicted as continuous. We shall examine some reliability-growth equations and review their properties and advantages along with the associated curves. The equations which are presented are those which may be the most practical and which are plotted most conveniently.

The first case is one where reliability growth is a function of the number of tests and design improvements. It is assumed that, as a result of each test, an improvement is made which reduces the failure rate by a fixed percentage of the *current* rate. This is defined by Eq. (12.1).*

$$R = e^{-k^n \lambda_0} \tag{12.1}$$

where λ_0 is the original failure rate, k is the percentage of the *existing* failure rate which remains after each test and resulting design change, and n is the number of tests and design changes. For example, if 10 percent of the existing failure rate is eliminated after each test and design change, then $k = 0.9$ and the failure rate after the first test λ_1 is $0.9 \lambda_0$; the failure rate after the second test λ_2 is $0.9 \lambda_1 = 0.9^2 \lambda_0 = 0.81 \lambda_0$; and the failure rate after the nth test λ_n is $0.9 \lambda_{n-1} = 0.9^n \lambda_0$. In this growth equation, the limit of reliability is unity.†

Figures 12.1 to 12.8 show some corresponding growth curves for eight different values of λ_0 and a number of values of k. Note that the curves are plotted on *inverted* semilogarithmic paper and that, when plotted in this manner, reliabilities greater than 0.9 are, for practical purposes, straight lines. This is an important advantage of this reliability-growth equation because the reliability-growth curve can be plotted very rapidly. Even when the reliability is less than 0.9, only two or three points are needed to plot the lower portion of the curves. Table 12.1 provides reliability numerics for many of these curves and for additional growth curves with other values of λ_0 and k.

The second growth curve is defined by Eq. (12.2).

$$R = 1 - ae^{-bx} \tag{12.2}$$

*Equation (12.1) is a modification of a more general growth curve $R = ce^{-\beta e^{-\gamma n}}$, known as the *Gompertz* curve.

†The limit may be lowered by multiplying the expression on the right of Eq. (12.1) by a constant between zero and one. The value of the constant will then define the limit. However, the resulting expression does not yield straight lines, and some of the value of the original equation's simplicity is lost.

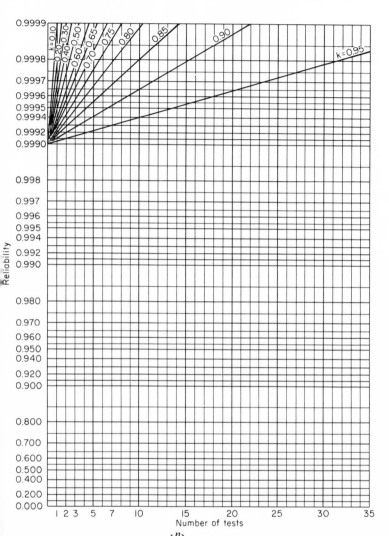

FIG. 12.1 Growth curves for $R = e^{-k^{n}\lambda_0}$ for different values of k, $\lambda_0 = 0.001$.

where a is a constant which locates the initial reliability point, b is a second constant which determines the rate of growth, and x is time or number of tests. When a is one, the curve goes through zero at x equal to zero. When a is between zero and one, the original reliability (at $x = 0$) is between one and zero; when a is greater than one, the reliability is undefined for some initial time period. The equation has no meaning for negative values of a. All positive values of b yield valid growth curves. It can be seen from Figs. 12.9 and 12.10 that, when plotted on *inverted* semilogarithmic paper, the curves are perfectly straight lines for all usable values of a and b. As in Eq. (12.1),

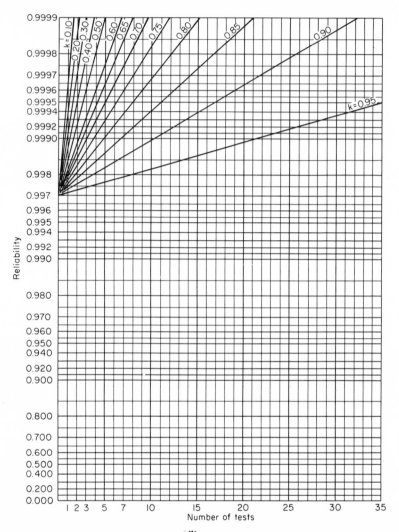

FIG. 12.2   Growth curves for $R = e^{-k^n \lambda_0}$ for different values of k, $\lambda_0 = 0.003$.

the reliability limit in Eq. (12.2) is unity, and it can be reduced by multiplying the right-side expression by a positive constant less than unity. Again, however, the lines will no longer be straight. This is true for any growth equation whose limit is less than unity.

A third reliability-growth equation, defined by Eq. (12.3), is a simplification of one developed by Sogorka and Peterson.*

---

*J. Sogorka and J. Peterson, Dynamic Characteristics of Reliability Growth and Its Implications, *Joint Mil. Ind. Symp. Guided Missile Res. Reliability, 5th, Chicago, 1959.*

$$R = 1 - \frac{a}{x^n} \qquad (12.3)$$

where a is a positive constant less than x and x is time or number of tests.*
Reliability is undefined when x is less than a. The curves for this equation are

*When $n = 1$, Eq. (12.3) is also a simplification of a reliability-growth function described in Chap. 11 of "Reliability: Management, Methods and Mathematics," by Lloyd and Lipow.

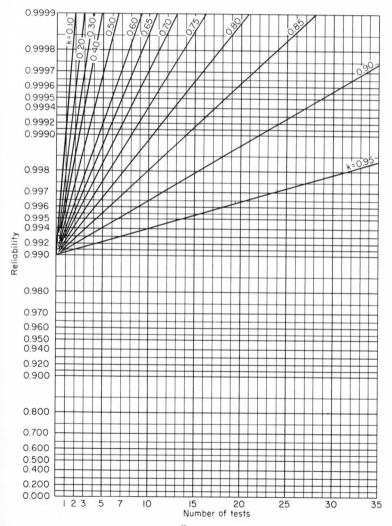

FIG. 12.3   Growth curves for $R = e^{-k^n \lambda_0}$ for different values of k, $\lambda_0 = 0.010$.

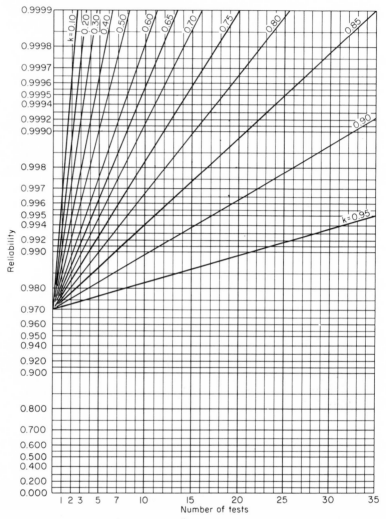

FIG. 12.4 Growth curves for $R = e^{-k^n\lambda_0}$ for different values of k, $\lambda_0 = 0.030$.

straight lines when plotted on logarithmic paper *rotated 90°*, as shown in Figs. 12.11 and 12.12. Again, the reliability limit is unity, but it may be reduced by multiplying the right-side expression by a positive constant less than 1. As before, the line will no longer be straight. (When the constant is used, the original Lloyd-Lipow equation results.)

These three growth equations yield curves which are all concave downward when plotted on standard (linear) graph paper. Curves for the first equation are concave downward everywhere when the initial failure rate λ_0 is less than one. The curve will turn upward initially if λ_0 is greater than one, but only as long as the product $k^n\lambda_0$ is greater than one. (Note that these are extreme cases and are rarely encountered.) Curves of the second and third equations are concave downward throughout their lengths for all usuable values of their parameters. The three equations define reliability-growth rates which are fast initially and which gradually become slower as actual reliability increases. That is, the reliability increases at a decreasing rate. The equations, therefore, represent realistic situations.

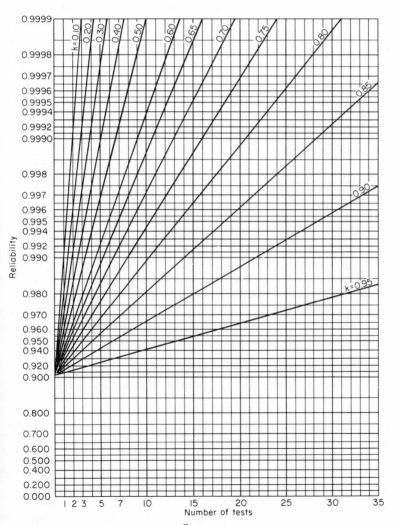

FIG. 12.5   Growth curves for $R = e^{-k^n\lambda_0}$ for different values of k, $\lambda_0 = 0.100$.

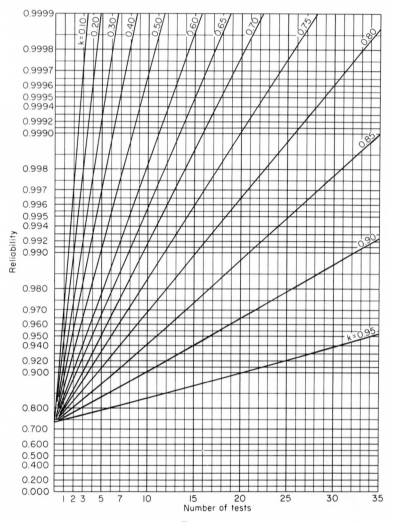

FIG. 12.6 Growth curves for $R = e^{-k^n \lambda_0}$ for different values of $k, \lambda_0 = 0.300$.

The advantage of these equations is that straight lines result when their curves are plotted on appropriate (rotated logarithmic or inverted semi-logarithmic) graph paper. This means that it is necessary to compute and plot only two values on this paper to define a curve. All other values can then be rapidly determined from the straight line and plotted on standard paper if desired. This eliminates a large number of tedious calculations. Furthermore, since only two values are needed to plot the equation, the *current reliability* and the *reliability objective* can be conveniently used for these two points.

Additional points for earlier reliabilities can then also be plotted. Curves for the three equations can be drawn, and that curve which passes through the points for the present reliability and the reliability goal and which most closely fits the earlier points can be used. The corresponding growth equation is then determined. This graphical method is much more practical than trying to mathematically select a growth equation and its particular parameters that will fit a diverse set of points. Also, it enables the selection of an equation that will both fit the present data and satisfy the objective, and the *required* reliability-growth rate can be found.

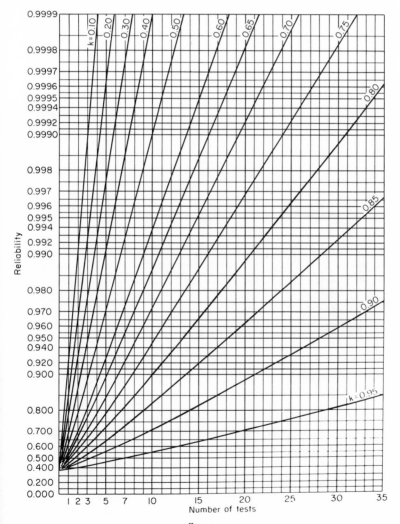

FIG. 12.7 Growth curves for $R = e^{-k^n \lambda_0}$ for different values of k, $\lambda_0 = 1.000$.

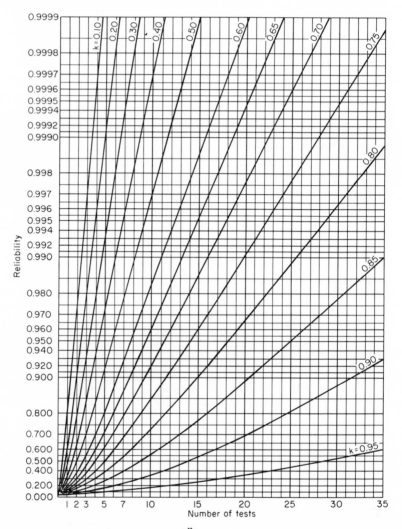

FIG. 12.8 Growth curves for $R = e^{-k^n \lambda_0}$ for different values of k, $\lambda_0 = 3.000$.

The only disadvantage of these equations is that they assume an ultimate reliability of one. However, this is a disadvantage in theory rather than in fact because the ultimate reliability is achieved only as the number of tests or hours becomes infinite. Since a program span covers a finite period of time or a limited number of design changes, the ultimate reliability does not affect the choice of growth equations as long as the objective is realistic.

As an example of the use of these curves, let us assume that the present reliability of an equipment is 0.940 and that the reliability requirement is

TABLE 12.1 Reliability after n tests for various initial failure rates λ_0 and different values of k;

$$R = e^{-k^n \lambda_0}$$

$n \backslash \lambda_0$	4.000	2.000	1.000	0.400	0.200	0.100	0.040	0.020	0.010	0.004	0.002	0.001
						$k = 0.95$						
1	0.0224	0.1496	0.3867	0.6839	0.8270	0.9094	0.9627	0.9812	0.9905$^+$	0.9962	0.9981	0.9991
2	0.0271	0.1645$^-$	0.4056	0.6970	0.8349	0.9137	0.9645$^+$	0.9821	0.9910	0.9964	0.9982	0.9991
3	0.0324	0.1800	0.4243	0.7097	0.8424	0.9178	0.9663	0.9830	0.9915$^-$	0.9966	0.9983	0.9991
5	0.0453	0.2128	0.4613	0.7338	0.8566	0.9255$^+$	0.9695$^+$	0.9846	0.9923	0.9969	0.9985$^-$	0.9992
7	0.0612	0.2474	0.4974	0.7563	0.8696	0.9325$^+$	0.9725$^-$	0.9861	0.9930	0.9972	0.9986	0.9993
10	0.0912	0.3020	0.5495$^+$	0.7870	0.8871	0.9419	0.9763	0.9881	0.9940	0.9976	0.9988	0.9994
15	0.1567	0.3959	0.6292	0.8308	0.9115$^+$	0.9547	0.9816	0.9908	0.9954	0.9981	0.9991	0.9995$^+$
20	0.2384	0.4882	0.6987	0.8664	0.9308	0.9648	0.9858	0.9929	0.9964	0.9986	0.9993	0.9996
30	0.4238	0.6510	0.8068	0.9177	0.9580	0.9788	0.9915$^-$	0.9957	0.9979	0.9991	0.9996	0.9998
50	0.7351	0.8574	0.9259	0.9697	0.9847	0.9923	0.9969	0.9985$^-$	0.9992	0.9997	0.9998	0.9999
						$k = 0.90$						
1	0.0273	0.1653	0.4066	0.6977	0.8353	0.9139	0.9646	0.9822	0.9910	0.9964	0.9982	0.9991
2	0.0392	0.1979	0.4449	0.7233	0.8504	0.9222	0.9681	0.9839	0.9919	0.9968	0.9984	0.9992
3	0.0541	0.2327	0.4824	0.7471	0.8643	0.9297	0.9713	0.9855$^+$	0.9927	0.9971	0.9985$^+$	0.9993
5	0.0942	0.3070	0.5541	0.7896	0.8886	0.9427	0.9767	0.9883	0.9941	0.9976	0.9988	0.9994
7	0.1476	0.3842	0.6198	0.8259	0.9088	0.9533	0.9810	0.9905$^-$	0.9952	0.9981	0.9990	0.9995$^+$
10	0.2479	0.4979	0.7056	0.8698	0.9326	0.9657	0.9861	0.9931	0.9965$^+$	0.9986	0.9993	0.9997
15	0.4389	0.6625$^-$	0.8139	0.9209	0.9597	0.9796	0.9918	0.9959	0.9979	0.9992	0.9996	0.9998
20	0.6149	0.7842	0.8855$^+$	0.9525$^+$	0.9760	0.9876	0.9951	0.9976	0.9988	0.9995$^+$	0.9998	0.9999
30	0.8440	0.9187	0.9585$^-$	0.9832	0.9916	0.9958	0.9983	0.9992	0.9996	0.9998	0.9999	1.0000
50	0.9796	0.9897	0.9949	0.9979	0.9990	0.9995$^-$	0.9998	0.9999	0.9999	1.0000	1.0000	1.0000

TABLE 12.1 Reliability after n tests for various initial failure rates λ_0 and different values of k;

$$R = e^{-k^n \lambda_0}$$ (Continued)

$n\backslash\lambda_0$	4.000	2.000	1.000	0.400	0.200	0.100	0.040	0.020	0.010	0.004	0.002	0.001
						$k = 0.80$						
1	0.0408	0.2019	0.4493	0.7261	0.8521	0.9231	0.9685+	0.9841	0.9920	0.9968	0.9984	0.9992
2	0.0773	0.2780	0.5273	0.7741	0.8799	0.9380	0.9747	0.9873	0.9936	0.9974	0.9987	0.9994
3	0.1290	0.3592	0.5993	0.8148	0.9027	0.9501	0.9797	0.9898	0.9949	0.9980	0.9990	0.9995−
5	0.2696	0.5193	0.7206	0.8772	0.9366	0.9678	0.9870	0.9935−	0.9967	0.9987	0.9993	0.9997
7	0.4322	0.6574	0.8108	0.9195+	0.9589	0.9792	0.9916	0.9958	0.9979	0.9992	0.9996	0.9998
10	0.6508	0.8067	0.8982	0.9580	0.9788	0.9893	0.9957	0.9979	0.9989	0.9996	0.9998	0.9999
15	0.8687	0.9320	0.9654	0.9860	0.9930	0.9965−	0.9986	0.9993	0.9996	0.9999	0.9999	1.0000
20	0.9549	0.9772	0.9885+	0.9954	0.9977	0.9988	0.9995+	0.9998	0.9999	1.0000	1.0000	1.0000
30	0.9951	0.9975+	0.9988	0.9995+	0.9998	0.9999	1.0000	1.0000	1.0000	1.0000	1.0000	1.0000
40	0.9995−	0.9997	0.9999	0.9999	1.0000	1.0000	1.0000	1.0000	1.0000	1.0000	1.0000	1.0000
						$k = 0.70$						
1	0.0608	0.2466	0.4966	0.7558	0.8694	0.9324	0.9724	0.9861	0.9930	0.9972	0.9986	0.9993
2	0.1409	0.3753	0.6126	0.8220	0.9066	0.9522	0.9806	0.9902	0.9951	0.9980	0.9990	0.9995+
3	0.2536	0.5036	0.7096	0.8718	0.9337	0.9663	0.9864	0.9932	0.9966	0.9986	0.9993	0.9997
5	0.5105+	0.7145+	0.8453	0.9350	0.9669	0.9833	0.9933	0.9966	0.9983	0.9993	0.9997	0.9998
7	0.7193	0.8481	0.9209	0.9676	0.9837	0.9918	0.9967	0.9984	0.9992	0.9997	0.9998	0.9999
10	0.8932	0.9451	0.9721	0.9888	0.9944	0.9972	0.9989	0.9994	0.9997	0.9999	0.9999	1.0000
15	0.9812	0.9905+	0.9953	0.9981	0.9991	0.9995+	0.9998	0.9999	1.0000	1.0000	1.0000	1.0000
20	0.9968	0.9984	0.9992	0.9997	0.9998	0.9999	1.0000	1.0000	1.0000	1.0000	1.0000	1.0000
25	0.9995−	0.9997	0.9999	0.9999	1.0000	1.0000	1.0000	1.0000	1.0000	1.0000	1.0000	1.0000
30	0.9999	1.0000	1.0000	1.0000	1.0000	1.0000	1.0000	1.0000	1.0000	1.0000	1.0000	1.0000

TABLE 12.1 Reliability after n tests for various initial failure rates λ_0 and different values of k; $R = e^{-k^n \lambda_0}$ (Continued)

$n \backslash \lambda_0$	4.000	2.000	1.000	0.400	0.200	0.100	0.040	0.020	0.010	0.004	0.002	0.001
						$k = 0.60$						
1	0.0907	0.3012	0.5488	0.7866	0.8869	0.9418	0.9763	0.9881	0.9940	0.9976	0.9988	0.9994
2	0.2369	0.4868	0.6977	0.8659	0.9305+	0.9646	0.9857	0.9928	0.9964	0.9986	0.9993	0.9996
3	0.4215-	0.6492	0.8057	0.9172	0.9577	0.9786	0.9914	0.9957	0.9978	0.9991	0.9995	0.9998
5	0.7327	0.8560	0.9252	0.9694	0.9846	0.9923	0.9969	0.9984	0.9992	0.9997	0.9998	0.9999
7	0.8941	0.9456	0.9724	0.9889	0.9944	0.9972	0.9989	0.9994	0.9997	0.9999	0.9999	1.0000
10	0.9761	0.9880	0.9940	0.9976	0.9988	0.9994	0.9998	0.9999	1.0000	1.0000	1.0000	1.0000
15	0.9981	0.9991	0.9995+	0.9998	0.9999	1.0000	1.0000	1.0000	1.0000	1.0000	1.0000	1.0000
20	0.9999	0.9999	1.0000	1.0000	1.0000	1.0000	1.0000	1.0000	1.0000	1.0000	1.0000	1.0000
						$k = 0.50$						
1	0.1353	0.3679	0.6065+	0.8187	0.9048	0.9512	0.9802	0.9900	0.9950	0.9980	0.9990	0.9995+
2	0.3679	0.6065+	0.7788	0.9048	0.9512	0.9753	0.9900	0.9950	0.9975	0.9990	0.9995+	0.9998
3	0.6065+	0.7788	0.8825-	0.9512	0.9753	0.9876	0.9950	0.9975	0.9988	0.9995+	0.9998	0.9999
5	0.8825-	0.9394	0.9692	0.9876	0.9938	0.9969	0.9988	0.9994	0.9997	0.9998	0.9999	1.0000
7	0.9692	0.9845-	0.9922	0.9969	0.9984	0.9992	0.9997	0.9998	0.9999	1.0000	1.0000	1.0000
10	0.9961	0.9980	0.9990	0.9996	0.9998	0.9999	1.0000	1.0000	1.0000	1.0000	1.0000	1.0000
15	0.9999	0.9999	1.0000	1.0000	1.0000	1.0000	1.0000	1.0000	1.0000	1.0000	1.0000	1.0000

TABLE 12.1 Reliability after n tests for various initial failure rates λ_0 and different values of k;
$R = e^{-k^n \lambda_0}$ (Continued)

n\\λ_0	4.000	2.000	1.000	0.400	0.200	0.100	0.040	0.020	0.010	0.004	0.002	0.001
						$k = 0.30$						
1	0.3012	0.5488	0.7408	0.8869	0.9418	0.9704	0.9881	0.9940	0.9970	0.9988	0.9994	0.9997
2	0.6977	0.8353	0.9139	0.9646	0.9822	0.9910	0.9964	0.9982	0.9991	0.9996	0.9998	0.9999
3	0.8976	0.9474	0.9734	0.9893	0.9946	0.9973	0.9989	0.9995−	0.9997	0.9999	0.9999	1.0000
4	0.9681	0.9839	0.9919	0.9968	0.9984	0.9992	0.9997	0.9998	0.9999	1.0000	1.0000	1.0000
5	0.9903	0.9952	0.9976	0.9990	0.9995+	0.9998	0.9999	1.0000	1.0000	1.0000	1.0000	1.0000
6	0.9971	0.9985+	0.9993	0.9997	0.9999	0.9999	1.0000	1.0000	1.0000	1.0000	1.0000	1.0000
7	0.9991	0.9996	0.9998	0.9999	1.0000	1.0000	1.0000	1.0000	1.0000	1.0000	1.0000	1.0000
8	0.9997	0.9999	0.9999	1.0000	1.0000	1.0000	1.0000	1.0000	1.0000	1.0000	1.0000	1.0000
9	0.9999	1.0000	1.0000	1.0000	1.0000	1.0000	1.0000	1.0000	1.0000	1.0000	1.0000	1.0000
10	1.0000	1.0000	1.0000	1.0000	1.0000	1.0000	1.0000	1.0000	1.0000	1.0000	1.0000	1.0000
						$k = 0.10$						
1	0.6703	0.8187	0.9048	0.9608	0.9802	0.9900	0.9960	0.9980	0.9990	0.9996	0.9998	0.9999
2	0.9608	0.9802	0.9900	0.9960	0.9980	0.9990	0.9996	0.9998	0.9999	1.0000	1.0000	1.0000
3	0.9960	0.9980	0.9990	0.9996	0.9998	0.9999	1.0000	1.0000	1.0000	1.0000	1.0000	1.0000
4	0.9996	0.9998	0.9999	1.0000	1.0000	1.0000	1.0000	1.0000	1.0000	1.0000	1.0000	1.0000
5	1.0000	1.0000	1.0000	1.0000	1.0000	1.0000	1.0000	1.0000	1.0000	1.0000	1.0000	1.0000

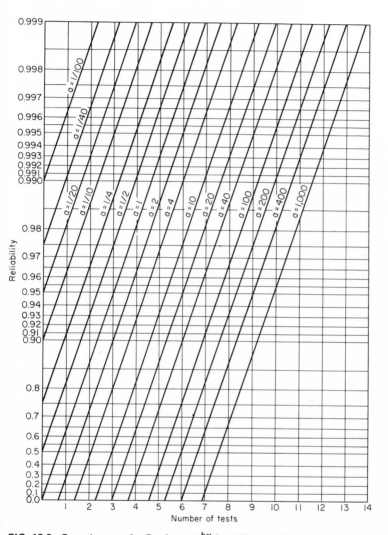

FIG. 12.9 Growth curves for $R = 1 - ae^{-bx}$ for different values of a; $b = 1$.

0.995, to be achieved within 1 year. Let us also assume that an average of six tests* per year are performed and that, as a result of analyses of the test data, design improvements are made. Finally, for purposes of this example, we shall assume that a total of six tests have already been made and that the reliability measured after each test is as given in Table 12.2.

*A test, as used in this example, is a set of operations used to evaluate the equipment and which provides sufficient data to enable estimation of the equipment's reliability.

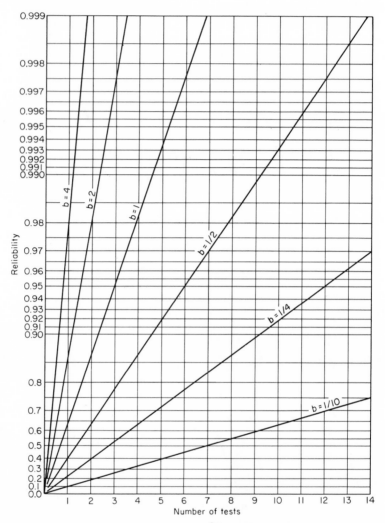

FIG. 12.10   Growth curves for $R = 1 - ae^{-bx}$ for different values of b; $a = 1$.

By comparing the achieved reliabilities with the curves in Figs. 12.1 to 12.12, we can find one or more specific growth equations which can be used to describe the reliability growth of this equipment. Using the first general growth equation $R = \exp(-k^n\lambda_0)$, Eq. (12.1), we select, after reviewing the various sets of curves, the curves in Fig. 12.6 (for $\lambda_0 = 0.300$)—specifically those for $k = 0.75$ and $k = 0.80$. Since our data fall between these two curves, we estimate that a curve for $k = 0.77$ will result in the best fit.

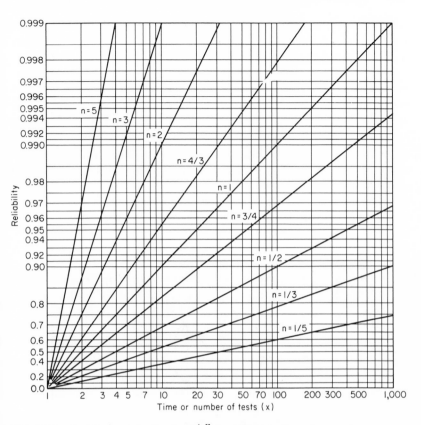

FIG. 12.11 Growth curves for $R = 1 - a/x^n$ for different values of n; $a = 1$.

Similarly, we can use Eq. (12.2), $R = 1 - ae^{-bx}$, and find a particular curve of best fit. As indicated earlier, all curves with the same value of a cross the Y axis at the same point, and all curves with the same value of b have the same

TABLE 12.2   Equipment reliability after each test

Test number	Reliability
1	0.80
2	0.85
3	0.88
4	0.905
5	0.925
6	0.940

slope. By comparing our data with the curves in Fig. 12.10, we find that the slope corresponds most closely to a value of $b = 1/4$. Using this value, we then find that the curve crosses the Y axis at a value of $a = 1/4$.

Either of the two equations $R = e^{-0.77^n \times 0.300}$ or $R = 1 - 0.25e^{-0.25x}$ will accurately describe the reliability growth experienced to date. We then extend the selected curve to n (or x) = 12 and determine what the reliability will be if the same rate of growth continues for the next year. This is shown in Fig. 12.13. We find that the reliability at the end of a year will be only about 0.987, and that it will not meet the requirement. Therefore, we must either increase the number of tests that will be performed during the next year and maintain the same rate of growth, or we must increase the rate of reliability growth per test, or we must attain some combination of these increases. The particular course of action will depend on many program factors.

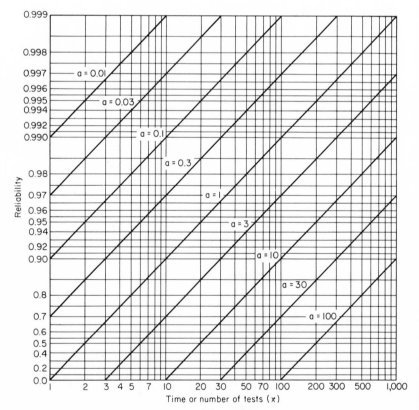

FIG. 12.12 Growth curves for $R = 1 - a/x^n$ for different values of a; $n = 1$.

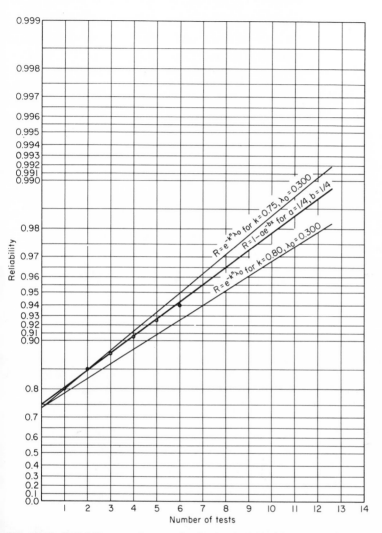

FIG. 12.13   Reliability-growth curves for the data in Table 12.2.

12.3   OTHER GROWTH EQUATIONS

We will briefly mention some other growth equations. As indicated earlier, these are more complex than the three we have just considered, and they usually require considerably more effort to plot and evaluate. When specific information about their parameters is available and their applicability is known, they can provide useful mathematical descriptions of the reliability-growth function.

$$R = k\left(1 - ae^{-bx - cx^2 - \cdots}\right) \tag{12.4}$$

These are extensions of the second growth equation and include higher-power terms in the exponent. They are much more general expressions. The number of terms in the exponent is generally limited to two. In this case the values of four parameters are required to completely describe the growth function.

$$R = \frac{k}{1 + ae^{-bx}} \tag{12.5}$$

When $k = 1$, this equation yields curves similar to those of Eq. (12.1). For reliabilities greater than 0.9, the curves are approximately straight when plotted on inverted semilogarithmic paper. Equation (12.1) is easier to use because the assumption which is made regarding the reduction in failure rate per test or design change can be directly inserted into the equation.

$$R = k\left(1 - \frac{b}{ax + b}\right) \tag{12.6}$$

This equation is similar to Eq. (12.3) when $k = 1$, but includes an additional term in the denominator. It can be simplified by dividing both numerator and denominator by b and letting $c = a/b$, leading to Eq. (12.6a), or by dividing both numerator and denominator by a and letting $c = b/a$, leading to Eq. (12.6b).

$$R = k\left(1 - \frac{1}{cx + 1}\right) \tag{12.6a}$$

$$R = k\left(1 - \frac{c}{x + c}\right) \tag{12.6b}$$

All these, however, are more complex than Eq. (12.3), and they do not result in straight lines.

$$R = ke^{-cx} \tag{12.7}$$

When $x = 1$, Eq. (12.7) provides lines which are almost straight when plotted on rotated logarithmic paper for values of R greater than 0.95. However, it is a simple equation to calculate and plot using a table of negative exponential values. Determining the appropriate value of c may be more difficult than establishing a value for k in Eq. (12.1).

There are other growth equations which have been developed, particularly in the field of psychology of learning. Some of these may be adaptable for use in reliability applications.

ADDITIONAL READINGS

Earles, D. R.: Reliability Growth Prediction during the Initial Design Analysis, *Proc. Nat. Symp. Reliability Quality Control, 7th, Philadelphia,* Institute of Electrical and Electronics Engineers, Inc., New York, 1961, pp. 380-393.

Lloyd, D. K., and M. Lipow: "Reliability: Management, Methods, and Mathematics," Prentice-Hall, Inc., Englewood Cliffs, N.J., 1962.

McDonald, P. R.: "The Theory and Practical Application of Improvement Curves," Procurement Associates, Covina, Calif., 1967.

Sogorka, J., and J. Peterson: Dynamic Characteristics of Reliability Growth and Its Implications, *Proc. Joint Mil.-Ind. Symp. Guided Missile Res. Technol., 5th, Chicago, 1959.*

13

ASSESSMENT METHODS: ATTRIBUTES

13.1 THE ASSESSMENT PROCESS

Reliability assessment is the process of utilizing actual test and performance data to estimate the achieved (inherent) reliability of the current equipment design. Assessments differ fundamentally from predictions in that assessment is an after-the-fact (a posteriori) activity utilizing actual data obtained from the equipment being evaluated, whereas prediction is a before-the-fact (a priori) function which employs data from previously existing equipment *already in service.* An assessment defines the *current* level of the equipment's reliability, regardless of the state of the equipment's development. A prediction forecasts the level of reliability which can be obtained in the field after development is completed and after those faults and deficiencies which have been experienced are corrected.

13.1.1 Data Applicability

The data which are used in an assessment are generated during the inspection, testing, and operation of the item of equipment being designed and evaluated or of other units of the same basic design and construction. When units have undergone design modifications and changes, specific criteria must be met to permit utilization of data from earlier models. These criteria are established by each individual program, sometimes for each subsystem or even for each item of hardware. A reliability-assessment guide prepared by North American-Rockwell Corporation devotes a complete section to criteria governing the inclusion or exclusion of test data when an assessment is made.

For the data to be truly applicable, the conditions and environments which exist during its generation should also correspond to or at least simulate those encountered in actual use. If an electronic assembly, to cite a simple example, is to be employed in a hot, humid atmosphere or in a missile, then favorable failure rates resulting from indoor ground testing at room temperature may have little or no relationship to those that will be experienced in use. The increased environmental severity has long been recognized, as evidenced by the qualification and acceptance test programs required for most items deliverable to government agencies. However, such test programs usually only verify the basic capability of the design to withstand the environments by testing one or at most a few units. Actual reliability demonstration on a quantitative basis is seldom undertaken.

It is not a lack of awareness of the requirements for quantitative assessments that creates this situation. Rather, the elements of cost, schedule, availability, environmental facilities required, and sometimes degradation of test units combine to render impractical a reliability test program carried out under conditions of use. Consequently, the data which become available must be screened and possibly modified before they are used for assessments. Perhaps an actual failure rate experienced in ground testing of a unit can be increased to account for the effects of a more severe environment. Or, if the relationship of failures at overstress conditions to failures at use conditions is known, a smaller number of units might be tested at off-design conditions. Sometimes it is possible to weight ground tests and count them only as partial tests; three ground tests, if successful, might be counted as the equivalent of one flight test. (On the other hand, one unsuccessful ground test might be the equivalent of three failures under actual conditions.) Some of the basic guidelines and requirements commonly used for determining data applicability were discussed in Chap. 6. The following paragraphs restate those criteria which relate to the utilization of data in reliability assessments.

Data from tests performed under normal operating conditions or conditions which closely simulate conditions of application should be included in the assessment. Such data are applicable regardless of whether the tests resulted in successes or failures. In fact, the retention of some results and deletion of others would invalidate the assessment.

Tests which are performed at off-design or accelerated conditions may also yield data which are useful in assessments. Only some of the characteristics, however, may display normal values; other parameters may exhibit significantly different performance than would normally be experienced. Design and test engineers should be consulted prior to such testing to distinguish between those parameters which provide useful data and those which will yield abnormal results. Values of those characteristics which are not affected should be used and the others, of course, should be discarded. However, the fact that a characteristic does not meet requirements is not, in itself, sufficient justification for deleting the data. There must be adequate engineering knowledge that the failure was actually the result of the abnormal test conditions.

All characteristics which bear on system performance or affect reliability should be evaluated. The fact that nine parameters all perform well within specification limits becomes of relatively little concern if the value of a tenth characteristic is such that the system fails. Even when the assessment is qualitative because there are insufficient data to yield realistic numerics, it is still necessary that all significant characteristics be assessed. Justification may be required for the omission of any characteristic which could affect the reliability.

Where obtainable, data derived from overall systems tests are very desirable. All component characteristics and interactions will affect the system performance, and since it is usually the system's characteristics which are most important, it is preferable to measure these directly. When system quantities are insufficient to provide realistic numerics, or when performance anomalies occur which must be traced to assemblies and components to effect a correction, then test data from lower levels, i.e., assemblies and components, should be utilized. When used in conjunction with system test data, information obtained from lower-level tests can provide insight into interaction effects which could not otherwise be determined. In many cases system data are not available until quite late in the development program, and data from component, assembly, and subsystem tests are both necessary and valuable for assessments. Assessments made early in the program provide needed information on design capabilities and indicate those areas where improvements are required. Therefore, system and lower-level assessments supplement each other and, where possible, both should be performed.

13.1.2 Quantitative Assessments

Once the validity of the data has been verified and any non-applicable data have been deleted, we can determine whether an assessment can be made on a quantitative basis. (A quantitative assessment is one which derives numerical probability values from statistical analysis of the data.) In order for a quantitative assessment to be meaningful, a sufficient number of units must have been tested, operating hours accumulated, cycles completed, etc., to provide probabilistic validity. Otherwise, a qualitative assessment should be performed in which pertinent information is summarized and evaluated and knowledgeable engineering statements are made regarding extrapolation of test results to evaluate operational capability.

The number of units to be tested, or the number of operating hours or cycles to be completed if a quantitative assessment is made, is a matter of judgment and compromise. Many of the older requirements have been revised in recent years. A few years ago a rough working rule when grouping data was to form about 20 groups or cells of equal width, with at least 5 measurements in most of the cells.* This rule can still apply when we are dealing with large numbers of small parts such as resistors, diodes, bolts, etc. However, those of us who are associated with the missile and aerospace industries seldom have 20 units, let alone five times that many! More realistic requirements are needed to permit quantitative measures of the reliability of spacecraft where it is considered very fortunate to have even 10 units on which to accumulate data.

The suggested minimum numbers of test units required for the various assessment methods are given below. Detailed explanations of the methods themselves and their applications follow in the remaining sections of this chapter and in Chap. 14.

1. When an assessment is made based only on the number of units tested and the number of failures experienced, at least 20 units should be tested if there are no failures, and at least 10 if there are one or more failures. If high levels of confidence are required, the minimum number of units to be tested should be appropriately increased.

2. When the equivalent-life method (Sec. 13.3) is used, the minimum suggested quantities are similar to those just defined. If there are no failures, 20 equivalent mission lives should be accrued; if one or more failures occur, 10 equivalent missions are recommended as a minimum. The accumulation of mission times should include testing of at least five units to provide some assurance of component homogeneity. If only one or two units were tested for long periods of time, we would have relatively little assurance that additional units would be similar. On the other hand, the testing of many units for relatively short times (less than the mission time) could be equally misleading, and would not demonstrate the capability of one unit for successfully completing a full mission.

3. Variables data (actual numerical values of characteristics) provide more information than do attributes data and, in general, fewer tests are required for the same degree of confidence than when the assessment is based on attributes. However, the recommended *minimum* quantities do not differ significantly. Assessments based on variables data, including wear-out times, may be made using data from as few as 10 units when the characteristics have

*A cell is an interval or subdivision of the scale of measurement.

normally distributed values. If a distribution is unknown or if it is known to be non-normal, the suggested minimum number of units is 20. Again, higher confidence levels require increased numbers of test units. Also, methods discussed in Chap. 5 should be considered for normalizing data from skewed distributions.

4. Weibull distributions, which are more complex than other distributions commonly used in assessments, should be used sparingly and only when the times to failure do not fit a simpler distribution. At least 10 applicable *failures* are required to enable determination of the statistical characteristics, even when the graphical method presented in the next chapter is used. If the particular Weibull distribution is already known from previous tests, then effects of design, processing, or other changes may be assessed from data on only five failures.

The foregoing guidelines are rules of thumb for minimum quantities required for each type of assessment. Additional comments pertinent to quantity relationships are made in subsequent sections where applicable to the particular assessment method and in Chap. 15 on reliability demonstration.

We have mentioned two types of data—attributes and variables—which are used in assessments. Attributes tests are those which are recorded as either successes or failures or as conforming or non-conforming. Two assessment methods use attributes data. The first considers only the number of tests and the number of failures, while the second uses number of operating hours/cycles (or equivalent missions) along with the failure quantity. Both of these methods and appropriate examples are discussed in this chapter.

The second type of assessment utilizes variables data—actual values of equipment characteristics. Values of significant characteristics are measured during or following the tests, as appropriate, and the distribution of these values is statistically determined and compared with specification requirements. A number of methods are available, and these are discussed in Chap. 14.

13.2 THE FAILURE-TEST RATIO METHOD

The first attributes method used for assessments consists merely of summing the number of applicable tests (trials, operations, etc.) and applicable successes (or failures). This method is employed when there is a large number of tests and when the data is of a go/no go (success/failure) nature, that is, when actual numerical measurements are not available. In this case the binomial is the basic statistical distribution utilized for obtaining the reliability-confidence values. Since the validity of the statistics is based on independence of individual tests or trials, care is required both in defining what constitutes a test and in ensuring that the outcome of one trial is not affected by previous results.

In terms of statistical efficiency, this is the least efficient assessment method, requiring a larger number of tests than other methods to obtain the same level of confidence. However, in terms of program efficiency, it may be less costly to obtain and analyze a large quantity of attributes data than a small quantity of variables data. In general, when testing on a success/failure basis is rapid, when the attributes test is inexpensive as compared with testing by variables, and when there are large quantities of units, this method may be the least expensive overall. On the other hand, when the cost of performing a test is high, when each test is lengthy, or when only a small number of units are available, other methods are preferred.

This method is the easiest to use and requires no calculations on the part of the user except possibly for linear interpolation between readily available table values. The number of tests and number of failures are determined. The reliability is found directly from the tables in Appendix C.2 using the confidence level desired. Tables for confidence levels from 50 to 99 percent are provided. They were prepared using the following equations.

1. The first equation (13.1) involves summing the first $f + 1$ terms of the binomial expansion to provide the reliability-confidence numerical relationship:

$$(R + Q)^n = R^n + nR^{n-1}Q + \frac{n(n-1)}{2!} R^{n-2}Q^2$$

$$+ \frac{n(n-1)(n-2)}{3!} R^{n-3}Q^3 + \cdots$$

$$+ \frac{n(n-1)(n-2)\cdots(n-f+1)}{f!} R^{n-f}Q^f = 1 - C$$

$$(13.1)$$

where R = the reliability
Q = the unreliability (= $1 - R$)
n = the number of tests
f = the number of failures
C = the confidence level expressed as a decimal

The use of this equation requires an iterative process. A reliability value is assumed and the confidence is computed. If the resulting confidence level is lower than desired, the original reliability choice was too high and a lower reliability value is chosen. If the confidence level is higher than desired, the original reliability choice was too low. A new, higher reliability value is then chosen. The process is repeated until the desired level of confidence is obtained. These pairs of reliability-confidence values are equivalent for the given number of tests and failures.

2. The second equation (13.2) enables a direct calculation to be made when tables of the F distribution are available.

$$R = \frac{1}{1 + [(f + 1)/(n - f)]F_{\alpha(2f+2, \, 2n-2f)}} \tag{13.2}$$

where R = the reliability
     n = the number of tests
     f = the number of failures
     α = the risk (= $1 - C$)
     $2f + 2$ = the number of degrees of freedom for the greater mean square
     $2n - 2f$ = the number of degrees of freedom for the lesser mean square*
Either of the above equations can be used to calculate reliability at any desired level of confidence since both yield the same pairs of reliability-confidence values for the same set of trial-success numbers.

Example 1. As a first example of the use of this method for assessing a component's reliability, assume that 100 tests have been performed under conditions comparable to actual operating conditions and that two failures have been experienced. Using Table C.2a, the reliability is determined to be 0.9733 at the 50 percent confidence level. If we desire more confidence in the reliability numeric, we might use Table C.2c and obtain a reliability value of 0.9476 at 90 percent confidence.

Example 2. A second example will serve to illustrate interpolation. When the exact number of *failures* is not found in the tables, direct linear interpolation should be used. However, when the number of *tests* is not found, linear interpolation is performed using the *reciprocals* of the test numbers. If 45 tests have been made and there have been 5 failures, the reliability (for 50 percent confidence) is found as follows. The reliability

N	$1/N$	R
40	0.02500	0.8594
45	0.02222	0.8750
50	0.02000	0.8874

for 40 tests and 5 failures is 0.8594 (from Table C.2a) and for 50 tests and 5 failures it is 0.8874. Linear interpolation using reciprocals of test quantities is employed.

---

*Since the F distribution depends on the number of degrees of freedom in the estimation of input values, the tables are arranged accordingly. For a more complete discussion of the F distribution and its utilization, the reader is referred to Sec. 3.9 and to Paul G. Hoel, "Introduction to Mathematical Statistics," John Wiley & Sons, Inc., New York, 1947. Some F tables use the terms "numerator" and "denominator." These correspond to "greater mean square" and "lesser mean square," respectively.

The reliability value found in this manner will usually not be in error by more than 0.0001. In this case, the actual value, calculated from the formulas, is 0.8749.

13.3 THE EQUIVALENT-LIFE METHOD

The second method is used when operating times or cycles are available (when equipment is tested for a period of time or number of cycles) and the number of mission hours or cycles is known or can be estimated. Although the exponential distribution is used to assess the reliability, the procedure is classed as an attributes method because only total operating times or cycles and number of failures are used. Actual times to failure are not considered. The basic assumption of this distribution is the existence of a constant failure rate—the distribution of failures is random, depending only on the failure rate and the number of units remaining operable. This method is known as the "equivalent-mission" method because the number of hours or cycles experienced is divided by the number of hours or cycles in a mission to determine the equivalent number of missions represented by the testing. Thus, if a mission is planned to last for 10 hr and 150 hr of applicable testing have been accumulated, the number of equivalent missions is 15.

When no failures have been experienced, the calculated reliability values are comparable to those obtained by the first method. However, when there have been failures, the equivalent-mission method yields higher reliability values than does the binomial method. This is because the equivalent-mission method, when applicable, can more fully utilize the data. For example, 5 failures in 10 trials yields a reliability of only 0.45 at 50 percent confidence because each trial is assumed to have terminated at the occurrence of the failure. However, 5 failures in 10 equivalent missions yields a reliability of 0.57 at 50 percent confidence when it is known that 10 full mission times have been accrued.

The number of failures experienced and the number of equivalent missions performed are the two input parameters. The number of equivalent missions E is determined by dividing the number of test hours (or cycles) T by the number of mission hours (or cycles) M: $E = T \div M$. The reliability is then found directly from Table C.3 for the desired level of confidence. These table values were derived by employing a chi-square table to obtain the value for the exponent to be used in the negative exponential distribution, using the following procedure (given here only for reference since Table C.3 provides the reliability numerics directly).

1. Determine the number of degrees of freedom d (to be used for obtaining the chi-square factor):
    a. If the testing was stopped after a failure, $d = 2f$, where f is the number of failures.
    b. If the testing was stopped after a given time, $d = 2f + 2$.

2. Obtain the chi-square factor χ^2 for d degrees of freedom [from ($1a$) or ($1b$)] and probability equal to 1 minus the desired confidence level; e.g., if $C = 0.90$, the probability used is 0.10.*

3. Determine the number of equivalent missions, as above: $E = T \div M$.

4. Obtain the value for the exponent x by dividing the chi-square value χ^2 from (2) by twice the number of equivalent missions—2 x E, from (3): $x = \chi^2 \div 2E$.

5. Using the value of x from (4) and a table of negative exponents, determine the reliability: $R = e^{-x}$.

Since the number of degrees of freedom depends on whether the testing was ended after a failure or after a given time, Table C.3 includes two columns for the number of failures. If the testing ended after a failure, column A is used; if testing stopped after a given time, column B is used.†

Example 3.   The third example employs the equivalent-mission method to arrive at a reliability estimate. Assume that a mission is planned for 3 days, that 360 hr of testing have been performed, that two failures have occurred, and that the testing was terminated after the second failure. It is desired to determine the reliability with 75 percent confidence. The number of equivalent missions is $360 \div 72$ (3 days) = 5. Using Table C.3b (75 percent confidence) for five equivalent missions, two failures, and column A (testing stopped after a failure), we obtain a reliability of 0.5836.

When the number of equivalent missions is greater than five and is not found in the tables, simple linear interpolation can be used. This is accurate to at least two places and, for large numbers of equivalent missions, the accuracy extends to the third decimal place. When there are fewer than five equivalent missions, the error resulting from linear interpolation becomes significant, and use of the chi-square and exponential tables is recommended, following the procedure described above.

ADDITIONAL READINGS

Amstadter, B. L., and T. A. Siciliano: Reliability Assessment Guides for Apollo Suppliers, North American-Rockwell Corporation, Downey, Calif., SID 64-1447A, 1965, IDEP no. 347.40.00.00-F1-28.

Bazovsky, I.: "Reliability Theory and Practice," Prentice-Hall, Inc., Englewood Cliffs, N.J., 1961.

*The probability value is based on the probability of exceeding the value given in the χ^2 table. Some references give the probability of *not* exceeding the table value. In the latter case, the probability is equal to the confidence level.

†This and associated methods were developed by Dr. B. Epstein, and they are discussed in detail in an article entitled Life Testing, *J. Am. Statis. Assoc.*, vol. 48, 1953. Other papers by Dr. Epstein on this subject are listed in the Additional Readings.

Bureau of Naval Weapons, Chief of: "Handbook: Reliability Engineering," U.S. Government Printing Office, 1964.

Calabro, S. R.: "Reliability Principles and Practices," McGraw-Hill Book Company, New York, 1962.

Chorafas, D. N.: "Statistical Processes and Reliability Engineering," D. Van Nostrand Company, Inc., Princeton, N.J., 1960.

Dummer, G. W., and N. B. Griffin: "Electronics Reliability: Calculation and Design," Pergamon Press, Ltd., London, 1966.

Epstein, B.: Life Testing, *J. Am. Statis. Assoc.,* vol. 48, 1953.

——: Statistical Techniques in Life Testing, *Wayne State Univ. Dept. of Math. Tech. Rept. 4,* Detroit, 1959.

——: "Tolerance Limits Based on Life Test Data Taken from an Exponential Distribution," Department of Mathematics, Wayne State University, Detroit, 1959.

—— and M. Sobel: "Some Tests Based on the First *r* Ordered Observations Drawn from an Exponential Population," Department of Mathematics, Wayne State University, Detroit, 1952.

Hiltz, P. A., and J. L. Gaffney: "Statistical Techniques for Reliability," North American-Rockwell Corporation, Downey, Calif., 1965.

Hoel, P. G.: "Introduction to Mathematical Statistics," John Wiley & Sons, Inc., New York, 1947.

Lloyd, D. K., and M. Lipow: "Reliability: Management, Methods, and Mathematics," Prentice-Hall, Inc., Englewood Cliffs, N.J., 1962.

14

ASSESSMENT METHODS: VARIABLES

14.1 COMPARISONS WITH SPECIFICATIONS

A large portion of reliability-assessment activity is concerned with the comparison of a parameter or characteristic with a specification limit or with a pair of limits. The distribution of the characteristic is estimated or determined and the percentage of the distribution which meets specification requirements is calculated. If one were fortunate enough to know the value of the characteristic for every unit, the task of reliability assessment could involve a very minimum of effort and the calculations would be exact. In fact, those units which did not meet requirements would be corrected or discarded. All units which were accepted would be satisfactory and their reliability would be unity, at least with respect to the measured characteristic and assuming that this did not change with time.

However, assessment rarely involves the measurement of every unit. Tests may be destructive or at least detrimental, it may be necessary to complete the mission before the necessary data are obtained, there may be so many units that pretesting them all is impractical, some units may be unavailable, or there may be other reasons which render 100 percent testing unfeasible. Reliability must be evaluated by comparing data from samples with requirements.

14.1.1 The Criterion of Adequate Performance

Comparison with specification limits has been introduced but may not always be a realistic measure of reliability. We recall that the reliability of a part or component is defined as its ability to *perform adequately,* and the specification requirement may not define the limit of adequacy. Sometimes a designer may set as a requirement a level of performance which he would like to obtain, but he will actually accept equipment whose capability does not quite meet this requirement if it meets the adequacy criterion. In other words, he is knowingly "pushing the state of the art" in an attempt to attain improved performance capability. The opposite could also be true. A specification might be set at a level below that which is considered 100 percent adequate because proper equipment just is not available. The widespread use of sampling plans, particularly in the manufacture of consumer products, which accept a small percentage of deficient items is tacit admission that for many items 100 percent reliability is not economically practical.

Another consideration is the actual application of the item. Sizes and ratings of component parts, for example, are standardized to reduce the number of different items and to facilitate design, purchasing, production control, and assembly even though the full capabilities are not needed in all applications. A 0.5-watt resistor may be used in different circuits where the power requirements range from almost nothing all the way to the maximum rating of 0.5 watt, although the common practice is to limit its use to some lower value, perhaps 0.25 watt. The circuit designer would not specify resistors of many different power ratings for all the various applications if an inexpensive 0.5-watt resistor would serve adequately everywhere. Also, if a vehicle were to be subjected to some maximum dynamic load, e.g., vibration, during operation, standard components that had the capability of withstanding this load would be employed, even though many of them would be used in assemblies that had isolation mounts. Since the reliability of an item depends on its application, units of the same type, even if they were inherently identical, would have different reliabilities because of differences in the applications.

If it is possible to determine a true criterion of adequacy, then the sample data can be compared with this requirement. This may well be the specification limit. Conscientious designers do set limits which are realistic from both economic and performance standpoints. The foregoing remarks were intended only to point out the possibility that the specification may differ from the criterion of adequacy and the reader should bear this in mind when making an assessment. The methods which follow assume the accuracy of the specification limit and discussions are in terms of this requirement.

14.1.2   Comparisons with a Single Specification Limit

The comparison of test data with requirements will be considered first when only one specification limit, either a minimum or a maximum, exists. The minimum octane rating of a fuel, the maximum forward voltage drop of a transistor, the maximum viscosity of a lubricant at a given temperature, and the minimum tensile strength of a welded joint are four examples of one-sided specification requirements.

Normal distributions and independence are assumed for the characteristics which are evaluated when the reliability is determined. When the distribution is not normal, the data should first be normalized by means of an applicable procedure described in Chap. 5. The purpose of normalizing the data is to enable direct comparisons of the sample data or distribution with the specification. Tables are readily available which provide reliability-confidence relationships for normally distributed variables. A set of curves based on these tables is included in Appendix C.

The errors which may be introduced by sampling have been discussed. Estimates of either the population mean or its standard deviation or both may be inaccurate. The error resulting from a small-size sample may be considerable. As the sample size increases the error generally decreases. The effects of these errors on reliability assessment will now be examined. Figure 14.1 shows a theoretical specification limit and the values obtained from measuring a sample of five units. Although the five sample units are satisfactory, there may be a much wider spread of values or a different

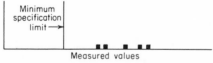

Measured values

FIG. 14.1   Specification limit and values of five sample units.

average if all units were measured. Figure 14.2 depicts some of the possibilities of the complete population distribution. Even though distribution c may be more probable than the other distributions, the possibility that the five sample units came from a population which had distribution a or b cannot be neglected. We can be fairly certain (confident) that the true distribution is no worse than distribution a, that is, that the percentage of the actual distribution lying below (to the left of) the minimum specification limit is no greater than the percentage of distribution *a* that is below the limit. We are somewhat less confident that the true distribution is no worse than distribution *b*, and even less confident with respect to distribution *c*.

The percent of the population distribution below the minimum specification limit is the probability of not meeting the specification requirement—the

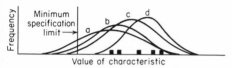

FIG. 14.2 Some possible normal population distributions associated with the data of Fig. 14.1.

unreliability. Conversely, the percentage above the limit is the reliability. It can be seen that when the confidence is high the reliability is low (distribution *a*), and as the reliability increases the confidence decreases (distributions *b* and *c*). The task then is to find a statistical parameter relating the distribution to confidence and reliability. This parameter is developed from the sample mean and standard deviation.

The difference between the sample mean and the specification limit could be called the safety margin of the mean. However, this in itself provides little useful information because, if the standard deviation is large, a considerable percentage of the distribution may be below (outside) the specification limit. Since definite percentages of a normal distribution lie beyond specific numbers of standard deviations, we are much more concerned with how many units of standard deviation there are between the sample mean and the specification limit. This, then, is the statistical parameter, and this comparison statistic is called $\hat{K}$. It is computed by dividing the absolute difference between the sample mean $\overline{X}$ and the specification limit L by the sample standard deviation s, as shown by

$$\hat{K} = \frac{|\overline{X} - L|}{s} \tag{14.1}$$

To determine the reliability, it is necessary merely to compute $\hat{K}$ from the data and to choose the appropriate table corresponding to the sample size and the desired level of confidence. The set of curves in Appendix C provides reliabilities and corresponding confidence levels for values of $\hat{K}$ from 0.0 to 7.0 for various sample sizes.

We shall now determine reliability-confidence values for the data shown pictorially in Fig. 14.1. Assume that the five values are 16, 17, 20, 22, and 23, and that the minimum specification limit is 10.0. First, the sample mean and standard deviation are calculated.

Calculation of Mean		*Calculation of Standard Deviation*
X_1	16	$16 - 19.6 = -3.6; (-3.6)^2 = 12.96$
X_2	17	$17 - 19.6 = -2.6; (-2.6)^2 = 6.76$
X_3	20	$20 - 19.6 = 0.4; \quad 0.4^2 = 0.16$
X_4	22	$22 - 19.6 = 2.4; \quad 2.4^2 = 5.76$
X_5	23	$23 - 19.6 = 3.4; \quad 3.4^2 = 11.56$
ΣX_i	98	$\Sigma(X_i - \bar{X})^2 = 37.20$

$$\bar{X} = \frac{\Sigma X_i}{N} = \frac{98}{5} = \underline{19.6} \qquad s = \sqrt{\frac{\Sigma(X_i - \bar{X})^2}{n-1}} = \sqrt{\frac{37.20}{4}}$$

$$= \sqrt{9.30} = \underline{3.05}$$

Next, the value of $\hat{K}$ is computed, using Eq. (14.1).

$$\hat{K} = \frac{|\bar{X} - L|}{s} = \frac{19.6 - 10.0}{3.05} = \frac{9.6}{3.05} = \underline{3.15}$$

Then, Table C.4c for sample size $N = 5$ is selected, and the approximate curve for $\hat{K} = 3.15$ is visualized. This curve is almost exactly three-fourths of the way between the curves for $\hat{K} = 3.0$ and $\hat{K} = 3.2$. Finally, the reliability-confidence values are found. For 90 percent confidence, the reliability is approximately 0.94; for 75 percent confidence, the corresponding reliability is about 0.991. The reliability value for any level of confidence from 50 percent to almost 99 percent can be found from this set of curves.

14.1.3 Comparisons with a Pair of Limits

Many specifications require that the item of equipment operate within a pair of limits. The specific requirement must be carefully reviewed to determine whether the probabilities are *interdependent* or are *independent* of each other. These two cases are treated in different ways. By interdependence is meant that if the equipment fails to meet one limit, it cannot also fail to meet the other: the conditions are mutually exclusive. For example, if an electrically detonated igniter cartridge must operate between two voltages, say 5 volts and 10 volts, then if a unit fails to meet one requirement, it cannot also fail to meet the other. If the igniter does not operate until 11 volts are applied, then it automatically meets the requirement that it must not detonate below 5 volts. The same unit cannot fail on both requirements: the failures are mutually exclusive.

On the other hand, if a temperature-actuated valve must be fully open when the temperature drops below 50°F and must be fully closed when the temperature rises above 75°F, a unit could fail to meet either or *both* requirements. A failure to open at a temperature below 50°F does not preclude a failure to close at a temperature above 75°F. In effect, there are two separate requirements and the reliability calculations are performed accordingly.

When there is interdependence as in the case of the cartridges, two values of $\hat{K}$ are calculated separately: $\hat{K}_u$ with reference to the upper limit, and $\hat{K}_l$ with respect to the lower.

$$\hat{K}_u = \frac{L_u - \overline{X}}{s} \tag{14.2a}$$

$$\hat{K}_l = \frac{\overline{X} - L_l}{s} \tag{14.2b}$$

where L_u and L_l are the upper and lower limits, respectively. The smaller of the two values $\hat{K}_u$ or $\hat{K}_l$ is used for calculating the reliability. Let us examine the reasons for this procedure.

Consider the representation of a sample distribution and its relationship to two specification limits in Fig. 14.3. Because of the possible errors in estimating μ and σ from $\overline{X}$ and s, the true relationship between the population distribution and the specification limits could be any of those illustrated in Fig. 14.4. Distributions with smaller spreads are also possible but are not shown in the figure. The worst case depicted in Fig. 14.4 is distribution h which has a mean closer to the upper limit and a larger standard deviation. But case h represents the same worst-case relationship that was considered for a single specification limit, namely, an error in the mean with the true mean in an adverse location, i.e., closer to the specification limit, and a larger distribution spread.

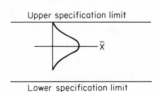

Upper specification limit

$\overline{X}$

Lower specification limit

FIG. 14.3 Sample distribution and specification limits.

Consequently, the use of the smaller of $\hat{K}_u$ or $\hat{K}_l$ will result in the most accurate reliability-confidence relationships.

For clarity, the relationship can be examined using the same approach that was used for a single limit. There is a high level of confidence that the true relationship is no worse than that shown by distribution h, less confidence with respect to distribution e, and still less with respect to distribution f. The unreliabilities of these distributions are all associated with the upper (closest) limit, which yields the smaller value of $\hat{K}$.

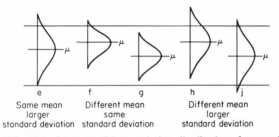

e f g h j
Same mean   Different mean   Different mean
larger   same   larger
standard deviation standard deviation standard deviation

FIG. 14.4 Some possible population distributions for sample distribution shown in Fig. 14.3.

This procedure is accurate for most cases of interdependence. It yields a slight error in the optimistic direction when the mean of the sample distribution is midway or very close to midway between the specification limits and when the distribution spread is as wide or wider than the limits, as in Fig. 14.5. Some possible population distributions are shown in Fig. 14.6. Populations with smaller standard deviations are again omitted for clarity.

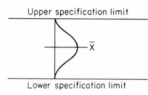

Upper specification limit

$\bar{X}$

Lower specification limit

FIG. 14.5 Sample distribution with mean midway between specification limits.

Because the sample mean in Fig. 14.5 is midway between the limits, distributions l and m are equivalent, and distributions n and p are also equivalent. The procedure is to use the closest limit (or either limit if $\hat{K}_u$ and $\hat{K}_l$ are identical). Distributions n and p yield the highest confidence, l and m less confidence, and k still less. By using only one limit, e.g., the upper limit, the small shaded area outside the opposite (i.e., lower) specification limit is neglected and this represents the slight optimism which results. Reviewing distribution n shows that, for high levels of confidence, the shaded area (error) is only a small percentage of the unreliability. (The percent

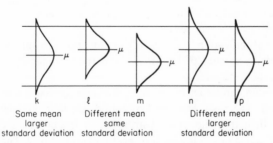

k l m n p
Same mean   Different mean   Different mean
larger   same   larger
standard deviation standard deviation standard deviation

FIG. 14.6 Some possible population distributions for $\bar{X}$ midway between limits.

error increases slightly at low confidence levels although the absolute error decreases.*)

The independent case is accurate because the relationship to one specification limit is not affected by the relationship to the other. The example of the temperature-actuated valve is shown pictorially in Fig. 14.7. The two distributions are largely independent. Even though the opening and closing temperatures for any one valve could be partially related, the distributions generally are not.†

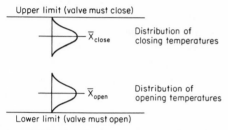

FIG. 14.7   Distributions of opening and closing temperatures.

The reliability for each function is calculated with respect to the appropriate specification limit. The overall valve reliability, assuming that these are the only considerations, is the product of the two reliabilities:

$$R_{valve} = R_{opening} \times R_{closing}$$

It should be mentioned that both reliabilities must be determined at the same level of confidence. The reader may wish to refer to the discussion in Sec. 14.4 on finding the product of independent reliabilities and the associated confidence.

14.2   WEAR-OUT

When the time to normal wear-out is equal to or less than twice the required mission time, wear-out often becomes the dominant failure mode. This also

---

*Some procedures and tables make the assumption of an accurate estimate of the mean and an error only in the standard deviation (distribution k). This results in a larger overall error than does the procedure presented here.

†If one group of valves were tested for opening temperature and another group were tested for closing temperature, the distributions would be completely independent. When the same valves are tested for both functions, the overall reliability could be slightly optimistic if the opening and closing temperatures were related.

applies when the wear-out life is primarily cycle or actuation dependent and requirements are defined in terms of number of cycles or actuations per mission. Wear-out can be a principal cause of failure even when the mean time or number of cycles to wear-out is several times the mission requirement if there are large variations in lifetime from unit to unit (if σ is a large percentage of μ). In fact, there have been cases of wear-out failure when the mean time to wear-out was many times greater than the requirement and the variability was not so great that wear-out failures were expected. In these cases, careful tracing of the component's history disclosed that the actual operating time which it had experienced was much greater than the mission time, the component having been subjected to many hours of operation during testing at the suppliers, during receiving inspection operations at the contractors, and at several points during subsystem and system assembly and field check-out. For this reason, even when the lifetime is known to be much greater than mission requirements, it is often necessary to keep accurate records of actual time that the equipment was operated prior to its intended use.

14.2.1 Use of the Normal Distribution

Failure from wear-out can be characterized by either the Weibull (which will be discussed shortly) or the gaussian distribution. The normal distribution is assumed unless there is definite knowledge that it is not applicable. The normal distribution also applies to cases when the wear-out distribution is lognormal or fits some other distribution which can be transformed to and analyzed as a gaussian distribution. Calculations of reliability-confidence numerics are carried out following the procedures already discussed. When the normal distribution is used, the mean and standard deviation are calculated in the usual manner, the value of $\hat{K}$ is determined using the required mission life as the lower one-sided specification limit, and the reliability is found from the curves in Table C.4 for the desired level of confidence. When the Weibull distribution is used, the procedure described later in this chapter is followed.

The probability of failure to meet specifications and the probability of failure due to wear-out may both apply to a particular component. In fact, the component may also be susceptible to random failures during the mission so that all three types of failures are applicable. As long as these *probabilities* are independent, the reliability of the component for the mission is the product of the individual reliabilities

$$R_{\text{mission}} = R_{\text{spec. limit}} \times R_{\text{random}} \times R_{\text{wear-out}} \tag{14.3}$$

It would be incorrect to simply add the probabilities of failure and then subtract the sum from unity to find the reliability because the *number* of failures occurring from wear-out, for example, depends on *both* the probability of wear-out failure and the probability of failure not occurring from the other causes.* If the probability of failing to meet specifications were 0.30, the probability of random failure were 0.40, and the probability of wear-out were 0.35, simple addition would yield a total of 1.05, which is obviously incorrect. The probability of meeting specifications would be 0.70 and 4/10 of the satisfactory units (28 percent of the total) would fail from random causes. The remaining 42 percent (100-30-28) of the total would be subject to wear-out. Of this 42 percent, 35/100 would fail and 65/100 would not, and about 27 percent of the total initial number would survive.

14.2.2 Time-dependent Characteristics

Some failures from wear-out are not of a catastrophic nature. Instead of operating for a period of time without evidence of impending failure and then experiencing a failure which makes it totally inoperative, the unit may gradually deteriorate with time. If the measurements are made periodically, the time to failure or wear-out can be predicted. If the tread thickness of an automobile tire is measured at regular intervals and is plotted as a function of mileage, the number of miles until the tire is worn out (assuming that a reasonable criterion of wear-out is defined) can be determined rather accurately. Similarly, if a boiler erodes due to the nature of the boiling fluid and the rate of erosion can be measured, then the length of time until the wall thickness is reduced to the minimum acceptable standard can be calculated. When rate of wear is predictable, it is not necessary to operate a large number of units to failure to make the assessment. The time to wear-out and the resulting reliability with respect to a required mission time can be computed before wear-out occurs. It is necessary to verify the wear rate over the entire lifetime of only a few units. Even only one unit may be sufficient to provide verification in some circumstances. This predictability is highly desirable when the failure of a component results in costly repairs to the system or in serious program delays. The boiler is replaced or reworked at a convenient maintenance time before a leak occurs which might, for example, cause a fire or explosion or damage other components. We shall evaluate the case when rate of wear is linear with time, although the methods described can be applied to the more general case of a variable wear rate if the rate can be described mathematically.

There are two standard deviations involved for each characteristic subject to such wear-out analysis rather than just one. The original value of the characteristic, e.g., wall thickness, has variation. We shall call its standard deviation s_c —standard deviation of the characteristic. Even though the wear

---

*If all reliabilities are high, e.g., greater than 0.99, the resulting error will be very small and simple addition of failure probabilities may be used.

rate is linear, it too varies, and the standard deviation of wear over the mission time is s_w. The total effect of two *independent* standard deviations s_t is found by Eq. (14.4).* This equation can be applied to any number of *independent* standard deviations by simply including the additional variances under the radical.

$$s_t = \sqrt{s_c^2 + s_w^2} \qquad (14.4)$$

Figure 14.8 depicts the relationship of the distribution of a characteristic which varies with time to the requirement. The original standard deviation s_c is shown on a vertical scale at time zero at the left-hand side of the figure. The total standard deviation over the mission life, s_t, is shown at the right side of the figure. Original and final values of the mean are also shown.

The analysis utilizes the $\hat{K}$ statistic and the curves in Table C.4 in the same manner in which they have already been used. The only additional steps are calculating s_w and combining it with s_c, and adding (or subtracting) the average change due to wear to the original mean value of the characteristic. We shall assume that linearity of the rate of wear has been verified.

A convenient time period, say one-fourth of the mission time, is used for operating a selected number of units. The amount of wear w is determined for each unit. The mean $\overline{w}$ and standard deviation s_w of wear (change) over this time period are calculated following the usual procedures:

$$\overline{w} = \frac{\Sigma w_i}{n} \qquad s_w = \sqrt{\frac{\Sigma (w_i - \overline{w})^2}{n - 1}}$$

Because they are for only one-fourth of the mission time, these values of $\overline{w}$ and s_w are multiplied by 4 to find their values for the entire mission.

The final value of $\overline{w}$ is added algebraically to the original mean $\overline{X}$ to find the final value of the mean $\overline{X}'$:

$$\overline{X}' = \overline{X} + \overline{w} \qquad (14.5)$$

The total standard deviation is found from Eq. (14.4), using the value of s_w for the total mission:

---

*If the wear rate depends on the original value of the characteristic, then the two standard deviations are not independent and s_c is multiplied by an appropriate constant to obtain s_t (assuming that the effect of the dependency is to increase the spread of the distribution).

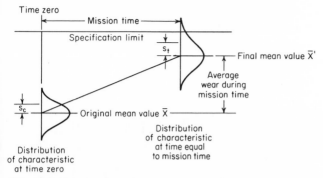

FIG. 14.8 Distribution shift with time and comparison with specifications.

$$s_t = \sqrt{s_c^2 + s_w^2}$$

The value of $\hat{K}$ is computed by Eq. (14.6) using $\overline{X}'$ and s_t :

$$\hat{K} = \frac{|\overline{X}' - L|}{s_t} \tag{14.6}$$

One example will serve to illustrate the method. A boiler is required to operate for 1,000 hr. Its principal mode of failure is leakage or rupture from wall thinning due to erosion which occurs lineally with time. The wall thickness is measured at the start of operation and again after 500 hr. Data are given in Table 14.1.

For the entire mission $\overline{w} = -7.4 \times 2$ (because data were taken at 500 hr and wear is linear) $= -14.8$ and $s_w = 4.6$.

$$\overline{X}' = \overline{X} + \overline{w} = 90 - 14.8 = 75.2$$

$$s_t = \sqrt{2.2^2 + 4.6^2} = 5.1$$

Suppose that a minimum wall thickness of 60 mils is required, and due to the severe consequences of a failure, a 95 percent confidence on reliability is specified.

$$\hat{K} = \frac{75.2 - 60}{5.1} = 3.0$$

From Table C.4h, for a sample size of 10, confidence of 95 percent, and $\hat{K} = 3.0$, the reliability is 0.96. If only 50 percent confidence were specified, the reliability would be almost 0.999.

TABLE 14.1 Boiler-wall-thickness data

Original wall thickness, mils	$(X_i - \bar{X})$, mils	$(X_i - \bar{X})^2$	Wall after 500 hr, mils	Net change, mils	$(w_i - \bar{w})$, mils	$(w_i - \bar{w})^2$
94	4	16	86	- 8	-0.6	0.36
89	-1	1	79	-10	-2.6	6.76
90	0	0	86	- 4	3.4	11.56
92	2	4	83	- 9	-1.6	2.56
86	-4	16	76	-10	-2.6	6.76
89	-1	1	83	- 6	1.4	1.96
90	0	0	84	- 6	1.4	1.96
91	1	1	87	- 4	3.4	11.56
91	1	1	82	- 9	-1.6	2.56
88	-2	4	80	- 8	-0.6	0.36
$\Sigma X_i = 900$		$\Sigma = 44$		$\Sigma = -74$		$\Sigma = 46.40$
$\bar{X} = 90$		$s_c = \sqrt{\dfrac{44}{9}} = 2.2$		$\bar{w} = -7.4$		$s_w = \sqrt{\dfrac{46.40}{9}} = 2.3$

14.2.3 Use of Weibull Distributions

The Weibull family of distributions—frequently called simply "the Weibull"—was introduced in Chap. 3 and its relative complexity was presented. These distributions are associated with times to failure and therefore supplement the exponential (which is a particular case of the Weibull) and normal (which can be closely approximated by a Weibull) distributions. Because of its increased complexity it is used only when there is prior knowledge that the failure distribution is best described by a Weibull or when a large number of failures (at least 10) have been experienced and the associated times do not fit a simpler distribution. Two examples of components with Weibull-distributed times to failure are ball bearings and fuel cells. Bearings in particular have had a great deal of testing over a number of years and relatively precise values of Weibull parameters have been determined for different bearing types.

Although two types of analytical solutions of the Weibull are available (method of moments and maximum likelihood),* neither is widely used in reliability applications because each involves extensive calculations. Instead, reliability utilization most often employs a graphical solution which requires determination of only the location parameter γ. Values of the scale parameter α and the shape parameter β are not necessary to find the reliability-confidence relationships when the graphical method is used.

Special graph paper, known as Weibull paper, makes this possible. A typical graph paper is shown in Fig. 14.9.† The paper does provide for finding the values of α and β if these are desired. The graphical procedure, while involving a number of steps and one or two iterations, is relatively straightforward and requires, at most, simple algebra. The steps are as follows:

1. Assign rank numbers. The times to test termination are arranged in rank order, $(1, 2, 3, \ldots, n - 1, n)$, according to the length of time of operation. The unit with the shortest operating time is ranked no. 1, the next shortest operating time is no. 2, and so on. The rules regarding repaired equipment, extraneous failures, and so on are

   a. If an item of equipment fails and is repaired so that it is returned to operable condition, the time of operation for ranking purposes is the time between failures. Thus, if a unit was repaired twice, it has three separate operating times, and each time will have its own rank number.

---

*W. Weibull, A Statistical Representation of Fatigue Failures in Solids, *J. Roy. Inst. Technol., Stockholm,* November, 1954.

†Reproduced by permission of Technical and Engineering Aids for Management, Lowell, Mass. This type of Weibull paper enables the determination of all three distribution parameters.

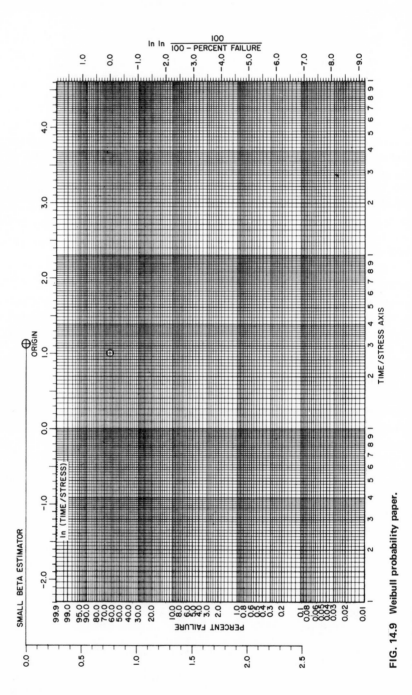

FIG. 14.9 Weibull probability paper.

266

b. Tests terminated for reasons other than failure of the test item, e.g., test equipment failure, operator error, end of test period, etc., are not chargeable and are called "suspended tests." These are placed in rank order along with other test times, but are treated separately in step 2.

c. When failures occur due to entirely separate basic causes, it is desirable that a separate Weibull analysis be made for each cause. There should be at least five failures for each separate Weibull plot; if there are not, all operating times are used in one analysis.

2. Determine order numbers. Order numbers (X_1, X_2, etc.) are assigned to those operating periods which terminated because of failure of the test item, that is, due to *chargeable* failures. Operating periods terminated as the result of other causes are not assigned order numbers.

a. When all tests are terminated due to failure of the test specimen, repaired or not, the order number X_i is the same as the rank number n_i assigned to the failure.

b. When suspended tests are included in the ranking, order numbers are assigned to the chargeable failures in the following manner.

(1) If a chargeable failure occurred first, its order number X_1 is 1.0. However, if one or more suspended tests have lower rank numbers than the chargeable failure, the order number of the chargeable failure is found from Eq. (14.7).

$$X_1 = \frac{n + 1}{n + 2 - n_1} \tag{14.7}$$

where n is the total number of operating periods regardless of the reason for termination and n_1 is the rank number of the first chargeable failure.

(2) The order number of the second chargeable failure is found from Eq. (14.8).

$$X_2 = X_1 + \frac{n + 1 - X_1}{n + 2 - n_2} \tag{14.8}$$

where n_2 is the rank number of the second chargeable failure.

(3) The order number of each succeeding chargeable failure is found in a similar manner from Eq. (14.9).

$$X_i = X_{i-1} + \frac{n + 1 - X_{i-1}}{n + 2 - n_i} \tag{14.9}$$

Before proceeding to step 3, we shall work out an example to illustrate the determination of order numbers. Assume that 8 units of an equipment item were tested and that a total of 20 operating times were accumulated as follows:

Unit 1: Operated 300 hr and test terminated because of failure of test equipment. After repair of test equipment and maintenance of the test item, it was operated an additional 240 hr and then failed. It was repaired satisfactorily and was run until a total of 1,000 hr were accumulated when testing was terminated.

Unit 2: Operated 380 hr and then failed. Unit was repaired and operated for an additional 520 hr when it failed again. There was no further operation.

Unit 3: Operated 40 hr when test was terminated because of an external power failure. No maintenance was performed and unit was returned to operation for an additional 730 hr when failure of the test item occurred. Unit was repaired and run until a total of 1,000 hr of test time was accrued, and test was then stopped.

Unit 4: Operated 1,000 hr without failure, and test was then stopped.

Unit 5: Unit failed after 275 operating hours. Unit repaired and failed again after an additional 410 hr. Unit was repaired again and failed a third time after a total of 885 hr. Unit repaired a third time and test was terminated after 1,000 hr of total operation.

Unit 6: Unit failed after 235 hr and was repaired. Operated an additional 395 hr when operator error caused a failure. Unit was repaired again and was run for a total of 1,000 hr when test was stopped.

Unit 7: Unit ran 505 hr and failed. Unit was repaired and was run an additional 355 hr and failed again. Unit ran for 110 hr more and test terminated when test equipment failed.

Unit 8: Unit was run 680 hr and failed. Repaired and ran 305 hr more when it failed again. Test stopped at this time.

The various times to failure are listed in Table 14.2, along with the rank order. Chargeable failures and suspended tests are indicated by the letters "Ch" and "S," respectively. Order numbers are calculated for each chargeable failure (nos. 3, 5, 6, 7, 9, 10, 12, 14, 16, 17, 18, and 19) according to the rules defined in step 2. The corresponding times to failure are given in parentheses following the order number.

$$X_1 = \frac{20 + 1}{20 + 2 - 3} = \frac{21}{19} = 1.11 \ (200 \text{ hr})$$

TABLE 14.2   Operating times (hours) and rank order of failures for eight test units

A. Operating Times							
Unit 1	Unit 2	Unit 3	Unit 4	Unit 5	Unit 6	Unit 7	Unit 8
300 S*	380 Ch	770 Ch†	1000 S	275 Ch	235 Ch	505 Ch	680 Ch
240 Ch	520 Ch	230 S		410 Ch	395 S‡	355 Ch	305 Ch
460 S				200 Ch	370 S	110 S§	
				115 S			

B. Rank Order			
1. 110	6. 240 Ch	11. 370	16. 505 Ch
2. 115	7. 275 Ch	12. 380 Ch	17. 520 Ch
3. 200 Ch	8. 300	13. 395	18. 680 Ch
4. 230	9. 305 Ch	14. 410 Ch	19. 770 Ch
5. 235 Ch	10. 355 Ch	15. 460	20. 1,000

*Because maintenance was performed at 300 hr, a new test period is counted from this time.

†Since no maintenance was performed after the power failure, the actual operating time is taken as the sum of the 40 hr before the power failure plus the 730 hr after the power failure.

‡Because failure was caused by operator error, it is not a chargeable failure, even though a repair was made.

§Suspended test by definition.

$$X_2 = 1.11 + \frac{20 + 1 - 1.11}{20 + 2 - 5} = 1.11 + \frac{19.89}{17}$$

$$= 1.11 + 1.17 = 2.28 \ (235 \ \text{hr})$$

$$X_3 = 2.28 + \frac{20 + 1 - 2.28}{20 + 2 - 6} = 2.28 + \frac{18.72}{16}$$

$$= 2.28 + 1.17 = 3.45 \ (240 \ \text{hr})$$

The process is continued until all order numbers for the 12 chargeable failures have been determined. The remaining nine order numbers are

$$X_4 = 4.62 \, (275 \ \text{hr}) \quad X_7 = 8.53 \, (380 \ \text{hr}) \quad X_{10} = 13.73 \, (520 \ \text{hr})$$

$$X_5 = 5.88 \, (305 \ \text{hr}) \quad X_8 = 10.09 \, (410 \ \text{hr}) \quad X_{11} = 15.55 \, (680 \ \text{hr})$$

$$X_6 = 7.14 \, (355 \ \text{hr}) \quad X_9 = 11.91 \, (505 \ \text{hr}) \quad X_{12} = 17.37 \, (770 \ \text{hr})$$

3. Determine probabilities. Associated with each order number X_i is a probability P_{X_i}. This number states the probability of a unit failing at or before the time corresponding to the X_i th failure. If a failure occurred at exactly 100 hr and the associated probability were 0.10, then the probability of a unit failing at or before 100 hr of operation is 10 percent. The probability numbers are found by using the binomial distribution (Table C.2) to find the reliability at two levels of confidence. The first confidence level is 50 percent. This level is needed because the associated reliabilities will be approximately on a straight line on Weibull paper when the next step in the procedure is carried out. The second confidence level is that level of confidence which is being utilized in the assessment. For purposes of this example, we shall use 90 percent.

   a. Using Table C.2 for n tests (in this example, $n = 20$), and $X_i - 1$ failures, determine the probability associated with each failure at 50 percent and 90 percent confidence. Interpolation will be required because the order numbers are not integers (unless all failures are chargeable). Linear interpolation can be used and will provide at least two-place and usually greater accuracy for all order numbers.*

   b. Subtract these probabilities from unity. This is required because the probabilities determined in step 3a were success probabilities and Weibull paper is designed for plotting probabilities of failure rather than of success. List the order numbers and corresponding failure probabilities for the two confidence levels being used. The numerics for the sample problem are given in detail in Table 14.3 to illustrate this step of the procedure.

4. Plot the probabilities.

   a. The failure probabilities at 50 percent confidence and corresponding times to failure are plotted on Weibull graph paper. The line of best fit through the points at 50 percent confidence will be approximately straight at the correct value of γ. Because the value of γ is unknown when the first plot is made, the first line will most likely be curved; only if γ is zero will it be straight. If the line for 50 percent confidence is concave upward, γ is negative; if it is concave downward, γ is positive. In most cases γ will be positive because, for those components which have Weibull failure distributions, there is some initial period of time which is failure free.

   Note: It is not necessary to find the value of γ unless desired reliabilities must be extrapolated: a straight line is necessary for accurate extrapolation. If extrapolation is not required, steps (b) and

---

*Very rarely, the second decimal place for the first order number may be in error by 0.01.

**TABLE 14.3   Failure numbers, order numbers, and
probabilities**

Failure number	Order number (X_i)	$X_i - 1$	Success probabilities		Failure probabilities	
			$c = 0.50$	$c = 0.90$	$c = 0.50$	$c = 0.90$
3	1.11	0.11	0.9606	0.8833	0.0394	0.1167
5	2.28	1.28	0.9038	0.8011	0.0962	0.1989
6	3.45	2.45	0.8464	0.7285	0.1536	0.2715
7	4.62	3.62	0.7890	0.6608	0.2110	0.3392
9	5.88	4.88	0.7271	0.5916	0.2729	0.4084
10	7.14	6.14	0.6651	0.5256	0.3349	0.4744
12	8.53	7.53	0.5968	0.4559	0.4032	0.5441
14	10.09	9.09	0.5202	0.3805	0.4798	0.6195
16	11.91	10.91	0.4307	0.2970	0.5693	0.7030
17	13.73	12.73	0.3413	0.2181	0.6587	0.7819
18	15.55	14.55	0.2518	0.1445	0.7482	0.8555
19	17.37	16.37	0.1624	0.0777	0.8376	0.9223

(c) may be omitted, and the probabilities at the required confidence level, for example, 90 percent, may be plotted at this time.

b. The value of γ is estimated and is subtracted algebraically from the time of each chargeable failure. An efficient first estimate of γ is the time to the first failure. Thus, the first estimate of γ in the example is + 200 hr, based on the downward curvature of the line and on the data. The probabilities are again plotted, using the revised times.

c. The process is repeated until an approximately straight line is obtained. The curve was only very slightly concave upward after the first revision. A second estimate of γ of 150 hr was made and, except for the first two points, the line was almost straight. A final estimate of 175 hr was made for γ and resulted in a reasonably straight line. It is difficult to distinguish from the graph whether 175 hr is the best choice or whether 170 or 180 hr might be slightly better, but for practical purposes, the graphical solution is quite satisfactory.

d. Probabilities at 90 percent confidence (in this example) are then plotted using the final revised times.* The resulting line for this confidence level (or any level of confidence other than 50 percent) will be curved and should not be changed. Table 14.4 shows the original and revised times and corresponding reliabilities, and Fig.

---

*Technically, what is plotted is the 10 percent confidence line for failure probability, which is the reason that X_{i-1} was used instead of X_i when probabilities were found from Table C.2. The final result provides 90 percent confidence for reliability for X_i.

TABLE 14.4   Reliabilities and original and revised operating times for Weibull Plot

Reliabilities		Operating times, hr			
$c = 0.50$	$c = 0.90$	Original	First revision	Second revision	Final
0.9606	0.8833	200	0	50	25
0.9038	0.8011	235	35	85	60
0.8464	0.7285	240	40	90	65
0.7890	0.6608	275	75	125	100
0.7271	0.5916	305	105	155	130
0.6651	0.5256	355	155	205	180
0.5968	0.4559	380	180	230	205
0.5202	0.3805	410	210	260	235
0.4307	0.2970	505	305	355	330
0.3413	0.2181	520	320	370	345
0.2518	0.1445	680	480	530	505
0.1624	0.0777	770	570	620	595

14.10 depicts the original 50 percent confidence line and the final 50 and 90 percent lines. The graph paper depicted in the figure is a simplified version of the Weibull paper shown in Fig. 14.9.

5. Determine equipment reliability. To determine the appropriate time on the graph, the value of γ is subtracted from the required operating time of the equipment. Using the line for 90 percent confidence (assuming that this is the desired confidence level for the assessment), the revised time and corresponding failure probability are read from the graph. If the original required time was 200 hr, the revised graph value is 200 less 175 or 25 hr. The failure probability is subtracted from one to give the corresponding reliability. On the graph paper shown in the example, a separate scale for the reliability has been added. The original graph paper does not include this scale but does include other scales and means for finding the values of α and β. The reliability at 90 percent confidence for 25 graph hours (200 actual hours) is approximately 0.89. From Fig. 14.10 it can be seen that, if the reliability for 200 hr had been desired at 50 percent confidence, an accurate estimate could not have been made from the original curve and it would have been necessary to determine γ. This is usually true at higher reliability values.

14.3 OVERSTRESS METHODS

Overstress or accelerated testing can sometimes be used to evaluate the reliability of units of a new lot more rapidly than testing under normal conditions. The restrictions which must necessarily be made limit such testing

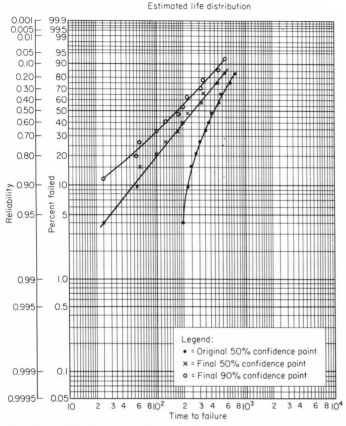

FIG. 14.10 Weibull plots for sample problem.

for assessment purposes to those parts and components for which there are known relationships between failure rates at normal and overstress conditions. A second major requirement is that no new failure modes may be introduced by the overstress conditions. Only when these requirements are met and are applicable from lot to lot can the reliability of a new lot be evaluated in this manner. When the relationships are accurately defined, accelerated tests can provide reliability assessment data at a fraction of the cost of regular testing and can be used to great advantage.

The relationship between accelerated tests and normal tests can be with respect to a random failure rate, rate of degradation or change of a characteristic, or time to wear-out. As long as the relationship is known, the overstress data can be reduced to data for normal operating conditions, usually by multiplying by an appropriate constant. A multiplier to convert overstress data to data at normal conditions will be less than one for a random failure rate or for a rate of degradation, while a multiplier for wear-out or mean time to failure will be greater than one. The multipliers must be known for all statistical parameters to be used in the assessment, e.g., for both the mean and standard deviation for a normal distribution applied to wear-out, but they are not necessarily the same for each parameter.

There is a second application which utilizes overstress testing more frequently than does the assessment function. Some types of overstress testing have the effect of weeding out potentially unreliable units without affecting good units. This is particularly true when the overstress test exceeds the requirement of the part or component but does not exceed its capability. A good example of this is the 20,000-g acceleration test imposed on semiconductors. (At least one manufacturer increases even this high g level by 50 percent to 30,000 g on some units without any measurable effect on the performance or longevity of the units which pass.) Such a test serves to weed out those units which have a potential mechanical weakness and a lower reliability. The test also may cause some units to fail whose reliability would otherwise have been satisfactory, but by imposing it on all units the resultant overall reliability of the lot after testing is considerably higher than it might have been if the test had not been performed. It is important, of course, that the remaining units not be degraded.

Many burn-in tests can be classed as overstress or accelerated because the equipment is operated in excess of its normal requirement. Even testing at 100 percent of full rating is, in this sense, an overstress test because some types of parts are seldom if ever operated at full rated conditions. This last fact iterates the statement made earlier that a specification limit is applied to all parts of a type even though many may be used in applications where the operating conditions are much less severe.

14.4 SYSTEM ASSESSMENTS

Assessed reliabilities of the individual equipment items can be combined to provide reliabilities of assemblies, subsystems, and systems. The methods of combining these reliabilities are the same as those used for combining predicted reliabilities when a system reliability prediction is made. One difference is that assessment numerics are used rather than prediction numerics. Redundancies, modes of failure, system effects, and other related aspects are considered, and the most applicable of the three methods described in Chap. 10 is employed.

However, there is a second important difference which must be considered. Most predictions are "point estimates"; that is, reliability values for each element are found from standard tables and are then appropriately combined to obtain the system prediction. Confidence levels are not given for the individual values, and consequently no particular level of confidence is associated with the system prediction. On the other hand, the assessed reliability value for each item does have a corresponding confidence level. We have seen that there are many equivalent reliability-confidence pairs applicable to a set of data. How, then, can we combine the reliability assessment values of many different items, perhaps at varying levels of confidence, into a realistic reliability-confidence pair for the system? Three methods are possible.

One method is to use only the point estimate for each component, thereby obtaining a point estimate of the system reliability. If there have been ten tests and two failures of a particular equipment item, then the point estimate of its reliability is 0.80. But what if there were no failures? Intuitively, we feel that the reliability is not 100 percent, but we cannot assign a particular reliability value in such cases. When all assessments are based on variables data, we can avoid this problem by using normal table values based on the mean and standard deviation of the sample (not the curves in Table C.4 which account for possible sampling errors in the mean and standard deviation, even at 50 percent confidence). But in the case of attributes data, point estimates of 100 percent for zero failures and n tests may lead to justifiable questions regarding the true reliability.

An alternate method, one which is somewhat conservative but which can provide reasonable reliability-confidence values for the system, utilizes the 50 percent confidence level for component reliabilities. The reliability of each component is assessed at 50 percent confidence, and these reliabilities are combined to obtain the 50 percent confidence value for the system. As shown by the tables in Appendix C, the 50 percent confidence level is more conservative than the point estimate, particularly for small sample sizes and few failures. (In the previous instance, for a sample size of ten with two failures which resulted in a point estimate of 0.80, the reliability value with 50 percent confidence is only 0.74.) Consequently, if component reliabilities are to be combined to provide a system reliability assessment, the minimum quantities suggested in Chap. 13 should be adhered to. In this way, an unduly conservative reliability value will be avoided.

Once the system reliability has been assessed at 50 percent confidence, it can be converted to a lower reliability value at a higher level of confidence. The conservative approach is to base the system reliability-confidence pair numerics on the smallest sample size associated with any of the individual component assessments. That is, the system reliability is first assessed at 50 percent confidence. The number of tests is assumed to be equal to the smallest number of tests used in the assessment of any of its components. Then, using the appropriate table in the Appendix and the sample size and reliability just determined, the new reliability-confidence pair is found.

The table used to convert reliability-confidence pairs depends on the type of component assessments which are made. If attributes methods are used for most of the component assessments, then the equivalent number of system failures is found by interpolation from Table C.2a or Table C.3a as appropriate. (Interpolation will be needed because the equivalent number of system failures will usually not be an integer.) This number of failures and tests (or equivalent missions) is then used to find the system reliability at the desired level of confidence, again using interpolation as required.

Suppose that the reliability of a system at 50 percent confidence, based on the reliability of its components also at 50 percent confidence, is 0.756. Suppose further that the smallest number of tests of any component was 25. Then, from Table C.2a the equivalent number of failures is 5.5. If we wished to find the system reliability at 90 percent confidence, we would use Table C.2c for 25 tests and 5.5 failures to obtain a system reliability of 0.638. If the testing were on a time (equivalent-mission) basis and the smallest number of equivalent missions had been 25, then Tables C.3a and C.3c would be used. From Table C.3a, for a reliability of 0.756 and 25 equivalent missions with testing stopped at the end of the time period, the equivalent number of failures is 6.33. At 90 percent confidence, with 25 equivalent missions and 6.33 failures, the reliability, from Table C.3c, is 0.645.

If variables methods were used, then the curves in Table C.4 would be utilized. For the same reliability of 0.756 and 25 tests, the $\hat{K}$ factor from Table C.4b would be about 0.70. Then, using a $\hat{K}$ of 0.70 with $N = 25$, at 90 percent confidence the reliability would be about 0.66.

The author prefers a third method for combining component reliability assessments into the system assessment which is a compromise of the two alternatives which we have just discussed. This method is not as conservative as the latter alternative nor as indeterminant as the former with respect to system confidence level. Point estimates are made of all component reliabilities. These are combined into an estimate of the system reliability. The number of equivalent tests is again determined by the smallest number of tests associated with any of the components. The equivalent number of failures or $\hat{K}$ value is then determined, as appropriate, from the reliability estimate and number of tests. The equivalent number of failures Q is found from the point estimate: $R = 1 - (Q/n)$, or $Q = n(1 - R)$, where R is the system reliability and n is the number of tests; the equivalent $\hat{K}$ is found from the *normal* tables, using only the value of R. These values of n and Q (or $\hat{K}$) are then used to obtain reliability-confidence pairs for the system as described above.

We should note that for system confidence levels above 50 percent, we must first change all assessed reliabilities to point estimates, find the system 50 percent level, and then change this reliability-confidence pair to the numerics at the desired level of confidence. If we were to find the component reliabilities at the final confidence level, for example, 90 percent, and then combine these reliabilities to obtain the system reliability value, the system confidence level could be considerably above 90 percent but the exact level would be unknown. We would be unnecessarily penalizing ourselves because the correct system reliability at 90 percent confidence might be significantly higher than the value calculated in this manner.

ADDITIONAL READINGS

Amstadter, B. L., and T. A. Siciliano: Reliability Assessment Guides for Apollo Suppliers, North American-Rockwell Corporation, Downey, Calif., SID 64-1447A, 1965, IDEP no. 347.40.00.00-F1-28.

Bazovsky, I.: "Reliability Theory and Practice," Prentice-Hall, Inc., Englewood Cliffs, N.J., 1961.

Bureau of Naval Weapons, Chief of: "Handbook: Reliability Engineering," U.S. Government Printing Office, 1964.

Calabro, S. R.: "Reliability Principles and Practices," McGraw-Hill Book Company, New York, 1962.

Chorafas, D. N.: "Statistical Processes and Reliability Engineering," D. Van Nostrand Company, Inc., Princeton, N.J., 1960.

Dummer, G. W., and N. B. Griffin: "Electronics Reliability: Calculation and Design," Pergamon Press, Ltd., London, 1966.

Eisenhart, C., M. W. Hastay, and W. A. Wallis (eds.): "Techniques of Statistical Analysis," McGraw-Hill Book Company, New York, 1947.

Hald, A.: "Statistical Theory with Engineering Applications," John Wiley & Sons, Inc., New York, 1952.

Hiltz, P. A., and J. L. Gaffney: "Statistical Techniques for Reliability," North American-Rockwell Corporation, Downey, Calif., 1965

Johnson, L. G.: "The Statistical Treatment of Fatigue Experiments," American Elsevier Publishing Company, Inc., New York, 1964.

Lloyd, D. K., and M. Lipow: "Reliability: Management, Methods, and Mathematics," Prentice-Hall, Inc., Englewood Cliffs, N.J., 1962.

Weibull, W.: A Statistical Representation of Fatigue Failures in Solids, *J. Roy. Inst. Technol., Stockholm,* November, 1954.

15

RELIABILITY
DEMONSTRATION

15.1 DEFINITION OF RELIABILITY DEMONSTRATION

We must distinguish between predicted reliability, inherent reliability, and reliability demonstration. Predicted reliability, as we have seen, is the estimated reliability of a fully developed, operable equipment. It is the value that is expected to be obtained after the infant-mortality period has passed and most if not all of the experienced failure modes have been eliminated through design modification, changes in procedures, and other corrective actions. Predicted reliability also assumes that the equipment will be operated within its design limitations prior to the onset of the wear-out period.

Inherent reliability is the actual reliability of the equipment resulting from its design, function, usage, and required operating period. If it does not degrade during operation (that is, if it is operated during the period when its failure rate is constant), it will retain this reliability throughout the period. Its inherent reliability with respect to a particular mission changes only as design improvements are made. As explained in Chap. 12, these occur as a series of steps in the upward direction if the design changes are improvements in fact as well as in intent. The inherent reliability exists regardless of the amount of time or number of cycles that are demonstrated or even if the equipment is never operated and no demonstration takes place.

Reliability demonstration is the evaluation, through operation, of the capability of an equipment for meeting a specified reliability value. It is the systematic, often mathematical, process by which the assessor gains confidence in this capability. It is, of course, dependent on the inherent reliability, but it can differ considerably. For example, an equipment may be 99.9 percent reliable, but this high a reliability can rarely be fully demonstrated because of budgetary and time constraints. Also, if no design changes are made, inherent reliability remains constant while demonstrated reliability, at a given level of confidence, hopefully increases as more operating time or cycles are accrued.* Consequently, when the inherent reliability is reasonably high, e.g., greater than 0.9, demonstrated reliability will approach actual reliability asymptotically as more and more time/cycles are accumulated. If no failures are experienced and the demonstrated reliability is time dependent, it is a continuous function. Figure 15.1*a* illustrates this relationship. When reliability is cycle dependent, a step function results. This is shown in Fig. 15.1*b*. Figure 15.1*c* shows the relationship when failures occur. Each dip in the demonstrated reliability value represents the occurrence of a failure.

We will consider two common types of reliability demonstration. The first type is demonstration in the statistical sense, whereby actual numerical values are obtained. The second considers capability for meeting certain levels of stress, functional performance, etc., without "proving" a mathematical probability.

15.2 STATISTICAL DEMONSTRATION

The statistical demonstration of achieved reliability utilizes many of the assessment methods which have been discussed. In fact, in the mathematical sense, assessment and demonstration are complementary activities. In an assessment, the data which have been obtained from testing are evaluated to estimate the corresponding reliability. In a demonstration, the desired reliability is known, and the quantity of data necessary to verify that this reliability exists is determined. The numerics have the same relationship.

For example, if 69 tests have been made and no failures have been experienced, using Table C.2a we find that the assessed reliability at 50

*Alternately, we can state that the demonstrated reliability remains fixed while the confidence level increases. We recall, from Chaps. 13 and 14, that there are many reliability-confidence pairs which are equivalent, and therefore we can determine either the change in demonstrated reliability at a constant confidence level or the change in confidence level at a fixed value of demonstrated reliability.

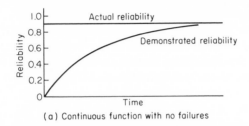

(a) Continuous function with no failures

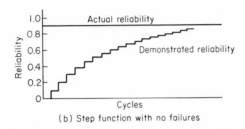

(b) Step function with no failures

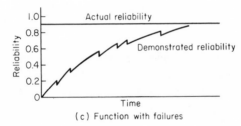

FIG. 15.1 Relationship between demonstrated and actual reliability with increasing operating time or cycles but no design change.

(c) Function with failures

percent confidence is very close to 0.99. If the reliability demonstration requirement were 0.99 with 50 percent confidence, 69 tests would have to be made with no failures to demonstrate compliance with the requirement. Similarly, the safety margin $\hat{K}$ is used with variables data for assessment of reliability. For demonstration of reliability, the goal is known and the appropriate required $\hat{K}$ is determined. If 10 tests are to be performed to demonstrate 99 percent reliability with 90 percent confidence, then using Table C.4h, a $\hat{K}$ value of about 3.45 must be obtained.

15.2.1 Equivalent Tests

Let us examine the statistical demonstration of reliability in more detail. Suppose that the requirement is to demonstrate a reliability of 0.85 at the 90 percent confidence level. From Table C.2c, 15 tests with no failures are required. Assuming that one failure has occurred, a course of action is necessary. Although the requirement for 90 percent confidence has not been met, we would not wish to reject the assumption of 85 percent reliability because of only one failure. (At 50 percent confidence, the demonstrated reliability is 89.1 percent, and a point estimate results in a reliability of 93.3 percent.) Instead, we might wish to test additional units. If the ratio of failures to total number of tests remained at 1/15, a total sample size of 45 units with three failures would demonstrate 85.7 percent reliability with 90 percent confidence. If the sample size were increased to 150 with 10 failures, the demonstrated reliability at 90 percent confidence would rise to 0.90.

Thus, when the actual reliability exceeds the requirement, statistical demonstration is possible at any desired level of confidence *if* enough units, time, and funding are available. The greater the margin between actual and required reliability, the sooner (in terms of number of units) will the requirement be demonstrated. Table 15.1 illustrates this point for a reliability demonstration requirement of 85 percent with 90 percent confidence. The demonstration requirement is met when the average number of failures is equal to or less than the number of failures permitted in the demonstration. The required sample size in Table 15.1 is taken from Table C.2c. It is assumed that the sample results will, on an average, reflect the true reliability. It can be seen from the table that when the actual reliability is close to the required reliability the average number of failures will exceed the permitted number of failures until a great number of units have been tested. Conversely, when the

TABLE 15.1   Relationship of demonstrated reliability to requirements for a required demonstration of 0.85 reliability at 90 percent confidence

Requirements		Inherent reliability					
Sample size	Failures permitted	0.88	0.91	0.94	0.97	0.99	1.00
		Average number of failures experienced					
15	0	1.80	1.35	0.90	0.45	0.15	0.00
25	1	3.00	2.25	1.50	0.75	0.25	
34	2	4.08	3.06	2.04			
43	3	5.16	3.87	2.58			
52	4	6.24	4.68				
60	5	7.20	5.40				
68	6	8.16	6.12				
76	7	9.12	6.84				
85	8	10.20					
258	31	30.96					

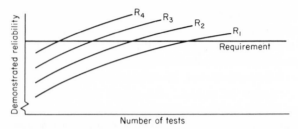

FIG. 15.2 Relation of demonstrated reliability to requirement for different actual reliabilities: $R_4 > R_3 > R_2 > R_1$ > requirement.

actual reliability greatly exceeds the requirement, the allowed number of failures exceeds the actual number after only a *relatively* small number of tests. Figure 15.2 graphically depicts the general relationship.

15.2.2 Validity of Sequential Testing

In a rigorous statistical sense, it is not 100 percent valid to test a small sample, for example, 15 units, and, if the demonstration requirement is not met, to test additional units until it is met. However, when high confidence levels are involved, the actual reliability may be better than the required reliability even though the demonstration is not passed. When this is the case, the practical approach is to test additional units. This may be thought of as sequential testing although, in a true sequential test procedure, the complete plan including all sample sizes and *both* acceptance and rejection criteria at each step is defined in advance. Chapter 10 of "Reliability: Management, Methods and Mathematics"* describes sequential plans in more detail.

Note that when the required demonstration is at low levels of confidence, i.e., at or just above 50 percent, this procedure should not be used. If the reliability were slightly below the requirement, it could easily happen due to chance that, at some stage during the testing, the actual number of failures would be less than the allowed number and the product would be accepted. Consequently, the testing of additional samples when the first sample does not meet demonstration requirements should be limited to confidence levels of 75 percent or higher and then only when the point estimate indicates that the actual reliability exceeds the requirement.

Just as additional units may be evaluated, additional operating time may be accrued if demonstration requirements are not met initially. Again, additional testing should be limited to those instances where the confidence requirement is at least 75 percent and the initial testing indicates that the actual reliability does, in fact, exceed the required reliability.

*D. K. Lloyd and M. Lipow, "Reliability: Management, Methods, and Mathematics," Prentice-Hall, Inc., Englewood Cliffs, N.J., 1962.

Similarly, when the $\hat{K}$ value obtained in a variables test does not meet the required value, the testing of a larger sample may result in acceptance. As we note by reviewing the curves in Table C.4, as the sample size increases the required $\hat{K}$ value for a given level of reliability and confidence decreases. It should also be noted that, as the sample size increases, the $\hat{K}$ value approaches a minimum constant value and this constant corresponds to a normal table z value for the particular required reliability. Unless the initial $\hat{K}$ value exceeds this minimum value, the testing of additional units should not be undertaken.

It frequently happens that the testing of additional samples (when the demonstration requirement is not met initially) is not permitted. In these cases it behooves the manufacturer to determine what risk he is willing to take that his product will not be accepted. If he wishes to reduce this risk, he can initially select a large enough sample so that the probability of meeting associated criteria is high. As the sample size increases the risk of making an error decreases so that, as is noted in the preceding example and in Table 15.1, the requirement becomes less severe; i.e., the ratio of permitted failures to the sample size increases or the required $\hat{K}$ value decreases. The probability that a product with a particular reliability will meet demonstration requirements can be determined by using the standard statistical procedures which have been discussed.

If the actual reliability is only slightly better than the required reliability, the number of tests which may be required to pass demonstration might well be excessive. The alternative to testing large numbers of products is to improve the inherent reliability. It is frequently less costly to make the improvement than to test additional units. In the long run, improved reliability not only increases the probability of demonstration being passed but also enhances the manufacturer's reputation and can lead to additional business.

15.2.3 Mission Time and Quantity Considerations

A frequently encountered requirement relates to the required mission time. A demonstration of the operation of one or more units for a time period equivalent to the mission time (including subsystem and system check-out time and on-site readiness inspection time) may be required. If a total time of 1,000 hr of failure-free operation is required for demonstrating capability of accomplishing a 100-hr mission, the time can be accrued in a number of ways. One hundred units could be tested for 10 hr each, ten units for 100 hr, five units for 200 hr, or one unit for 1,000 hr. In a strict statistical sense these demonstrations are all equivalent, amounting to ten equivalent mission lives. At 50 percent confidence, this would result in a demonstrated reliability of 93.3 percent. However, from a practical point of view, they can be vastly different. The testing of 100 units for 10 hr does not demonstrate the capability of even one unit operating for 100 hr. Wear-out might well occur in the period between 10 and 100 hr. (The fact that 200 automobiles all ran for a minimum of 50,000 miles does not mean that 50 automobiles will all run for 200,000 miles although the total mileage—10 million—is the same.) Also, the fact that one unit operates for 1,000 hr does not, in itself, prove that other units can operate satisfactorily for 100 hr. There may easily be failure modes appearing randomly on some units which limit their useful lives.

The operation of ten units for 100 hr each would be a more meaningful test and we would be more confident (at least subjectively) in our test results. However, testing of five units for 200 hr each could be preferable, particularly if early wear-out were suspected. If the mean wear-out life were 150 hr and the standard deviation were 25 hr, then assuming a normal distribution and no failure modes other than wear-out, the probability of one unit operating for 100 hr would be 0.97725 and of ten units would be 0.7944. In this case, the testing of ten units for the required mission time would probably fail to disclose any mode of failure, whereas the testing of five units for up to 200 hr would show the marginal wear-out capability.

An even more desirable test procedure would be the testing of ten units for 200 hr. This would provide good assurance that no unidentifiable failure modes existed. In general, a minimum sample size of ten is recommended, with five units as the lower limit. Also, a test time of at least twice the mission time is suggested. This would provide a minimum of 20 equivalent mission lives with no chargeable failures. If one or more chargeable failures occur, ten equivalent mission lives are suggested as a lower limit for demonstration testing. These should be distributed among a minimum of five units with a minimum operating time for each unit of at least the length of the mission.

Sometimes a contract will require the demonstration of reliability without stating the method of demonstration. This may be interpreted by the reliability, purchasing, engineering, or other cognizant organization of the purchaser to mean statistical demonstration by attributes, that is, the testing and evaluation of a large number of units on a success/failure basis. This can easily lead to the imposition of unrealistic requirements on the supplier if the implications are not thoroughly understood. If the contractual reliability demonstration requirement is 0.90 at 90 percent confidence for a 1,000-hr mission, from Table C.3c, more than 22,000 test hours must be accrued without failure. This may well be an impossible task. In fact, in most aerospace programs, the statistical demonstration of reliability is not possible regardless of whether attributes or variables methods are used. The use of an attributes method would result in testing of more units than are associated with the entire program. The use of variables methods would require the testing of an excessive number of units (from a cost-or-schedule standpoint) to destruction or at least to the point where the units would be considered unusable for an actual mission.

15.2.4 Cost Considerations

If reliability demonstration is not possible or practical at the system level, it may be feasible to demonstrate the reliabilities of the components and subsystems and to use the appropriate mathematical model to combine the results and obtain a value for the reliability of the system. When reliability demonstration is carried out at the component or subsystem level, the quantity of each type of component or subsystem to be tested must be determined. In order to obtain maximum demonstration per dollar expended, criteria should be developed which will provide the greatest efficiency in the overall demonstration program. Two such criteria relating to *series* systems are presented here. For a more detailed discussion of these criteria, including proofs of the relationships, the reader is referred to the *Proceedings of the 1963 Aerospace Reliability and Maintainability Conference.**

Three ground rules are required before the actual criteria are presented. Should the system conditions result in a change in these ground rules, the criteria should also be appropriately modified. The ground rules are the following: (1) The reliability requirements are sufficiently high that no failures are permitted during demonstration. This assumption is quite applicable to many critical programs in the aerospace as well as in other industries. (2) All reliabilities are demonstrated at the same level of confidence. It would be invalid with respect to demonstration optimization to test one equipment at one confidence level and another equipment at a different level. (3) Demonstration is performed on an attributes basis.

The criteria which provide the greatest demonstration efficiency are

1. The number of units of the component or subsystem types to be tested should be *directly* proportional to the square roots of their respective quantities in the system. Thus, if there are four times as many components of type A as there are of type B in the system, then the ratio of demonstration test quantities should be 2:1.

2. The number of units to be tested should be *inversely* proportional to the square roots of the respective direct testing costs. If the direct costs with respect to type A are twenty-five times as high as the costs for type B, then only $1/\sqrt{25}$ or one-fifth as many of type A should be tested. Note that since the *unit test cost* for type A is twenty-five times as high and the *test quantity* is one-fifth, the *total direct expenditure* for type A will be five times as great. Therefore, we may restate this criterion as follows: The total test expenditures should be *directly* proportional to the square roots of the respective

*B. L. Amstadter, Maximizing Reliability Demonstration for a Given Expenditure, *Proc. Aerospace Reliability Maintainability Conf., Washington, 1963,* American Institute of Aeronautics and Astronautics, New York, 1963, pp. 38-42.

direct unit test costs. The two criteria can be combined into a single equation:

$$n \approx \sqrt{\frac{k}{C}} \tag{15.1}$$

where n = the demonstration test quantity
$\quad k$ = the number in the system
$\quad C$ = the direct test cost per unit
When there are two equipment types in a series system, Eq. (15.1) can be written

$$n_1 : n_2 = \sqrt{\frac{k_1}{C_1}} : \sqrt{\frac{k_2}{C_2}} \tag{15.2}$$

where the subscripts refer to the two different equipments. These equations can be extended to cover any number of different equipment types as long as they are in series in the logic diagram (i.e., the failure of any one equipment item fails the system).

$$n_1 : n_2 : n_3 : \cdots : n_i = \sqrt{\frac{k_1}{C_1}} : \sqrt{\frac{k_2}{C_2}} : \sqrt{\frac{k_3}{C_3}} : \cdots : \sqrt{\frac{k_i}{C_i}} \tag{15.3}$$

15.2.5 Overstress Methods in Demonstration

While statistical demonstration of component reliability at actual system conditions is, in many cases, not practical, it can sometimes be accomplished by testing the component at conditions whose severity is greater than that of the intended mission. The failure rates of some types of components can be greatly accelerated by increasing the stress levels. Failure rates of some capacitors, for example, are increased by the sixth or seventh power of the ratio of the dc voltage levels when tested above rated voltage. A group of these capacitors, when tested at twice their operating voltage, may exhibit a failure rate from 64 to 128 times as high as will be experienced in use. Therefore, if a value of 100 is used, the failure rate at operating conditions can be estimated as 1 percent of the failure rate at test conditions, and the operating mean time to failure would be estimated as 100 times as long as the MTTF experienced during testing.

As a direct result of these relationships, it may be possible to demonstrate the required reliability by testing a relatively small number of units at overstress or accelerated conditions. The procedure is to multiply the number of tests or test hours accumulated at overstress conditions by the appropriate constant. The number of experienced failures remains the same. The demonstrated reliability is then determined at the desired level of confidence using the converted number of hours or tests and the actual number of failures. Accurate knowledge of the relationship of tests at accelerated conditions to tests at normal operating conditions is the principal requisite.

As an example, if 500 hr of testing have been accumulated and two failures have occurred during the test period and if a multiplying constant of 100 is appropriate, then the demonstrated reliability is determined using 50,000 hr and two failures. Assuming that the mission time is 2,500 hr, the reliability is determined from Table C.3 for 20 equivalent missions and two failures. At 50 percent confidence this is 0.8748. The corresponding reliability at 90 percent confidence is 0.7672.

15.3 CAPABILITY DEMONSTRATION

The second type of reliability demonstration refers to those instances when the equipment is subjected to increasing levels of environmental and operational stresses equal to and in excess of those that will be experienced in the intended mission. During and/or after subjection to these stresses the equipment is operated at normal conditions and its functional capability is ascertained. If the equipment does not fail until the stresses are far in excess of expected stress levels, we gain confidence in its capability.

Most government contracts require performance of a set of tests called "qualification tests." These include subjection to such environments as vibration, acceleration, shock, thermal cycling, humidity, salt spray, etc., in accordance with prescribed procedures. Procedures detailed in MIL-E-5272 are frequently specified. Section 6 of the "Reliability Handbook"* includes detailed lists of tests, specifications, and related information. The particular procedures to follow are specified by the contract.

While qualification tests require subjection of the equipment to environmental stresses considered applicable to the particular equipment and mission, they may not include testing at higher levels. Such higher-level testing may be called out separately in the contractual specification or may be proposed by the contractor as proof of equipment capability, particularly when statistical demonstration is not practical. Overstress testing is usually

---

*W. G. Ireson, "Reliability Handbook," McGraw-Hill Book Company, New York, 1966.

performed in stages or steps. The equipment is first operationally tested at the design level and is then subjected to the first overstress level. It is again operationally tested at the design level and is then subjected to the next higher level of overstress testing. The number of overstress levels is usually limited to two or three. In a two-level test program, these might be 150 and 200 percent of the qualification test level.

When the equipment is not operating during subjection to these stresses, e.g., most spacecraft equipment does not operate during launch, testing at operational levels after each stress step test is mandatory. However, when the equipment is operated during the time when the stresses are being applied, the intermediate tests of operational levels are sometimes omitted if the operation is satisfactory at the overstress conditions. A final test at normal conditions is always performed to check for any possible change in operating characteristics.

In addition to overstress levels of environments, equipment may be subjected to higher-than-expected levels of operational parameters. This aspect has already been discussed in conjunction with statistical demonstration, where the number of operating hours at accelerated conditions is converted to an equivalent number of hours at nominal conditions, using a known relationship. Often, however, the relationship is not known and a conversion is not possible. In these cases, testing at overstress operating parameters (voltage, current, power, flow rate, pressure, etc.) is used to gain qualitative confidence in the capability of the equipment rather than to obtain quantitative reliability numerics.

15.4 ANALYSIS OF DEMONSTRATION REQUIREMENTS

Whenever demonstration is required or contemplated, the requirement should be thoroughly evaluated to determine whether statistical or overstress testing is specified. When statistical demonstration is suggested, the particular numerical requirement should be reviewed so that the optimum plan can be chosen with respect to inherent reliability, risk, cost of testing, and cost of rejection. When high reliability and high confidence levels are required, statistical demonstration may be very costly and time-consuming and may warrant consideration of non-statistical procedures such as the use of accelerated or overstress levels.

ADDITIONAL READINGS

Amstadter, B. L.: Maximizing Reliability Demonstration for a Given Expenditure, *Proc. Aerospace Reliability Maintainability Conf., Washington, 1963,* American Institute of Aeronautics and Astronautics, New York, 1963, pp. 38-42.

ARINC Research Corporation: "Reliability Engineering," Prentice-Hall, Inc., Englewood Cliffs, N.J., 1964.

Bazovsky, I.: "Reliability Theory and Practice," Prentice-Hall, Inc., Englewood Cliffs, N.J., 1961.

Bureau of Naval Weapons, Chief of: "Handbook: Reliability Engineering," U.S. Government Printing Office, 1964.

Calabro, S. R.: "Reliability Principles and Practices," McGraw-Hill Book Company, New York, 1962.

Chorafas, D. N.: "Statistical Processes and Reliability Engineering," D. Van Nostrand Company, Inc., New York, 1960.

Ireson, W. G.:"Reliability Handbook," McGraw-Hill Book Company, New York, 1966.

Lloyd, D. K., and M. Lipow: "Reliability: Management, Methods, and Mathematics," Prentice-Hall, Inc., Englewood Cliffs, N.J., 1962.

Mood, A. M.: "Introduction to the Theory of Statistics," McGraw-Hill Book Company, New York, 1950.

16

SYSTEM RELIABILITY
CONSIDERATIONS

16.1 INTRODUCTION

Some important aspects of system reliability have already been discussed, in Chaps. 7 and 8 on logic diagrams and mathematical models and in Chaps. 10 and 11 on prediction and apportionment methods. A brief review of these chapters is suggested before we further examine the facets of system reliability.

The principal difference between system reliability considerations and component reliability considerations is one of emphasis rather than one of mathematics. System reliability per se emphasizes selection among alternatives rather than mathematical methods of evaluation which have been discussed. This is not to say that the mathematics are unimportant. They are at least as necessary as they are for other phases of reliability activity. However, the assumption of prior mathematical capability is made and the discussion will focus on those aspects which themselves are most pertinent to systems.

Perhaps the principal contribution of reliability to overall program economy is the selection of the best overall system from among several candidates. The selection usually involves considerations of cost, performance, availability, maintainability, and weight as well as those of reliability, and it requires the formulation of criteria of *effectiveness*. In fact, the formulation of these criteria must precede the calculations. For example, if a municipal power generating system were being considered, several measures of effectiveness might be possible. If the principal purpose were the maintenance of continuous power with a minimum number of outages, cost could be a secondary consideration. On the other hand, if cost were a primary consideration, the effectiveness measure might be average kilowatthours of energy per dollar spent.

16.2 INITIAL CONSIDERATIONS

Therefore, before we discuss examples of possible trade-offs, we will discuss some of the considerations related to a system. First among these is mission analysis. We will also review system analyses and then consider auxiliary functions such as support equipment. In effect, we will take an *integrated system* approach.

16.2.1 Mission Requirements

The first of these preliminary considerations—evaluation of mission requirements—is necessary if criteria are to be formulated for analyzing alternatives and evaluating trade-offs. Without specific knowledge of the requirements of a mission, realistic decisions on redundancy, design changes, and other aspects of reliability improvement cannot be reached. These requirements include both the functional operations for which the mission was designed and the conditions which will be encountered by the various systems and subsystems during preparation, transportation, storage, and readiness phases, and actual operation. A mission need not involve mobility from place to place—a fixed installation such as the power generating plant or a complex radar site has a mission. The radar installation may emphasize accuracy rather than range, or it may be designed for longevity or minimum repair time at the expense of the first two characteristics. Similarly, an automobile or home appliance may be designed for durability rather than for performance, or vice versa. These requirements and characteristics require definition and clarification before criteria of reliability can be established, and, hence, before reliability itself can be evaluated.

There can be several requirements. In Chap. 11, in the discussion of the relationship of criticality to the apportionment of allowances, an example was cited in which there were both primary and secondary functions of a mission. Their relative importance will have significant effects on product improvement, redundancy, and other decisions affecting reliability. The relative importance may even change from mission to mission for the same system. Therefore, careful consideration must be given to evaluating each potential mission for all the various pertinent requirements.

In addition to the operational requirements, each mission will have an associated complex of environments which are sometimes referred to as the *environmental envelope*. The environmental envelope includes all phases of the mission, encompassing transportation, storage, field operation, and so on. Sometimes transportation may induce a more severe environment than usage, e.g., a ground-based computer system. If delivery is considered to include all phases up to actual deployment, then delivery (launching) of a satellite imposes much greater loads than will be experienced during deployment (orbit). The equipment, whether in an operating or a non-operating state, must be capable of successfully withstanding subjection to all applicable environments. Normally, when an equipment is not operating, the effects of an otherwise severe environment are not as detrimental as when the equipment is in operation. If capability for withstanding a relatively short subjection (such as a missile launch) has been fully demonstrated, then the assumption is usually made that the operating failure rate is unaffected. Of course, if the subjection to a severe environment is prolonged, it must be taken into account by increasing either the non-operating failure rate or the operating rate or both.

The effects of the total environment must be considered. In recent years, new environments of zero gravity, absolute vacuum, and solar and other forms of radiation have been added to the list of existing environments. New modes of failure have arisen, and materials and components which may be excellent in one set of environments may be very poor in another. Knowledge of the environments and their potential effects is therefore a very necessary aspect of mission evaluation.

16.2.2 System Analysis

The most important of these preliminary considerations before trade-offs can be studied and optimum configurations selected is the analysis of the system itself. When evaluating components, their functions are usually well defined and criteria for determining and improving reliabilities can be readily formulated. Conversely, systems and major subsystems must first be studied with respect to their functions and purposes in a particular mission. These directly affect the criteria of adequate performance and thus the reliability.

As in the case of a mission, most systems and subsystems have both primary and secondary functions. A liquid fueled propulsion system on a spacecraft might be used for both attitude control and orbital translations. The first purpose requires extreme accuracy and fast response time with respect to short power impulses; the second purpose requires that the system attain some minimum thrust per pound of propellant burned or within a limited time from ignition. These two functions might be in opposition: large engines could generate the necessary thrust for translation but could have difficulty in generating short-duration impulses needed for precise attitude control. Small engines could have the inverse problem. Similarly, an automobile owner would like to have both performance and economy of operation. His choice of automobiles, and especially of engines, reflects the relative emphasis he places on these two characteristics since an automobile usually cannot fulfill both objectives to the same degree. Consequently, not only should the system purposes and functions be defined, but their relative importance should be specified.

The criteria by which reliability will be measured depend, therefore, on the system purposes and their relative importance. Furthermore, the nature of a non-conformance and its effects on system performance should be considered. If a spacecraft environmental control system is required to maintain cabin air temperature between 65 and 80°F, should a rise in temperature to 82°F be considered to be as serious as a rise to 100°F or as a complete breakdown of the temperature control subsystem? The method of accounting for partial failures or non-conformances should be defined so that realistic reliability numerics can be computed. These considerations of performance deficiencies depend on adequate definition of system functional requirements.

System analysis should be directed toward obtaining knowledge of such subjects as the system's reactions to various internal and external stimuli, the interactions among its components and subsystems, and the expected average and maximum stress levels which each component and subassembly will experience. Knowledge of the functions and duties of each component with respect to the system and mission is essential. Failure-modes-and-effects analyses (FMEAs) of all components, which delineate each mode of failure, possible causes, and effects on the component *and* on the system, are valuable aids in a systematic definition and understanding of component functions. These are best prepared jointly by Reliability, Design, and Systems Engineering personnel, and they may reveal system anomalies and contradictions as well as disclosing how functions are performed and which redundancies are real and which are only apparent. In large projects such as space explorations these three organizations are sometimes so far removed from each other organizationally and physically that a coordinated effort is necessary to ensure system integrity and optimization.

16.2.3 Auxiliary Functions

Auxiliary functions make up the third set of considerations. An integrated system consists not only of the system actually performing the mission but also includes the auxiliary equipment required for installation, check-out, maintenance, repair, and sometimes verification of mission accomplishment. The effects of changes of an equipment must consider both the operation of the immediate system and these other aspects. For example, an automobile could be made to run many more thousands of miles without oil leakage if all joints were welded—but how would one change the oil or make certain kinds of repairs?

Ease of switching, removal, replacement, etc., must be taken into account when redundant systems are evaluated. A system in a radioactive environment requires a different design for redundancy than does one which permits human access. A ground-based communication system differs considerably from one which is on a submarine or in a satellite. A change in the size or weight of the land-based system may have far less severe consequences than a change in the size or weight of the submarine or satellite communication system. All these factors influence the studies which we will now discuss.

16.3 SYSTEM IMPROVEMENT

Once the mission requirements have been evaluated, the system functions analyzed, and auxiliary factors reviewed, the primary task of improving the system can be started. From a broad viewpoint, the task is that of optimizing *system effectiveness.* It is for this reason that the prior considerations were needed. With them we can formulate quantitative measures of effectiveness, and we can compare alternatives with known objectives. If we were to attempt to weigh the advantages of different candidate systems without these measures, each organization would be using a different scale—probably oriented toward its own functions. Each organizational entity employs its own yardstick of effectiveness; Design Engineering stresses performance, Reliability seeks longevity of operation, Human Engineering emphasizes ease of operation, and so on. Program management may still direct each group to consider alternatives with respect only to its own principal area of responsibility, but guidelines are available for the final overall evaluation and decision process. Even when each group considers only its own function, when mission objectives are known the decision will have a higher degree of optimization. Reliability Engineering, for example, will know the relative importance of reliability, availability, maintainability, and mean time to failure, and can then be completely objective in its analyses and recommendations.

16.3.1 Redundancy

Redundancy is a principal method of improving system reliability. In some cases it is the only practical means of meeting system requirements within the scheduled time. Redundancy is, in itself, a very complex subject and this text treats only the more important aspects. Among these are numerical evaluations of active and sequential configurations; numbers of units needed to meet requirements; causes, modes, and effects of failure; maintenance considerations; and performance aspects.

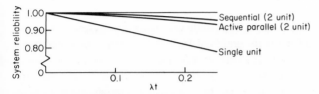

FIG. 16.1 System reliability for small λt as a function of time for three configurations.

We shall first look at the numerical relationships affecting the choice of active or sequential redundancy and then examine other factors which influence this choice. *If* the choice is not affected by other factors and *if* switching is perfect, then sequential redundancy is clearly preferable. Figures 16.1 and 16.2 illustrate the relationship for these conditions between the reliabilities of a single unit and two-unit active and sequential configurations. A constant operating failure rate is assumed for each unit.

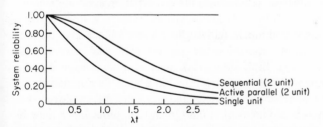

FIG. 16.2 System reliability for large λt as a function of time for three configurations.

The equations for the curves in Figs. 16.1 and 16.2 were indicated in Chaps. 3 and 8 and are repeated here in slightly modified form.

Single unit: $R = e^{-\lambda t}$ (8.2a)

Active redundancy: $R = 1 - (1 - e^{-\lambda t})^2$ (8.17a)

Sequential redundancy: $R = e^{-\lambda t}(1 + \lambda t)$ (8.34)

However, it is recognized that failure detection and switching are not perfect. Failure-detection devices (sensors, transducers, power conditioners, etc.) and switching equipment (valves, relays, etc.) often are less reliable than the components. Therefore, Eq. (8.34) should be rewritten as follows:

$$R = e^{-\lambda t}(1 + R_{ds}\lambda t) \tag{8.48}$$

When either type of redundancy is possible with respect to system operation, active redundancy is frequently preferred from the numerical considerations. Using simple calculations it can easily be shown that when the unreliability of detection and switching, Q_{ds} $(= 1 - R_{ds})$, is more than *one-half* the unreliability of the component Q, then an active configuration is more reliable than a sequential one. For example, if $R = 0.90$ so that $Q = 0.10$ and $R_{ds} = 0.95$ and $Q_{ds} = 0.05$, the reliability of an active system is 0.9900 and the reliability of a sequential one is also 0.9900. If Q_{ds} exceeds 0.05, then the reliability of the sequential system falls below 99 percent.

While a standby failure rate by itself does not affect the advantage of sequential over active redundancy (unless it is close to the operating failure rate), it does further reduce the break-even point with respect to Q_{ds}. That is, if standby reliability is less than perfect, then Q_{ds} must be *less* than 50 percent of Q for the sequential system to be as reliable as the active one. Consequently, the statement that active redundancy is frequently preferred is confirmed.

The *minimum* number of redundant units can often be easily determined from the numerical reliability relationships. Any system is made up of a number of series functional blocks (although each series block may be quite complex in terms of its internal series—parallel logic and may contain considerable redundancy). Each series block must be more reliable than the overall system requirement. If the elements within a block do not individually possess this degree of reliability, then redundancy must be employed. For example, if the system requirement is 0.80, then each series block must be better than 0.80. If the reliability of a component is only 0.50, then at least three of these components in active parallel $[1 - (1 - e^{-\lambda t})^3 = 0.875]$ or two in sequential redundancy $[e^{-\lambda t}(1 + \lambda t) = 0.85]$ are needed. These are only minimum requirements, and they are based on an assumed high reliability of other series blocks. However, if the reliabilities of other blocks are also relatively low, then this block may require a reliability approaching unity and thus will require four or even five redundant components. It is here that trade-off evaluation can be effective, determining which redundancy can best meet *all* system requirements such as cost, performance, and maintainability.

Causes and effects of the various modes of failure must be considered in the choice of redundancy—in fact, in the determination of whether redundancy will even be significantly effective. If the principal cause of failure of a valve, for example, is externally induced vibration, then the use of a redundant unit does not provide the desired benefits of redundancy and may even be detrimental. The cause of failure of the first valve is also the cause of failure of the second and the two units are not independent. Similarly, if two diodes are installed to provide sufficient current-carrying capacity because one diode alone is marginal, the failure of one unit results in imposing conditions on the second which significantly reduce its reliability. Such considerations are sometimes accounted for in the logic diagram. In the first case, only one valve may be depicted (with appropriate notation); in the second case, the diodes may be shown in series rather than in parallel because both are really needed in the system.

There are other instances when the environments or operating conditions are considerably different for redundant systems than for non-redundant ones. This does not necessarily contradict the statement made earlier in the book that failure rates of unfailed units in a redundant system are considered to be constant even after other units have failed. This is the usual case. However, there are cases when the failure rates do change and, in these cases, the choice of redundancy can be affected. If cooling is marginal and if each unit generates a large amount of heat, then the number of units which are operating at one time affects the operating temperature, perhaps critically; and sequential redundancy is preferable. Power requirements may also give preference to sequential operation. System power may be marginal or even insufficient to operate redundant components simultaneously.

Performance aspects are frequently the most important considerations in the choice of redundancy. If the system operating characteristics are significantly affected by components performing the same function together, active redundancy may not be possible. For example, if the pressure rise or increased flow rate resulting from two pumps working together causes unacceptable system conditions, then, when redundant pumps are employed, they must be sequential. Conversely, if the system can accept these different conditions but cannot tolerate a loss or sizable reduction of fluid flow for even a short time period, then sequential pumps should not be used and active redundancy is required.

Maintainability considerations can also influence the redundancy choice and configuration. Depending on the particular system, it may be possible to repair or replace a failed unit without shutting down a system which is sequentially redundant but not one which has an active parallel configuration. In other systems, when the system is operating, some active configurations are repairable while others are not. The fact that a system is or is not maintained can affect the particular configuration design when active redundancy is used. The configuration shown in Fig. 16.3 might be desirable for a non-maintained system where the failure of an element reduces the reliability of that element's function but does not affect the reliability of other functions which are still redundant. If element a fails, function A is still performed by element a', and the reliabilities of functions B and C are unaffected.

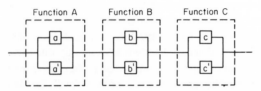

FIG. 16.3   Three parallel redundant functions in series.

If maintenance and repair are performed and require the shutdown of the function, then any failure in this configuration causes at least temporary system shutdown. But if all three functions are in series in one unit and the units are in parallel, as in Fig. 16.4, it may well be possible to replace an entire unit or leg without shutting down the system. In this case, with maintenance, both reliability and availability are improved.

In addition to purely system considerations, the location of manufacture and assembly can affect the choice of redundancy. If one supplier provides components which perform function A while components from other suppliers perform functions B and C, then each supplier may design his components or "black box" with internal redundancy, as shown in Fig. 16.3. But if one source provides all three functions, the configuration shown in Fig. 16.4 may be both feasible and desirable.

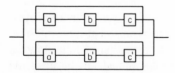

FIG. 16.4   Three series functions in parallel.

One further aspect of redundancy relates to the modes of failure. One redundant configuration can provide more protection against one mode of failure while another configuration with the same number, type, and relative position of parts can provide more protection against the opposite mode. Two simple parts which illustrate this point are diodes and check valves. For a diode, the opposite failure modes are open and short; the opposite modes for a check valve are failure to open and failure to close. Figures 7.15 and 7.16

and the related discussions in the chapter on logic diagrams illustrate the two possible redundant configurations and explain their relative advantages.

Another situation relates to fusing. While fuses provide protection against one mode of failure, they increase the probability of another failure mode. The choice of whether or not to employ fuses depends on the relative probabilities of the various failure modes, on the effects on the system, and in some cases on whether space and weight limitations permit fusing if it is desirable. The same considerations hold true for many redundant parts. The use of two components in series (or in parallel) can increase the reliability with respect to one mode of failure and decrease it with respect to another. System considerations will often determine whether redundancy is possible and desirable, and if so they will also indicate the best configuration.

16.3.2 Component Improvement

The second principal means of improving system reliability is component improvement. Both redundancy and component improvement should be explored, and usually some effort and funding should be allotted to each. Redundancy appears to be the most economical solution, but sometimes appearances are deceiving. For example, if a component has a reliability of 0.95, then two in parallel would have a reliability of 0.9975, other considerations being equal. The cost of a product-improvement program to reduce the failure rate by 95 percent would normally far exceed the cost of a second component. However, if the component is in a missile or spacecraft, the means of compensating for the increase in weight, space, power, and so on can easily reverse the advantage.

If other factors are the same, the cost of product improvement exceeds the cost of redundancy when only one unit is involved. However, when there are hundreds or perhaps thousands of the same type of part used in the system, the comparison of product-improvement costs to redundancy costs favors the former. Thus, such parts as capacitors, resistors, and semiconductors are considerably benefited by product-improvement programs.

Some components such as valves cannot be made truly redundant economically. These parts are ready candidates for these programs, and some government-sponsored projects, e.g., Minuteman, have had specific funds allocated to parts improvement. Participating contractors have had independent programs whose sole purpose was the reliability evaluation, improvement, and verification of the parts which they supplied. These programs turned out to be the keys to the high reliabilities achieved in the overall projects.

A third argument in favor of product improvement stems from the nature of the failure. We stated earlier that if the failures are caused by the imposition of stresses beyond the capacity of the equipment, for example, excessive shock or vibration, then adding redundant parts does not necessarily greatly upgrade the reliability. Some slight improvement will no doubt occur due to random variation among parts (some parts having more capability than others of the same design due to chance). However, the excessive stresses which caused one unit to fail will be imposed on all units and will cause them to fail also. If one of the stresses is high temperature, the use of more than one component may actually reduce the reliability by increasing the temperature and imposing even higher heat stresses. Careful analysis is required to determine the basic causes of the failure and to take appropriate action, where possible, to eliminate these causes. When failures are truly random, redundancy can greatly improve the reliability, but when failure results from excessive stress, redundancy may lose most of its effectiveness.

A good product-improvement program includes many elements. First among these are the failure modes, causes, and effects analyses mentioned earlier. The basic causes of failure are first determined so that later modifications and changes will be effective. A very basic study program in the fundamental physics of failure has been carried out jointly by the Illinois Institute of Technology Research Institute and the Rome Air Development Center. In addition to periodic reports of their findings, yearly symposia are held on the I.I.T. campus and are open to interested contractors and to the public.

After the basic causes and mechanisms of failure are determined, the manufacturing process is studied in great detail to pinpoint those variables that lead to failures and to find means of improving the process. Every step of the manufacturing is thoroughly examined. Quality control and inspection activities are also scrutinized for, while the adage that "quality cannot be inspected into a product" is true, unfortunately the inspection process itself can sometimes result in deficiencies which lead to later failures. Process specifications, test plans and procedures, and inspection instructions should all be thoroughly reviewed with the objective of verifying both their completeness and their accuracy.

An integral part of a product-improvement program is the evaluation of environmental capability. Parts and components are subjected to environmental stresses equal to and exceeding anticipated mission stresses to ensure their capability for satisfactory operation after subjection. Environmental testing is, of course, necessary for those components which were originally deficient in this area, and where the product-improvement program was specifically designed to overcome these deficiencies. It is equally important for those components which were deficient in other respects to ensure that corrections of deficiencies in these other areas do not result in adversely affecting environmental capability.

Another important consideration is the expected amount of increase in system reliability per dollar expended for component improvement. This can lead to interesting trade-off comparisons. The following greatly simplified example will illustrate such a trade-off analysis.

Suppose that a system consists of only two subsystems in series, for example a radar detection and guidance subsystem and a missile subsystem, and that it has a required reliability of 0.95. Further suppose that, after development, each radar subsystem has a cost C_1 of $1,000,000 and a reliability R_1 of 0.95 and that its reliability can be upgraded according to the incremental costs per unit given in Table 16.1. That is, in order to upgrade the reliability of each radar from 0.95 to 0.99, an additional expenditure of $1,100,000 is required for each subsystem. Suppose, further, that the missile subsystem has an initial cost C_2 of $2,000,000 and an initial reliability R_2 of 0.90 and that the expected costs per subsystem of upgrading the reliability are as given in Table 16.2. Since a series system was assumed, its reliability $R_s = R_1 \times R_2$. Also, since R_s must be 0.95, it can be achieved if $R_1 = 0.96$ and $R_2 = 0.99$, if $R_1 = 0.97$ and $R_2 = 0.98$, or if $R_1 = 0.99$ and $R_2 = 0.96$. Table 16.3 delineates the reliability and associated costs. From Table 16.3, case C appears to be least costly and, if the costs of upgrading reliability were estimated accurately, funds would be allocated in the manner indicated.

However, there are other alternatives. If the system operational requirements (weight, space, performance, and so on) permitted active redundancy or if sequential redundancy could be employed without necessitating large switching costs, then the following alternatives involving both product improvement and redundancy could be considered.

1. Two radars in parallel and in series with the missile subsystem:

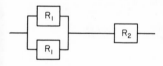

The overall system reliability would be $[1 - (1 - R_1)^2] \times R_2$. The reliabilities would be $R_1 = 0.95$, $R_2 = 0.96$, and $R_s = 0.9576$. The associated system cost is $2($1,000,000) + $3,100,000 = $5,100,000$.

2. One radar in series with two parallel missile subsystems:

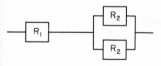

The reliabilities are $R_1 = 0.96$, $R_2 = 0.90$, and $R_s = 0.9504$, and the system cost is $1,100,000 + 2($2,000,000)$, also equal to $5,100,000.

TABLE 16.1 Costs of upgrading the radar system

Incremental reliability	Final reliability	Incremental radar cost	Total radar cost
0.01	0.96	$100,000	$1,100,000
0.01	0.97	200,000	1,300,000
0.01	0.98	300,000	1,600,000
0.01	0.99	500,000	2,100,000

TABLE 16.2 Costs of upgrading the missile subsystem

Incremental reliability	Final reliability	Incremental missile cost	Total missile cost
0.02	0.92	$ 200,000	$2,200,000
0.02	0.94	300,000	2,500,000
0.02	0.96	600,000	3,100,000
0.02	0.98	900,000	4,000,000
0.01	0.99	1,200,000	5,200,000

TABLE 16.3 Costs of achieving a system reliability of 0.95

Case	R_1	R_2	R_s	C_1	C_2	C_s
A	0.96 × 0.99	0.9504	$1,100,000	$5,200,000	$6,300,000
B	0.97 × 0.98	0.9506	1,300,000	4,000,000	5,300,000
C	0.99 × 0.96	0.9504	2,100,000	3,100,000	5,200,000

3. If possible, either (1) or (2) could be utilized with the redundant subsystems used sequentially, and a slight overall increase in system reliability would result if the switching were highly reliable.

Other potential advantages of redundancy could also be considered in combination with product improvement. Availability, which will be considered next, can add further possibilities. Final choices will be affected by mission and system considerations. Thus, the need for adequate study of the preliminary considerations is again confirmed.

16.4 OTHER CRITERIA OF EFFECTIVENESS

Because our principal criterion in trade-off analysis is that of system effectiveness, we must consider other measures of effectiveness. *Availability* is one of these. It may be defined as the ratio of the time that the system is usable (uptime) to the total amount of time that it is or may be needed. The total time is the sum of the usable time and the downtime for maintenance/repair, as shown in Eq. (16.1).

$$\text{Availability} = \frac{\text{uptime}}{\text{uptime} + \text{downtime}} \qquad (16.1)$$

where the downtime is the product of the number of maintenance/repair actions requiring shutdown and the average time for each action. The average time is usually called the *mean time to repair* (MTTR), and it includes both scheduled and unscheduled downtimes. Availability can also be expressed as the ratio of the mean time between failures (MTBF) divided by the sum of the MTBF plus the MTTR:

$$\text{Availability} = \frac{\text{MTBF}}{\text{MTBF} + \text{MTTR}} \qquad (16.2)$$

A trade-off can be developed between the MTBF, which is a measure of reliability, and the MTTR, which is a measure of maintainability. As the MTBF (and the reliability) increases, the MTTR can also increase; if the MTBF decreases (and the reliability goes down), the average time to accomplish a repair must also go down in order for the same availability to be retained. This is shown by Eq. (16.2). Table 16.4 shows some MTBF/MTTR pairs, each of which results in an availability of 99 percent. The time units can be any appropriate period, e.g., hours, days, etc.

Sometimes it is not necessary that a system be as reliable as possible but rather that it be usable the maximum possible percent of the time. At first

TABLE 16.4 Some MTBF/MTTR pairs which provide 99 percent availability

MTBF	MTTR	Availability
2,500	25.25	0.99
2,000	20.20	0.99
1,500	15.15	0.99
1,000	10.10	0.99
500	5.05	0.99
100	1.01	0.99

this statement may appear to be contradictory, but reflection will reveal that ease of maintenance, repair, or replacement might compensate for reliability. If we interpret the reliability definition of probability of adequate perform- ance for a given period of time to mean *without shutdown* for maintenance/ repair, then we can evaluate as an alternative the total percent of time that a system will be operable when it is required. An automobile is serviced at periodic intervals to ensure that it will be in good operating condition when it is needed. If it were operated for extended time periods without servicing, its performance would be less than desired and probably less than adequate. With servicing, the mean time between maintenance/repair actions is reduced, but the availability of adequate performance is increased.

It is often possible to design a system so that it can be serviced or repaired easily. We have seen an example when reliability and availability were both increased (Fig. 16.4). In other cases, one might be reduced in favor of the other. A trade-off might develop whereby ease of maintenance/repair is enhanced at the "expense" of reliability. A redundant configuration might be designed for module replacement rather than for part or component replacement. Such a design can lead to more frequent downtimes but they will be of shorter duration and may result in greater availability. If both configurations shown in Figs. 16.3 and 16.4 require shutdown for mainte- nance/repair action, configuration 16.4 would be less reliable than configura- tion 16.3 but it could well have a higher availability factor if an entire leg could be easily removed and replaced.

When switching requires system shutdown and active parallel redundancy does not, an active system will usually have greater availability than a sequentially redundant one, regardless of whether or not repair or replace- ment is accomplished. Without repair, the sequential system is more reliable, and it may be more reliable when maintenance/repair actions are infrequent. When there is frequent maintenance/repair activity, the active parallel system can have equal or greater reliability as well as having greater availability. Each case requires individual evaluation.

Systems which are required to operate for extended time periods may have very low reliabilities, even approaching zero, and yet still have very high availabilities. Suppose that the failure rate of a subsystem is 0.01 per hour and that the mission time is 10,000 hr. Further assume that two subsystems are in sequential configuration, that either can be maintained or repaired without system shutdown, that switching reliability is perfect, and that the average maintenance/repair time is 5 hr. The system operates with one subsystem until it fails, at which time the second subsystem is automatically switched in without delay, and repair of the failed system is initiated. A failure rate of 0.01 per hour results in an MTBF of 100 hr and an average of 100 subsystem failures in 10,000 hr. Since the average repair time is 5 hr, the system will be operating for 100 periods of 5 hr or a total of 500 hr with no spare. The reliability of one subsystem for this time period is $e^{-0.01 \times 500} = e^{-5.00}$ which is 0.0067. The system reliability for 10,000 hr is thus less than 1 percent.

However, the availability is very high. We know that the system must be inoperative less than 500 hr, which gives an availability of at least 95 percent. The actual availability is very close to unity and is found by considering the number of failures which can be expected in the second subsystem when there is no spare. In 500 hr, an average of only about five failures will occur. Thus the system will be without an operable subsystem only five times during the 10,000-hr period. This is a maximum of only 25 hr since the first system will be repaired in 5 hr. Furthermore, since on an average, the second failure will occur about halfway through the 5-hr repair period, the system will be inoperable approximately only 12 1/2 hr. (There is a very small probability that the first subsystem will fail again in the few hours before the second is repaired, but this probability is too small to have a noticeable effect.) Consequently, the overall availability will be 99.87 percent.

Maintainability has been discussed in conjunction with its effects on redundancy and on availability. We shall now briefly consider types of maintainability and other factors related to this aspect of system effectiveness. Maintenance is usually divided into two types: preventive maintenance whereby actions are taken on a scheduled periodic basis or on the basis of operating age and state of system degradation, and corrective maintenance, i.e., repair, whereby a failed component or system is restored to operating condition.

Preventive maintenance includes such actions as lubrication, replacement of age-dependent parts such as O-rings, replacement of filters, etc. These actions are taken before a failure occurs and usually before a significant change in performance is noted. It is frequently a very necessary aspect of system availability and longevity. In some instances, notably commercial aircraft operation, preventive maintenance is a major cost item and maintenance practices are very thoroughly monitored to ensure maximum safety and longevity with minimum overhaul time. In these cases, maintainability is a very real measure of system effectiveness.

Corrective maintenance, usually called repair, can also be a significant measure of effectiveness. The choice of repairing a redundant component as it fails or waiting until the system fails before making the repair depends on costs (the cost of system shutdown, the cost of a complete failure, the cost of repair crews, and so on). Quantitative measures of these costs must be established, even though many assumptions may have to be made. When the costs are quantized, the technical aspects of the maintenance policy can be separated from the assumptions, and objective decisions can be made from the technical information. Questions arising from the assumptions can then be correctly directed at the ground rules rather than at the technical accuracy.

Based on the relative importance of maintainability, availability, and reliability, maintenance policies can be formulated with respect to the optimum size of the maintenance/repair crew, the type and number of spare parts, and the frequency and schedule for preventive-maintenance activity. (Failures are unscheduled and unless there is a large constant backlog, repair cannot be scheduled in advance.) The correctness of the maintenance policy will depend on how well the operating time, downtime, and maintenance costs are estimated and on the accuracy of the failure and repair rates. Fortunately, this policy may not be too sensitive because usually the optimum maintenance policy is optimum for a wide range of costs and rates, and there can be some error in our estimates without significantly affecting the policy.

Performance is perhaps the most commonly used measure of system effectiveness. Certainly, in many situations it is the deciding factor in the choice among alternatives. In the development phases of most programs, the designers concentrate their efforts on designing a system that will meet operational requirements. Concepts which will not meet performance requirements are discarded regardless of their other merits, and rightly so. However, this approach is sometimes carried far beyond the point where system requirements are met, and higher and higher performance goals are set regardless of the effects on reliability, maintainability, and other aspects of the system. Cost becomes the only limiting factor.

When performance requirements are met, the value of additional capability can be compared with longevity, availability, and other factors to arrive at an optimum combination. Reliability may be improved by adding redundancy, but this may adversely affect performance. When the relative values are established, trade-offs can be compared objectively and rational decisions will result. Maintainability, in particular, may be affected by the choice of designs. A design with greater performance may be chosen at the expense of excessive repair times when failures do occur. If failure is very rare, this may be of little importance, but if frequent maintenance/repair actions are needed, excessive repair times with their associated costs may more than outweigh the performance advantage.

Finally, weight must frequently be considered. Weight is of primary concern in the aerospace industry but it can also affect decisions on consumer products. Certainly, manufacturers of such items as electric irons, razors, etc., are conscious of the weight of their products since sales may be directly related. Trade-off of weight can be considered with any one or more of the above measures of system effectiveness. Frequently, reliability is a direct function of weight, so that a trade of weight for performance adversely affects reliability. In space programs, weight considerations are multiplied many times over in terms of booster requirements per pound of payload, and reliability engineers must be constantly alert for trade-offs that, if implemented, could significantly degrade reliability.

16.5 SUMMARY

This chapter presented some of the important considerations affecting system reliability. The two principal ways of upgrading reliability—redundancy and product improvement—were discussed, and various choices were qualitatively analyzed. The reader should particularly note the comparison of active and sequential redundancy. Many comparisons were indicated based on development, maintenance, downtimes, and other costs. It is recognized that considerable difficulty may be encountered in obtaining quantitative information on which to base some trade-offs, but these data are necessary for objective decisions. It is hoped that the chapter pointed out areas of further investigation and encourages the reader to continue his endeavors in the area of reliability.

ADDITIONAL READINGS

ARINC Research Corporation: "Reliability Engineering," Prentice-Hall, Inc., Englewood Cliffs, N.J., 1964.

Barlow, R. E., and F. Proschan: "Statistical Theory of Reliability," John Wiley & Sons, Inc., New York, 1965.

Bazovsky, I.: "Reliability Theory and Practice," Prentice-Hall, Inc., Englewood Cliffs, N.J., 1961.

Bureau of Naval Weapons, Chief of: "Handbook: Reliability Engineering," U.S. Government Printing Office, 1964.

Calabro, S. R.: "Reliability Principles and Practices," McGraw-Hill Book Company, New York, 1962.

Chorafas, D. N.: "Statistical Processes and Reliability Engineering," D. Van Nostrand Company, Inc., Princeton, N.J., 1960.

Ireson, W. G.: "Reliability Handbook," McGraw-Hill Book Company, New York, 1966.

Shooman, M. L.: "Probabilistic Reliability: An Engineering Approach," McGraw-Hill Book Company, New York, 1968.

Appendix **A**

MATHEMATICAL
DERIVATIONS

A.1 DERIVATION OF EQS. (5.3) AND (5.4)

To minimize $\Sigma(Y_{oi} - bX_i - a)^2$, partial derivatives are taken with respect to a and b and are set equal to zero.

$$\frac{\partial \Sigma(Y_{oi} - bX_i - a)^2}{\partial a} = \Sigma 2(Y_{oi} - bX_i - a)(-1) = 0$$

$$\Sigma(Y_{oi} - bX_i - a) = 0$$

$$\Sigma Y_{oi} - \Sigma bX_i - \Sigma a = 0$$

Since $\Sigma Y_{oi} = n\overline{Y}$ and $\Sigma X_i = n\overline{X}$, this may be written:

$$n\overline{Y} - bn\overline{X} - na = 0$$
$$\overline{Y} - b\overline{X} - a = 0$$
$$a = \overline{Y} - b\overline{X} \tag{5.3}$$

$$\frac{\partial \Sigma(Y_{oi} - bX_i - a)^2}{\partial b} = \Sigma 2(Y_{oi} - bX_i - a)(-X_i) = 0$$

$$\Sigma[X_i(Y_{oi} - bX_i - a)] = 0$$

$$\Sigma(X_i Y_{oi}) - \Sigma bX_i^2 - \Sigma aX_i = 0$$

Since $a = \overline{Y} - b\overline{X}$ from (5.3) and $\Sigma X_i = n\overline{X}$

$$\Sigma(X_i Y_{oi}) - b\Sigma(X_i^2) - (\overline{Y} - b\overline{X})n\overline{X} = 0$$

$$\Sigma(X_i Y_{oi}) - b\Sigma(X_i^2) - n\overline{X}\overline{Y} + bn\overline{X}^2 = 0$$

$$\Sigma(X_i Y_{oi}) - n\overline{X}\overline{Y} = b[\Sigma(X_i^2) - n\overline{X}^2]$$

$$b[\Sigma(X_i^2) - n\overline{X}^2 - n\overline{X}^2 + n\overline{X}^2] = \Sigma(X_i Y_{oi}) - n\overline{X}\overline{Y} - n\overline{X}\overline{Y} + n\overline{X}\overline{Y}$$

$$b[\Sigma(X_i^2) - 2\overline{X}\Sigma X_i + \Sigma\overline{X}^2] = \Sigma(X_i Y_{oi}) - \overline{X}\Sigma Y_{oi} - \overline{Y}\Sigma X_i + \Sigma\overline{X}\overline{Y}$$

$$b\Sigma(X_i^2 - 2\overline{X}X_i + \overline{X}^2) = \Sigma(X_i Y_{oi} - \overline{X}Y_{oi} - \overline{Y}X_i + \overline{X}\overline{Y})$$

$$b\Sigma[(X_i - \overline{X})^2] = \Sigma[(X_i - \overline{X})(Y_{oi} - \overline{Y})]$$

$$b = \frac{\Sigma[(X_i - \overline{X})(Y_{oi} - \overline{Y})]}{\Sigma[(X_i - \overline{X})^2]} \tag{5.4}$$

A.2 PROOF OF EQUIVALENCE OF EQS. (5.6), (5.7a), AND (5.7b)

$$r = \sqrt{1 - \frac{\Sigma(Y - Y_R)^2}{\Sigma(Y - \overline{Y})^2}} \overset{(1)*}{=} \sqrt{1 - \frac{\Sigma(Y - bX - \overline{Y} + b\overline{X})^2}{\Sigma(Y - \overline{Y})^2}}$$

$$\overset{(2)}{=} \sqrt{\frac{\Sigma(Y - \overline{Y})^2 - \Sigma[(Y - \overline{Y}) - b(X - \overline{X})]^2}{\Sigma(Y - \overline{Y})^2}} \overset{(3)}{=} \sqrt{\frac{\Sigma(y^2) - \Sigma(y - bx)^2}{\Sigma(y^2)}}$$

$$\overset{(4)}{=} \sqrt{\frac{\Sigma(y^2) - \Sigma(y^2 - 2bxy + b^2x^2)}{\Sigma(y^2)}} \overset{(5)}{=} \sqrt{\frac{2b\Sigma(xy) - b^2\Sigma(x^2)}{\Sigma(y^2)}}$$

$$\overset{(6)}{=} \sqrt{\frac{2(\Sigma xy/\Sigma x^2)\Sigma xy - (\Sigma xy/\Sigma x^2)^2 \Sigma x^2}{\Sigma y^2}}$$

$$\overset{(7)}{=} \sqrt{\frac{2(\Sigma xy)^2/\Sigma x^2 - (\Sigma xy)^2/\Sigma x^2}{\Sigma y^2}} \overset{(8)}{=} \sqrt{\frac{(\Sigma xy)^2}{\Sigma x^2 \, \Sigma y^2}}$$

$$\overset{(9)}{=} \sqrt{\frac{(\Sigma xy)^2/n^2}{(\Sigma x^2/n)(\Sigma y^2/n)}} \overset{(10)}{=} \frac{\Sigma xy/n}{\sigma_X \sigma_Y} \overset{(11)}{=} \frac{\Sigma[(X - \overline{X})(Y - \overline{Y})]/n}{\sigma_X \sigma_Y}$$

$$\overset{(12)}{=} \frac{\Sigma(XY - \overline{X}Y - X\overline{Y} + \overline{X}\overline{Y})/n}{\sigma_X \sigma_Y} \overset{(13)}{=} \frac{\Sigma XY/n - \Sigma\overline{X}Y/n - \Sigma X\overline{Y}/n + \Sigma\overline{X}\overline{Y}/n}{\sigma_X \sigma_Y}$$

$$\overset{(14)}{=} \frac{\Sigma XY/n - n\overline{X}\overline{Y}/n - n\overline{X}\overline{Y}/n + n\overline{X}\overline{Y}/n}{\sigma_X \sigma_Y}$$

$$\overset{(15)}{=} \frac{\Sigma XY/n - 2\overline{X}\overline{Y} + \overline{X}\overline{Y}}{\sigma_X \sigma_Y} \overset{(16)}{=} \frac{\Sigma XY/n - \overline{X}\overline{Y}}{\sigma_X \sigma_Y}$$

*Numbers in parentheses refer to the following mathematical operations:

1. Substituting $Y_R = bX + \overline{Y} - b\overline{X}$
2. Obtaining common denominator and combining terms
3. Substituting $x = X - \overline{X}$; $y = Y - \overline{Y}$
4. By expansion
5. Taking individual sums and combining terms
6. Substituting value of b from Eq. (5.4)
7. Simplifying
8. Simplifying further
9. dividing by n^2
10. Simplifying and by definition
11. Substituting $x = X - \overline{X}$; $y = Y - \overline{Y}$
12. Expanding
13. Expanding further
14. $\Sigma X = n\overline{X}$; $\Sigma Y = n\overline{Y}$; $\Sigma \overline{X}\overline{Y} = n\overline{X}\overline{Y}$
15. Simplifying and combining
16. Final simplification

A.3 DERIVATION OF EQ. (8.12) AND FURTHER APPLICATION OF L'HOSPITAL'S RULE

In Eq. (8.6), $(1/N_T)(dN_F/dt) = -dR/dt$, the term dN_F/dt is the frequency at which failures occur when the total number of units N_T remains constant. Multiplying both sides of Eq. (8.6) by N_T/N_S, we obtain

$$\frac{1}{N_S} \frac{dN_F}{dt} = -\frac{N_T}{N_S} \frac{dR}{dt}$$

When dN_F/dt is divided by the number of units still operating (i.e., multiplied by $1/N_S$), we obtain the *instantaneous* rate of failure for *one* component which, by definition, is λ. Also, since $R = N_S/N_T$, then $N_T/N_S = 1/R$. The next equation reflects these substitutions.

$$\lambda = -\frac{1}{R} \frac{dR}{dt}$$

or

$$\lambda dt = \frac{-dR}{R}$$

Integrating both sides, we have (since at $t = 0$, $R = 1$)

$$\int_0^t \lambda dt = -\int_1^R \frac{dR}{R} = -[\ln R]_1^R = -\ln R$$

Solving for R yields

$$R = \exp\left(-\int_0^t \lambda dt\right)$$

This equation is a general function and is not limited to the constant-failure-rate case. Having developed the general expression for R, we are now ready to show that the first term of Eq. (8.11) is zero.

$$[tR]_0^\infty = \left[t \exp\left(-\int_0^t \lambda dt\right)\right]_0^\infty = \left[\frac{t}{\exp\left(+\int_0^t \lambda dt\right)}\right]_0^\infty$$

Since λ is always greater than zero, even though it may not be constant, we may take it outside the integral sign without invalidating the proof.

$$\left[\frac{t}{\exp\left(\int_0^t \lambda dt\right)}\right]_0^\infty = \left[\frac{t}{\exp\left(\lambda \int_0^t dt\right)}\right]_0^\infty = \left[\frac{t}{e^{\lambda t}}\right]_0^\infty = \left[\frac{t}{e^{\lambda t}}\right]_\infty - \left[\frac{t}{e^{\lambda t}}\right]_0$$

At $t = 0$, the second part is $-0/e^0 = 0/1 \equiv 0$. At $t = \infty$, the first part is indeterminate but can be evaluated by L'Hospital's rule, which states that the limit of the ratio of two continuous functions is equal to the limit of the ratio of their derivatives:

$$\lim_{t \to \infty} \frac{t}{e^{\lambda t}} = \lim_{t \to \infty} \frac{dt/dt}{de^{\lambda t}/dt} = \lim_{t \to \infty} \frac{1}{\lambda e^{\lambda t}} = 0$$

Since both parts of the previous expression are zero, the difference is zero, and $-[tR]_0^\infty = 0$.

While we are using L'Hospital's rule, we shall show its application to the proof of other equations in Chap. 8, for example, Eq. (8.38). The rule states that if the limit of the ratio of two derivatives is indeterminate, we may take *their* derivatives, and so on as long as the resulting functions remain continuous. Thus,

$$\lim_{t \to \infty} \frac{\lambda t^2}{2} e^{-\lambda t} = \lim_{t \to \infty} \frac{\lambda t^2}{2e^{+\lambda t}} = \lim_{t \to \infty} \frac{d^2(\lambda t^2)/dt^2}{d^2(2e^{\lambda t})dt^2}$$

$$= \lim_{t \to \infty} \frac{2\lambda}{2\lambda^2 e^{\lambda t}} = \lim_{t \to \infty} \frac{1}{\lambda e^{\lambda t}} = 0$$

Similarly, $\lim_{t \to \infty} \dfrac{\lambda^2 t^3}{6} e^{-\lambda t} = 0$, and so on.

A.4 DERIVATION OF EQ. (8.40)

In the derivation of the expression for the mean time to failure, Eq. (8.7) was given as $f(t) = -dR/dt$. This equation applies to all possible failure density functions. Since the rate of change of failure has the same magnitude and the opposite sign as the rate of change of success or reliability ($dQ = -dR$), we may rewrite Eq. (8.7) as

$$f(t) = \frac{dQ}{dt}$$

By integration, $Q = \int f(t)\, dt$. When we integrate the failure density function over all time from time = 0 to time = ∞ we obtain the total probability of failure which is, of course, unity, and which is true for any failure density function.

For the specific case of a constant-failure-rate component ($R = e^{-\lambda t}$):

$$f(t) = \frac{-dR}{dt} = \frac{-de^{-\lambda t}}{dt} = -e^{-\lambda t} d(-\lambda t) = -(-\lambda) e^{-\lambda t} = \lambda e^{-\lambda t}$$

$$\int_0^\infty \lambda e^{-\lambda t}\, dt = \lambda \int_0^\infty e^{-\lambda t}\, dt = \lambda \int_0^\infty e^{-\lambda t} \frac{-\lambda}{-\lambda}\, dt$$

$$= -1 \int_0^\infty e^{-\lambda t} d(-\lambda t) = -\left[e^{-\lambda t} \right]_0^\infty = -(0 - 1) \equiv 1$$

Similarly, when we integrate from time 0 to time t, we obtain the probability of failure up to time t. Since the probability of success plus the probability of failure must equal unity, the remainder, from time t to infinity, is the probability that the component does not fail up to time t which is its reliability.

$$\int_0^\infty f(t)\, dt \equiv \int_0^t f(t)\, dt + \int_t^\infty f(t)\, dt \quad \text{or} \quad 1 \equiv Q + R$$

We can now use the facts that the probability of failure $Q = \int_0^t f(t)\,dt$ and the probability of success $R = \int_t^\infty f(t)\,dt$ to find the reliability of a two-unit sequential system (one unit operating and one on standby). The system will succeed if the first unit operates (does not fail) for the required time t *or* if the first unit fails at some time t_1 prior to t *and* the second unit operates (does not fail) from time t_1 to time t. We can evaluate the total probability as the sum of two terms

$$R_T = R_1 + (Q_1 \times R_2)$$

remembering that the reliability of the second unit, R_2, is for the time period from t_1 to t. This can be written

$$R_T = \int f_1(t)\,dt + \int \left[f_1(t) \times \int f_2(t)\,dt \right] dt_1$$

integrating the appropriate variables over their corresponding time intervals.

The interval of integration for the first term (to obtain the reliability of the first unit) is from time t to ∞ as discussed above and the variable is merely t. The failure density function is $\lambda_1 \exp(-\lambda_1 t)$.

$$R_1 = \int_t^\infty \lambda_1 e^{-\lambda_1 t}\,dt = \lambda_1 \int_t^\infty e^{-\lambda_1 t} \frac{-\lambda_1}{-\lambda_1}\,dt$$

$$= -1 \int_t^\infty e^{-\lambda_1 t}\,d(-\lambda_1 t)$$

$$= -1 \left[e^{-\lambda_1 t} \right]_t^\infty = -1 \left[0 - e^{-\lambda_1 t} \right] = \underline{e^{-\lambda_1 t}}$$

The interval of the left integration in the second term (to obtain the probability of failure of the first unit) is from time $= 0$ to time $= t$, but the variable of integration is now t_1 since unit one fails at time t_1. This is shown by the variable t_1 outside the brackets.

The last interval is somewhat more difficult to determine. The second unit must operate from time t_1 (when the first unit fails) to time t. The interval used for finding its probability of failure is time $= t_1$ to time $= t$, $[Q_2 = \int_{t_1}^{t} f_2(t)\, dt]$. But since the integral from time $=$ zero to ∞ is identically unity, it is convenient to appropriately change the time scale. When we consider the following figure:

and make appropriate changes in the abscissa to reflect that time zero for the second unit is actually located at t_1 we find that the new value of t for the second unit becomes $t - t_1$. The revised scale is shown in the next figure:

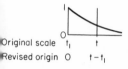

Using the revised scale, we can now redefine the integration interval for failure of the second unit as $\int_{0}^{t-t_1}$. The corresponding interval for success becomes $\int_{t-t_1}^{\infty}$. Since the second unit operates until actual time t, its variable of integration is t.

Integration of the second term follows. Note that, since the probability of failure of the first unit in each time increment is multiplied by the probability of the second unit succeeding, the right-side integration is performed first. This integration is shown in the next six lines.

The failure density function for the second unit, $f_2(t)$, is $\lambda_2 \exp(-\lambda_2 t)$.

$$\int_0^t \left\{ f_1(t) \int_{t-t_1}^{\infty} f_2(t)\, dt \right\} dt_1$$

$$= \int_0^t \left\{ \lambda_1 e^{-\lambda_1 t_1} \int_{t-t_1}^{\infty} \lambda_2 e^{-\lambda_2 t}\, dt \right\} dt_1$$

$$= \int_0^t \left\{ \lambda_1 e^{-\lambda_1 t_1} (-1) \int_{t-t_1}^{\infty} e^{-\lambda_2 t}\, d(-\lambda_2 t) \right\} dt_1$$

$$= \int_0^t \left\{ -\lambda_1 e^{-\lambda_1 t_1} \left[e^{-\lambda_2 t} \right]_{t-t_1}^{\infty} \right\} dt_1$$

$$= \int_0^t \left\{ -\lambda_1 e^{-\lambda_1 t_1} \left[0 - e^{-\lambda_2 (t-t_1)} \right] \right\} dt_1$$

The result of this integration is next combined with the first density function and the expression is simplified.

$$= \int_0^t \left\{ \lambda_1 e^{-\lambda_1 t_1} e^{-\lambda_2 (t-t_1)} \right\} dt_1$$

$$= \lambda_1 \int_0^t \left\{ e^{-\lambda_1 t_1} e^{-\lambda_2 (t-t_1)} \right\} dt_1$$

$$= \lambda_1 \int_0^t e^{-\lambda_1 t_1} e^{-\lambda_2 t} e^{\lambda_2 t_1}\, dt_1$$

$$= \lambda_1 \int_0^t \left\{ e^{-\lambda_1 t_1 + \lambda_2 t_1} e^{-\lambda_2 t} \right\} dt_1$$

The final integration is with respect to the probability of failure of the first unit from time zero to time t. In this integration, the total time t is a constant and t_1 is the variable since the unit fails at time t_1 as stated earlier. Here $\exp(-\lambda_2 t)$ is a constant and may be taken outside the integral sign as part of the coefficient. This last integration and simplification follows.

$$= \lambda_1 e^{-\lambda_2 t} \int_0^t \left\{ e^{-\lambda_1 t_1 + \lambda_2 t_1} \right\} dt_1$$

$$= \lambda_1 e^{-\lambda_2 t} \int_0^t \left\{ e^{-(\lambda_1 - \lambda_2)t_1} \right\} dt_1$$

$$= \lambda_1 e^{-\lambda_2 t} \int_0^t \left\{ \frac{1}{-(\lambda_1 - \lambda_2)} e^{-(\lambda_1 - \lambda_2)t_1} \right\} d - (\lambda_1 - \lambda_2)t_1$$

$$= -\frac{\lambda_1 e^{-\lambda_2 t}}{\lambda_1 - \lambda_2} \int_0^t e^{-(\lambda_1 - \lambda_2)t_1} d - (\lambda_1 - \lambda_2)t_1$$

$$= \frac{-\lambda_1 e^{-\lambda_2 t}}{\lambda_1 - \lambda_2} \left[e^{-(\lambda_1 - \lambda_2)t_1} \right]_0^t = \frac{\lambda_1 e^{-\lambda_2 t}}{\lambda_2 - \lambda_1} \left[e^{-(\lambda_1 - \lambda_2)t} - 1 \right]$$

$$= \frac{\lambda_1 e^{-\lambda_2 t} e^{-(\lambda_1 - \lambda_2)t} - \lambda_1 e^{-\lambda_2 t}}{\lambda_2 - \lambda_1} = \frac{\lambda_1 e^{-\lambda_1 t} - \lambda_1 e^{-\lambda_2 t}}{\lambda_2 - \lambda_1}$$

$$= \frac{\lambda_1}{\lambda_2 - \lambda_1} \left(e^{-\lambda_1 t} - e^{-\lambda_2 t} \right)$$

This term is then added to the probability of the first unit succeeding.

$$R_T = e^{-\lambda_1 t} + \frac{\lambda_1}{\lambda_2 - \lambda_1} \left(e^{-\lambda_1 t} - e^{-\lambda_2 t} \right)$$

A.5 PROOF OF EQ. (8.44)

$$
m = \int_0^\infty \left[\frac{\lambda_2\lambda_3\, e^{-\lambda_1 t}}{(\lambda_2 - \lambda_1)(\lambda_3 - \lambda_1)} + \frac{\lambda_1\lambda_3\, e^{-\lambda_2 t}}{(\lambda_1 - \lambda_2)(\lambda_3 - \lambda_2)} \right.
$$

$$
\left. + \frac{\lambda_1\lambda_2\, e^{-\lambda_1 t}}{(\lambda_1 - \lambda_3)(\lambda_2 - \lambda_3)} \right] dt
$$

$$
= \frac{\lambda_2\lambda_3}{(\lambda_2 - \lambda_1)(\lambda_3 - \lambda_1)\lambda_1} + \frac{\lambda_1\lambda_3}{(\lambda_1 - \lambda_2)(\lambda_3 - \lambda_2)\lambda_2}
$$

$$
+ \frac{\lambda_1\lambda_2}{(\lambda_1 - \lambda_3)(\lambda_2 - \lambda_3)\lambda_3}
$$

In order to simplify the above expression, it is necessary to first add some additional terms whose sum is zero and then to combine all terms. These additional terms are

$$
\frac{(\lambda_1\lambda_3 - \lambda_2\lambda_3 - \lambda_3^2 - \lambda_1\lambda_2 + \lambda_2^2 + \lambda_2\lambda_3)}{(\lambda_3 - \lambda_2)(\lambda_3 - \lambda_1)(\lambda_2 - \lambda_1)}
$$

$$
+ \frac{(\lambda_1\lambda_2 - \lambda_1^2 - \lambda_1\lambda_3 - \lambda_2\lambda_3 + \lambda_1\lambda_3 + \lambda_3^2)}{(\lambda_3 - \lambda_2)(\lambda_3 - \lambda_1)(\lambda_2 - \lambda_1)}
$$

$$
+ \frac{(\lambda_2\lambda_3 - \lambda_1\lambda_2 - \lambda_2^2 - \lambda_1\lambda_3 + \lambda_1^2 + \lambda_1\lambda_2)}{(\lambda_3 - \lambda_2)(\lambda_3 - \lambda_1)(\lambda_2 - \lambda_1)}
$$

$$
= \frac{(\lambda_3 - \lambda_2)(\lambda_1 - \lambda_2 - \lambda_3)}{(\lambda_3 - \lambda_2)(\lambda_3 - \lambda_1)(\lambda_2 - \lambda_1)} + \frac{(\lambda_1 - \lambda_3)(\lambda_2 - \lambda_1 - \lambda_3)}{(\lambda_3 - \lambda_2)(\lambda_1 - \lambda_3)(\lambda_1 - \lambda_2)}
$$

$$
+ \frac{(\lambda_2 - \lambda_1)(\lambda_3 - \lambda_1 - \lambda_2)}{(\lambda_2 - \lambda_3)(\lambda_1 - \lambda_3)(\lambda_2 - \lambda_1)}
$$

$$
= \frac{(\lambda_1 - \lambda_2 - \lambda_3)}{(\lambda_3 - \lambda_1)(\lambda_2 - \lambda_1)} + \frac{(\lambda_2 - \lambda_1 - \lambda_3)}{(\lambda_3 - \lambda_2)(\lambda_1 - \lambda_2)} + \frac{(\lambda_3 - \lambda_1 - \lambda_2)}{(\lambda_2 - \lambda_3)(\lambda_1 - \lambda_3)}
$$

$$= \frac{\lambda_1(\lambda_1 - \lambda_2 - \lambda_3)}{\lambda_1(\lambda_3 - \lambda_1)(\lambda_2 - \lambda_1)} + \frac{\lambda_2(\lambda_2 - \lambda_1 - \lambda_3)}{\lambda_2(\lambda_3 - \lambda_2)(\lambda_1 - \lambda_2)}$$

$$+ \frac{\lambda_3(\lambda_3 - \lambda_1 - \lambda_2)}{\lambda_3(\lambda_2 - \lambda_3)(\lambda_1 - \lambda_3)}$$

$$= \frac{\lambda_1^2 - \lambda_1\lambda_2 - \lambda_1\lambda_3}{\lambda_1(\lambda_2 - \lambda_1)(\lambda_3 - \lambda_1)} + \frac{\lambda_2^2 - \lambda_1\lambda_2 - \lambda_2\lambda_3}{\lambda_2(\lambda_1 - \lambda_2)(\lambda_3 - \lambda_2)}$$

$$+ \frac{\lambda_3^2 - \lambda_1\lambda_3 - \lambda_2\lambda_3}{\lambda_3(\lambda_2 - \lambda_3)(\lambda_1 - \lambda_3)}$$

We now add in the original terms and combine.

$$= \frac{\lambda_2\lambda_3}{\lambda_1(\lambda_2 - \lambda_1)(\lambda_3 - \lambda_1)} + \frac{\lambda_1^2 - \lambda_1\lambda_2 - \lambda_1\lambda_3}{\lambda_1(\lambda_2 - \lambda_1)(\lambda_3 - \lambda_1)}$$

$$+ \frac{\lambda_1\lambda_3}{\lambda_2(\lambda_1 - \lambda_2)(\lambda_3 - \lambda_2)} + \frac{\lambda_2^2 - \lambda_1\lambda_2 - \lambda_2\lambda_3}{\lambda_2(\lambda_1 - \lambda_2)(\lambda_3 - \lambda_2)}$$

$$+ \frac{\lambda_1\lambda_2}{\lambda_3(\lambda_1 - \lambda_3)(\lambda_2 - \lambda_3)} + \frac{\lambda_3^2 - \lambda_1\lambda_3 - \lambda_2\lambda_3}{\lambda_3(\lambda_1 - \lambda_3)(\lambda_2 - \lambda_3)}$$

$$= \frac{\lambda_2\lambda_3 - \lambda_1\lambda_2 - \lambda_1\lambda_3 + \lambda_1^2}{\lambda_1(\lambda_2 - \lambda_1)(\lambda_3 - \lambda_1)} + \frac{\lambda_1\lambda_3 - \lambda_1\lambda_2 - \lambda_2\lambda_3 + \lambda_2^2}{\lambda_2(\lambda_1 - \lambda_2)(\lambda_3 - \lambda_2)}$$

$$+ \frac{\lambda_1\lambda_2 - \lambda_1\lambda_3 - \lambda_2\lambda_3 + \lambda_3^2}{\lambda_3(\lambda_1 - \lambda_3)(\lambda_2 - \lambda_3)}$$

$$= \frac{(\lambda_2 - \lambda_1)(\lambda_3 - \lambda_1)}{\lambda_1(\lambda_2 - \lambda_1)(\lambda_3 - \lambda_1)} + \frac{(\lambda_1 - \lambda_2)(\lambda_3 - \lambda_2)}{\lambda_2(\lambda_1 - \lambda_2)(\lambda_3 - \lambda_2)}$$

$$+ \frac{(\lambda_1 - \lambda_3)(\lambda_2 - \lambda_3)}{\lambda_3(\lambda_1 - \lambda_3)(\lambda_2 - \lambda_3)} = \frac{1}{\lambda_1} + \frac{1}{\lambda_2} + \frac{1}{\lambda_3}$$

A.6   TRANSFORMATION OF EQUATION FOR THREE-UNIT SEQUENTIAL SYSTEMS TO FORM SHOWING RELIABILITY CONTRIBUTED BY EACH UNIT AS IN EQ. (8.40)

$$
\left(\frac{\lambda_2}{\lambda_2 - \lambda_1} \frac{\lambda_3}{\lambda_3 - \lambda_1} e^{-\lambda_1 t} \right) + \left(\frac{\lambda_1}{\lambda_1 - \lambda_2} \frac{\lambda_3}{\lambda_3 - \lambda_2} e^{-\lambda_2 t} \right)
$$

$$
+ \left(\frac{\lambda_1}{\lambda_1 - \lambda_3} \frac{\lambda_2}{\lambda_2 - \lambda_3} e^{-\lambda_3 t} \right)
$$

$$
= \frac{\lambda_2 \lambda_3 (\lambda_3 - \lambda_2)}{(\lambda_2 - \lambda_1)(\lambda_3 - \lambda_1)(\lambda_3 - \lambda_2)} e^{-\lambda_1 t}
$$

$$
+ \frac{\lambda_1 \lambda_3 (\lambda_1 - \lambda_3)}{(\lambda_1 - \lambda_2)(\lambda_3 - \lambda_2)(\lambda_1 - \lambda_3)} e^{-\lambda_2 t}
$$

$$
+ \frac{\lambda_1 \lambda_2 (\lambda_2 - \lambda_1)}{(\lambda_1 - \lambda_3)(\lambda_2 - \lambda_3)(\lambda_2 - \lambda_1)} e^{-\lambda_3 t}
$$

(8.41)

$$
= \left[\frac{\lambda_1^2\lambda_3 - \lambda_1^2\lambda_2 - \lambda_1\lambda_2\lambda_3 + \lambda_1\lambda_2^2 - \lambda_1\lambda_3^2 + \lambda_1\lambda_2\lambda_3 + \lambda_2\lambda_3(\lambda_3 - \lambda_2)}{(\lambda_2 - \lambda_1)(\lambda_3 - \lambda_1)(\lambda_3 - \lambda_2)} \right.
$$

$$
\left. \frac{-\lambda_1^2\lambda_3 + \lambda_1\lambda_3^2 - \lambda_1\lambda_2^2 + \lambda_1^2\lambda_2}{(\lambda_2 - \lambda_1)(\lambda_3 - \lambda_1)(\lambda_3 - \lambda_2)} \right] e^{-\lambda_1 t}
$$

$$
+ \frac{\lambda_1\lambda_3(\lambda_1 - \lambda_3)}{(\lambda_1 - \lambda_2)(\lambda_3 - \lambda_2)(\lambda_1 - \lambda_3)} e^{-\lambda_2 t}
$$

$$
+ \frac{\lambda_1\lambda_2(\lambda_2 - \lambda_1)}{(\lambda_1 - \lambda_3)(\lambda_2 - \lambda_3)(\lambda_2 - \lambda_1)} e^{-\lambda_3 t}
$$

$$
= \left[\frac{(\lambda_1^2 - \lambda_1\lambda_2 - \lambda_1\lambda_3 + \lambda_2\lambda_3)(\lambda_3 - \lambda_2)}{(\lambda_2 - \lambda_1)(\lambda_3 - \lambda_1)(\lambda_3 - \lambda_2)} \right.
$$

$$
\left. \frac{-\lambda_1\lambda_3(\lambda_1 - \lambda_3) - \lambda_1\lambda_2(\lambda_2 - \lambda_1)}{(\lambda_2 - \lambda_1)(\lambda_3 - \lambda_1)(\lambda_3 - \lambda_2)} \right] e^{-\lambda_1 t}
$$

$$
+ \frac{\lambda_1\lambda_3(\lambda_1 - \lambda_3)}{(\lambda_1 - \lambda_2)(\lambda_3 - \lambda_2)(\lambda_1 - \lambda_3)} e^{-\lambda_2 t}
$$

$$
+ \frac{\lambda_1\lambda_2(\lambda_2 - \lambda_1)}{(\lambda_1 - \lambda_3)(\lambda_2 - \lambda_3)(\lambda_2 - \lambda_1)} e^{-\lambda_3 t}
$$

$$
= \frac{(\lambda_1^2 - \lambda_1\lambda_2 - \lambda_1\lambda_3 + \lambda_2\lambda_3)(\lambda_3 - \lambda_2)}{(\lambda_2 - \lambda_1)(\lambda_3 - \lambda_1)(\lambda_3 - \lambda_2)} e^{-\lambda_1 t}
$$

$$
- \frac{\lambda_1\lambda_3(\lambda_1 - \lambda_3)}{(\lambda_2 - \lambda_1)(\lambda_3 - \lambda_2)(\lambda_3 - \lambda_1)} e^{-\lambda_1 t}
$$

$$
- \frac{\lambda_1\lambda_2(\lambda_2 - \lambda_1)}{(\lambda_3 - \lambda_1)(\lambda_3 - \lambda_2)(\lambda_2 - \lambda_1)} e^{-\lambda_1 t}
$$

$$
+ \frac{\lambda_1\lambda_3(\lambda_1 - \lambda_3)}{(\lambda_2 - \lambda_1)(\lambda_3 - \lambda_2)(\lambda_3 - \lambda_1)} e^{-\lambda_2 t}
$$

$$+ \frac{\lambda_1 \lambda_2 (\lambda_2 - \lambda_1)}{(\lambda_3 - \lambda_1)(\lambda_3 - \lambda_2)(\lambda_2 - \lambda_1)} e^{-\lambda_3 t}$$

$$= \frac{(\lambda_1 - \lambda_2)(\lambda_1 - \lambda_3)(\lambda_3 - \lambda_2)}{(\lambda_2 - \lambda_1)(\lambda_3 - \lambda_1)(\lambda_3 - \lambda_2)} e^{-\lambda_1 t}$$

$$+ \frac{\lambda_1 \lambda_3 (\lambda_1 - \lambda_3)}{(\lambda_2 - \lambda_1)(\lambda_3 - \lambda_2)(\lambda_3 - \lambda_1)} \left(e^{-\lambda_2 t} - e^{-\lambda_1 t} \right)$$

$$+ \frac{\lambda_1 \lambda_2 (\lambda_2 - \lambda_1)}{(\lambda_3 - \lambda_1)(\lambda_3 - \lambda_2)(\lambda_2 - \lambda_1)} \left(e^{-\lambda_3 t} - e^{-\lambda_1 t} \right)$$

$$= e^{-\lambda_1 t} + \frac{\lambda_1 \lambda_3}{(\lambda_1 - \lambda_2)(\lambda_3 - \lambda_2)} \left(e^{-\lambda_2 t} - e^{-\lambda_1 t} \right)$$

$$+ \frac{\lambda_1 \lambda_2}{(\lambda_3 - \lambda_1)(\lambda_3 - \lambda_2)} \left(e^{-\lambda_3 t} - e^{-\lambda_1 t} \right)$$

$$= e^{-\lambda_1 t} + \frac{\lambda_3}{\lambda_3 - \lambda_2} \frac{\lambda_1}{\lambda_1 - \lambda_2} \left(e^{-\lambda_2 t} - e^{-\lambda_1 t} \right)$$

$$+ \frac{\lambda_1 \lambda_2}{(\lambda_3 - \lambda_1)(\lambda_3 - \lambda_2)} \left(e^{-\lambda_3 t} - e^{-\lambda_1 t} \right)$$

$$= e^{-\lambda_1 t} + \frac{\lambda_3 - \lambda_2 + \lambda_2}{\lambda_3 - \lambda_2} \frac{\lambda_1}{\lambda_1 - \lambda_2} \left(e^{-\lambda_2 t} - e^{-\lambda_1 t} \right)$$

$$+ \frac{\lambda_1 \lambda_2}{(\lambda_3 - \lambda_1)(\lambda_3 - \lambda_2)} \left(e^{-\lambda_3 t} - e^{-\lambda_1 t} \right)$$

$$= e^{-\lambda_1 t} + \left(\frac{\lambda_3 - \lambda_2}{\lambda_3 - \lambda_2} + \frac{\lambda_2}{\lambda_3 - \lambda_2} \right) \frac{\lambda_1}{\lambda_1 - \lambda_2} \left(e^{-\lambda_2 t} - e^{-\lambda_1 t} \right)$$

$$+ \frac{\lambda_1 \lambda_2}{(\lambda_3 - \lambda_1)(\lambda_3 - \lambda_2)} \left(e^{-\lambda_3 t} - e^{-\lambda_1 t} \right)$$

$$= e^{-\lambda_1 t} + \frac{\lambda_1}{\lambda_1 - \lambda_2}\left(e^{-\lambda_2 t} - e^{-\lambda_1 t}\right)$$

$$+ \frac{\lambda_1 \lambda_2}{(\lambda_3 - \lambda_2)(\lambda_1 - \lambda_2)}\left(e^{-\lambda_2 t} - e^{-\lambda_1 t}\right)$$

$$+ \frac{\lambda_1 \lambda_2}{(\lambda_3 - \lambda_1)(\lambda_3 - \lambda_2)}\left(e^{-\lambda_3 t} - e^{-\lambda_1 t}\right)$$

$$= e^{-\lambda_1 t} + \frac{\lambda_1}{\lambda_1 - \lambda_2}\left(e^{-\lambda_2 t} - e^{-\lambda_1 t}\right)$$

$$+ \frac{\lambda_1 \lambda_2}{(\lambda_3 - \lambda_2)(\lambda_3 - \lambda_1)}\left(e^{-\lambda_3 t} - e^{-\lambda_1 t}\right)$$

$$- \frac{\lambda_1 \lambda_2}{(\lambda_3 - \lambda_2)(\lambda_2 - \lambda_1)}\left(e^{-\lambda_2 t} - e^{-\lambda_1 t}\right)$$

$$= \underbrace{e^{-\lambda_1 t}}_{\substack{\text{Reliability} \\ \text{contributed by} \\ \text{first unit}}} + \underbrace{\frac{\lambda_1}{\lambda_1 - \lambda_2}\left(e^{-\lambda_2 t} - e^{-\lambda_1 t}\right)}_{\substack{\text{Reliability} \\ \text{contributed by} \\ \text{second unit}}}$$

$$+ \underbrace{\frac{\lambda_1 \lambda_2}{\lambda_2 - \lambda_3}\left[\frac{\left(e^{-\lambda_3 t} - e^{-\lambda_1 t}\right)}{\lambda_1 - \lambda_3} - \frac{\left(e^{-\lambda_2 t} - e^{-\lambda_1 t}\right)}{\lambda_1 - \lambda_2}\right]}_{\substack{\text{Reliability} \\ \text{contributed by} \\ \text{third unit}}}$$

Appendix **B**

STATISTICAL TABLES

TABLE B.1   Values of the negative exponential $R = e^{-F}$ where $F = \lambda t$

F	0.0000*	0.0001	0.0002	0.0003	0.0004	0.0005	0.0006	0.0007	0.0008	0.0009
0.000	1.00000	0.99990	0.99980	0.99970	0.99960	0.99950	0.99940	0.99930	0.99920	0.99910
0.001	0.99900	0.99890	0.99880	0.99870	0.99860	0.99850	0.99840	0.99830	0.99820	0.99810
0.002	0.99800	0.99790	0.99780	0.99770	0.99760	0.99750	0.99740	0.99730	0.99720	0.99710
0.003	0.99700	0.99690	0.99681	0.99671	0.99661	0.99651	0.99641	0.99631	0.99621	0.99611
0.004	0.99600	0.99591	0.99581	0.99571	0.99561	0.99551	0.99541	0.99531	0.99521	0.99511
0.005	0.99501	0.99491	0.99481	0.99471	0.99461	0.99452	0.99442	0.99432	0.99422	0.99412
0.006	0.99402	0.99392	0.99382	0.99372	0.99362	0.99352	0.99342	0.99332	0.99322	0.99312
0.007	0.99302	0.99293	0.99283	0.99273	0.99263	0.99253	0.99243	0.99233	0.99223	0.99213
0.008	0.99203	0.99193	0.99183	0.99173	0.99164	0.99154	0.99144	0.99134	0.99124	0.99114
0.009	0.99104	0.99094	0.99084	0.99074	0.99064	0.99054	0.99045	0.99035	0.99025	0.99015
0.010	0.99005	0.98995	0.98985	0.98975	0.98965	0.98955	0.98946	0.98936	0.98926	0.98916
0.011	0.98906	0.98896	0.98886	0.98876	0.98866	0.98857	0.98847	0.98837	0.98827	0.98817
0.012	0.98807	0.98797	0.98787	0.98778	0.98768	0.98758	0.98748	0.98738	0.98728	0.98718
0.013	0.98708	0.98699	0.98689	0.98679	0.98669	0.98659	0.98649	0.98639	0.98629	0.98620
0.014	0.98610	0.98600	0.98590	0.98580	0.98570	0.98560	0.98551	0.98541	0.98531	0.98521
0.015	0.98511	0.98501	0.98491	0.98482	0.98472	0.98462	0.98452	0.98442	0.98432	0.98423
0.016	0.98413	0.98403	0.98393	0.98383	0.98373	0.98364	0.98354	0.98344	0.98334	0.98324
0.017	0.98314	0.98305	0.98295	0.98285	0.98275	0.98265	0.98255	0.98246	0.98236	0.98226
0.018	0.98216	0.98206	0.98196	0.98187	0.98177	0.98167	0.98157	0.98147	0.98138	0.98128
0.019	0.98118	0.98108	0.98098	0.98089	0.98079	0.98069	0.98059	0.98049	0.98039	0.98030
0.020	0.98020	0.98010	0.98000	0.97990	0.97981	0.97971	0.97961	0.97951	0.97941	0.97932
0.021	0.97922	0.97912	0.97902	0.97893	0.97883	0.97873	0.97863	0.97853	0.97844	0.97834
0.022	0.97824	0.97814	0.97804	0.97795	0.97785	0.97775	0.97765	0.97756	0.97746	0.97736
0.023	0.97726	0.97716	0.97707	0.97697	0.97687	0.97677	0.97668	0.97658	0.97648	0.97638
0.024	0.97629	0.97619	0.97609	0.97599	0.97590	0.97580	0.97570	0.97560	0.97550	0.97541
0.025	0.97531	0.97521	0.97511	0.97502	0.97492	0.97482	0.97472	0.97463	0.97453	0.97443
0.026	0.97434	0.97424	0.97414	0.97404	0.97395	0.97385	0.97375	0.97365	0.97356	0.97346
0.027	0.97336	0.97326	0.97317	0.97307	0.97297	0.97287	0.97278	0.97268	0.97258	0.97249
0.028	0.97239	0.97229	0.97219	0.97210	0.97200	0.97190	0.97181	0.97171	0.97161	0.97151
0.029	0.97142	0.97132	0.97122	0.97113	0.97103	0.97093	0.97083	0.97074	0.97064	0.97054
0.030	0.97045	0.97035	0.97025	0.97015	0.97006	0.96996	0.96986	0.96977	0.96967	0.96957
0.031	0.96948	0.96938	0.96928	0.96918	0.96909	0.96899	0.96889	0.96880	0.96870	0.96860
0.032	0.96851	0.96841	0.96831	0.96822	0.96812	0.96802	0.96793	0.96783	0.96773	0.96764
0.033	0.96754	0.96744	0.96735	0.96725	0.96715	0.96705	0.96696	0.96686	0.96676	0.96667
0.034	0.96657	0.96647	0.96638	0.96628	0.96618	0.96609	0.96599	0.96590	0.96580	0.96570
0.035	0.96561	0.96551	0.96541	0.96532	0.96522	0.96512	0.96503	0.96493	0.96483	0.96474
0.036	0.96464	0.96454	0.96445	0.96435	0.96425	0.96416	0.96406	0.96397	0.96387	0.96377
0.037	0.96368	0.96358	0.96348	0.96339	0.96329	0.96319	0.96310	0.96300	0.96291	0.96281
0.038	0.96271	0.96262	0.96252	0.96242	0.96233	0.96223	0.96214	0.96204	0.96194	0.96185
0.039	0.96175	0.96165	0.96156	0.96146	0.96137	0.96127	0.96117	0.96108	0.96098	0.96089
0.040	0.96079	0.96069	0.96060	0.96050	0.96041	0.96031	0.96021	0.96012	0.96002	0.95993
0.041	0.95983	0.95973	0.95964	0.95954	0.95945	0.95935	0.95925	0.95916	0.95906	0.95897
0.042	0.95887	0.95877	0.95868	0.95858	0.95849	0.95839	0.95829	0.95820	0.95810	0.95801
0.043	0.95791	0.95782	0.95772	0.95762	0.95753	0.95743	0.95734	0.95724	0.95715	0.95705
0.044	0.95695	0.95686	0.95676	0.95667	0.95657	0.95648	0.95638	0.95628	0.95619	0.95609
0.045	0.95600	0.95590	0.95581	0.95571	0.95562	0.95552	0.95542	0.95533	0.95523	0.95514
0.046	0.95504	0.95495	0.95485	0.95476	0.95466	0.95456	0.95447	0.95437	0.95428	0.95418
0.047	0.95409	0.95399	0.95390	0.95380	0.95371	0.95361	0.95352	0.95342	0.95332	0.95323
0.048	0.95313	0.95304	0.95294	0.95285	0.95275	0.95266	0.95256	0.95247	0.95237	0.95228
0.049	0.95218	0.95209	0.95199	0.95190	0.95180	0.95171	0.95161	0.95151	0.95142	0.95132

*Column headings provide fourth decimal place for value of F.

TABLE B.1   Values of the negative exponential $R = e^{-F}$ where $F = \lambda t$ (Continued)

F	0.0000*	0.0001	0.0002	0.0003	0.0004	0.0005	0.0006	0.0007	0.0008	0.0009
0.050	0.95123	0.95113	0.95104	0.95094	0.95085	0.95075	0.95066	0.95056	0.95047	0.95037
0.051	0.95028	0.95018	0.95009	0.94999	0.94990	0.94980	0.94971	0.94961	0.94952	0.94942
0.052	0.94933	0.94923	0.94914	0.94904	0.94895	0.94885	0.94876	0.94866	0.94857	0.94847
0.053	0.94838	0.94829	0.94819	0.94810	0.94800	0.94791	0.94781	0.94772	0.94762	0.94753
0.054	0.94743	0.94734	0.94724	0.94715	0.94705	0.94696	0.94686	0.94677	0.94667	0.94658
0.055	0.94649	0.94639	0.94630	0.94620	0.94611	0.94601	0.94592	0.94582	0.94573	0.94563
0.056	0.94554	0.94544	0.94535	0.94526	0.94516	0.94507	0.94497	0.94488	0.94478	0.94469
0.057	0.94459	0.94450	0.94441	0.94431	0.94422	0.94412	0.94403	0.94393	0.94384	0.94374
0.058	0.94365	0.94356	0.94346	0.94337	0.94327	0.94318	0.94308	0.94299	0.94290	0.94280
0.059	0.94271	0.94261	0.94252	0.94242	0.94233	0.94224	0.94214	0.94205	0.94195	0.94186
0.060	0.94176	0.94167	0.94158	0.94148	0.94139	0.94129	0.94120	0.94111	0.94101	0.94092
0.061	0.94082	0.94073	0.94064	0.94054	0.94045	0.94035	0.94026	0.94016	0.94007	0.93998
0.062	0.93988	0.93979	0.93969	0.93960	0.93951	0.93941	0.93932	0.93923	0.93913	0.93904
0.063	0.93894	0.93885	0.93876	0.93866	0.93857	0.93847	0.93838	0.93829	0.93819	0.93810
0.064	0.03800	0.93791	0.93782	0.93772	0.93763	0.93754	0.93744	0.93735	0.93725	0.93716
0.065	0.93707	0.93697	0.93688	0.93679	0.93669	0.93660	0.93651	0.93641	0.93632	0.93622
0.066	0.93613	0.93604	0.93594	0.93585	0.93576	0.93566	0.93557	0.93548	0.93538	0.93529
0.067	0.93520	0.93510	0.93501	0.93491	0.93482	0.93473	0.93463	0.93454	0.93445	0.93435
0.068	0.93426	0.93417	0.93407	0.93398	0.93389	0.93379	0.93370	0.93361	0.93351	0.93342
0.069	0.93333	0.93323	0.93314	0.93305	0.93295	0.93286	0.93277	0.93267	0.93258	0.93249
0.070	0.93239	0.93230	0.93221	0.93211	0.93202	0.93193	0.93183	0.93174	0.93165	0.93156
0.071	0.93146	0.93137	0.93128	0.93118	0.93109	0.93100	0.93090	0.93081	0.93072	0.93062
0.072	0.93053	0.93044	0.93034	0.93025	0.93016	0.93007	0.92997	0.92988	0.92979	0.92969
0.073	0.92960	0.92951	0.92941	0.92932	0.92923	0.92914	0.92904	0.92895	0.92886	0.92876
0.074	0.92867	0.92858	0.92849	0.92839	0.92830	0.92821	0.92811	0.92802	0.92793	0.92784
0.075	0.92774	0.92765	0.92756	0.92747	0.92737	0.92728	0.92719	0.92709	0.92700	0.92691
0.076	0.92682	0.92672	0.92663	0.92654	0.92645	0.92635	0.92626	0.92617	0.92608	0.92598
0.077	0.92589	0.92580	0.92570	0.92561	0.92552	0.92543	0.92533	0.92524	0.92515	0.92506
0.078	0.92496	0.92487	0.92478	0.92469	0.92459	0.92450	0.92441	0.92432	0.92422	0.92413
0.079	0.92404	0.92395	0.92386	0.92376	0.92367	0.92358	0.92349	0.92339	0.92330	0.92321
0.080	0.92312	0.92302	0.92293	0.92284	0.92275	0.92265	0.92256	0.92247	0.92238	0.92229
0.081	0.92219	0.92210	0.92201	0.92192	0.92182	0.92173	0.92164	0.92155	0.92146	0.92136
0.082	0.92127	0.92118	0.92109	0.92100	0.92090	0.92081	0.92072	0.92063	0.92054	0.92044
0.083	0.92035	0.92026	0.92017	0.92008	0.91998	0.91989	0.91980	0.91971	0.91962	0.91952
0.084	0.91943	0.91934	0.91925	0.91916	0.91906	0.91897	0.91888	0.91879	0.91870	0.91860
0.085	0.91851	0.91842	0.91833	0.91824	0.91814	0.91805	0.91796	0.91787	0.91778	0.91769
0.086	0.91759	0.91750	0.91741	0.91732	0.91723	0.91714	0.91704	0.91695	0.91686	0.91677
0.087	0.91668	0.91659	0.91649	0.91640	0.91631	0.91622	0.91613	0.91604	0.91594	0.91585
0.088	0.91576	0.91567	0.91558	0.91549	0.91539	0.91530	0.91521	0.91512	0.91503	0.91494
0.089	0.91485	0.91475	0.91466	0.91457	0.91448	0.91439	0.91430	0.91421	0.91411	0.91402
0.090	0.91393	0.91384	0.91375	0.91366	0.91357	0.91347	0.91338	0.91329	0.91320	0.91311
0.091	0.91302	0.91293	0.91284	0.91274	0.91265	0.91256	0.91247	0.91238	0.91229	0.91220
0.092	0.91211	0.91201	0.91192	0.91183	0.91174	0.91165	0.91156	0.91147	0.91138	0.91128
0.093	0.91119	0.91110	0.91101	0.91092	0.91083	0.91074	0.91065	0.91056	0.91046	0.91037
0.094	0.91028	0.91019	0.91010	0.91001	0.90992	0.90983	0.90974	0.90965	0.90955	0.90946
0.095	0.90937	0.90928	0.90919	0.90910	0.90901	0.90892	0.90883	0.90874	0.90865	0.90855
0.096	0.90846	0.90837	0.90828	0.90819	0.90810	0.90801	0.00792	0.90783	0.90774	0.90765
0.097	0.90756	0.90747	0.90737	0.90728	0.90719	0.90710	0.90701	0.90692	0.90683	0.90674
0.098	0.90665	0.90656	0.90647	0.90638	0.90629	0.90620	0.90611	0.90601	0.90592	0.90583
0.099	0.90574	0.90565	0.90556	0.90547	0.90538	0.90529	0.90520	0.90511	0.90502	0.90493

*Column headings provide fourth decimal place for value of F.

F	0.000*	0.001	0.002	0.003	0.004	0.005	0.006	0.007	0.008	0.009
0.10	0.90484	0.90393	0.90303	0.90213	0.90123	0.90032	0.89942	0.89853	0.89763	0.89673
0.11	0.89583	0.89494	0.89404	0.89315	0.89226	0.89137	0.89048	0.88959	0.88870	0.88781
0.12	0.88692	0.88603	0.88515	0.88426	0.88338	0.88250	0.88161	0.88073	0.87985	0.87897
0.13	0.87810	0.87722	0.87634	0.87547	0.87459	0.87372	0.87284	0.87197	0.87110	0.87023
0.14	0.86936	0.86849	0.86762	0.86675	0.86589	0.86502	0.86416	0.86329	0.86243	0.86157
0.15	0.86071	0.85985	0.85899	0.85813	0.85727	0.85642	0.85556	0.85470	0.85385	0.85300
0.16	0.85214	0.85129	0.85044	0.84959	0.84874	0.84789	0.84705	0.84620	0.84535	0.84451
0.17	0.84366	0.84282	0.84198	0.84114	0.84030	0.83946	0.83862	0.83778	0.83694	0.83611
0.18	0.83527	0.83444	0.83360	0.83277	0.83194	0.83110	0.83027	0.82944	0.82861	0.82779
0.19	0.82696	0.82613	0.82531	0.82448	0.82366	0.82283	0.82201	0.82119	0.82037	0.81955
0.20	0.81873	0.81791	0.81709	0.81628	0.81546	0.81465	0.81383	0.81302	0.81221	0.81140
0.21	0.81058	0.80977	0.80896	0.80816	0.80735	0.80654	0.80574	0.80493	0.80413	0.80332
0.22	0.80252	0.80172	0.80092	0.80011	0.79932	0.79852	0.79772	0.79692	0.79612	0.79533
0.23	0.79453	0.79374	0.79295	0.79215	0.79136	0.79057	0.78978	0.78899	0.78820	0.78741
0.24	0.78663	0.78584	0.78506	0.78427	0.78349	0.78270	0.78192	0.78114	0.78036	0.77958
0.25	0.77880	0.77802	0.77724	0.77647	0.77569	0.77492	0.77414	0.77337	0.77260	0.77182
0.26	0.77105	0.77028	0.76951	0.76874	0.76797	0.76721	0.76644	0.76567	0.76491	0.76414
0.27	0.76338	0.76262	0.76185	0.76109	0.76033	0.75957	0.75881	0.75805	0.75730	0.75654
0.28	0.75578	0.75503	0.75427	0.75352	0.75277	0.75201	0.75126	0.75051	0.74976	0.74901
0.29	0.74826	0.74752	0.74677	0.74602	0.74528	0.74453	0.74379	0.74304	0.74230	0.74156
0.30	0.74082	0.74008	0.73934	0.73860	0.73786	0.73712	0.73639	0.73565	0.73492	0.73418
0.31	0.73345	0.73271	0.73198	0.73125	0.73052	0.72979	0.72906	0.72833	0.72760	0.72688
0.32	0.72615	0.72542	0.72470	0.72397	0.72325	0.72253	0.72181	0.72108	0.72036	0.71964
0.33	0.71892	0.71821	0.71749	0.71677	0.71605	0.71534	0.71462	0.71391	0.71320	0.71248
0.34	0.71177	0.71106	0.71035	0.70964	0.70893	0.70822	0.70751	0.70681	0.70610	0.70539
0.35	0.70469	0.70398	0.70328	0.70258	0.70187	0.70117	0.70047	0.69977	0.69907	0.69837
0.36	0.69768	0.69698	0.69628	0.69559	0.69489	0.69420	0.69350	0.69281	0.69212	0.69143
0.37	0.69073	0.69004	0.68935	0.68867	0.68798	0.68729	0.68660	0.68592	0.68523	0.68455
0.38	0.68386	0.68318	0.68250	0.68181	0.68113	0.68045	0.67977	0.67909	0.67841	0.67773
0.39	0.67706	0.67638	0.67570	0.67503	0.67435	0.67368	0.67301	0.67233	0.67166	0.67099
0.40	0.67032	0.66965	0.66898	0.66831	0.66764	0.66698	0.66631	0.66564	0.66498	0.66431
0.41	0.66365	0.66299	0.66232	0.66166	0.66100	0.66034	0.65968	0.65902	0.65836	0.65770
0.42	0.65705	0.65639	0.65573	0.65508	0.65442	0.65377	0.65312	0.65246	0.65181	0.65116
0.43	0.65051	0.64986	0.64921	0.64856	0.64791	0.64726	0.64662	0.64597	0.64533	0.64468
0.44	0.64404	0.64339	0.64275	0.64211	0.64147	0.64082	0.64018	0.63954	0.63890	0.63827
0.45	0.63763	0.63699	0.63635	0.63572	0.63508	0.63445	0.63381	0.63318	0.63255	0.63192
0.46	0.63128	0.63065	0.63002	0.62939	0.62876	0.62814	0.62751	0.62688	0.62625	0.62563
0.47	0.62500	0.62438	0.62375	0.62313	0.62251	0.62189	0.62126	0.62064	0.62002	0.61940
0.48	0.61878	0.61816	0.61755	0.61693	0.61631	0.61570	0.61508	0.61447	0.61385	0.61324
0.49	0.61263	0.61201	0.61140	0.61079	0.61018	0.60957	0.60896	0.60835	0.60774	0.60714
0.50	0.60653	0.60592	0.60531	0.60471	0.60411	0.60351	0.60290	0.60230	0.60170	0.60110
0.51	0.60050	0.59990	0.59930	0.59870	0.59810	0.59750	0.59690	0.59631	0.59571	0.59512
0.52	0.59452	0.59393	0.59333	0.59274	0.59215	0.59156	0.59096	0.59037	0.58978	0.58919
0.53	0.58860	0.58802	0.58743	0.58684	0.58626	0.58567	0.58508	0.58450	0.58391	0.58333
0.54	0.58275	0.58217	0.58158	0.58100	0.58042	0.57984	0.57926	0.57868	0.57810	0.57753
0.55	0.57695	0.57637	0.57580	0.57522	0.57465	0.57407	0.57350	0.57293	0.57235	0.57178
0.56	0.57121	0.57064	0.57007	0.56950	0.56893	0.56836	0.56779	0.56722	0.56666	0.56609
0.57	0.56553	0.56496	0.56440	0.56383	0.56327	0.56270	0.56214	0.56158	0.56102	0.46046
0.58	0.55990	0.55934	0.55878	0.55822	0.55766	0.55711	0.55655	0.55599	0.55544	0.55488
0.59	0.55433	0.55377	0.55322	0.55267	0.55211	0.55156	0.55101	0.55046	0.54991	0.54936

*Column headings provide third decimal place for value of F.

TABLE B.1   Values of the negative exponential $R = e^{-F}$ where $F = \lambda t$
(Continued)

F	0.000*	0.001	0.002	0.003	0.004	0.005	0.006	0.007	0.008	0.009
0.60	0.54881	0.54826	0.54772	0.54718	0.54662	0.54607	0.54553	0.54498	0.54444	0.54389
0.61	0.54335	0.54281	0.54227	0.54172	0.54118	0.54064	0.54010	0.53956	0.53902	0.53848
0.62	0.53794	0.53741	0.53687	0.53633	0.53580	0.53526	0.53473	0.53419	0.53366	0.53312
0.63	0.53259	0.53206	0.53153	0.53100	0.53047	0.52994	0.52941	0.52888	0.52835	0.52782
0.64	0.52729	0.52677	0.52624	0.52571	0.52519	0.52466	0.52414	0.52361	0.52309	0.52257
0.65	0.52205	0.52152	0.52100	0.52048	0.51996	0.51944	0.51892	0.51840	0.51789	0.51737
0.66	0.51685	0.51633	0.51582	0.51530	0.51479	0.51427	0.51376	0.51325	0.51273	0.51222
0.67	0.51171	0.51120	0.51069	0.51018	0.50967	0.50916	0.50865	0.50814	0.50763	0.50712
0.68	0.50662	0.50611	0.50560	0.50510	0.50459	0.50409	0.50359	0.50308	0.50258	0.50208
0.69	0.50158	0.50107	0.50057	0.50007	0.49957	0.49907	0.49858	0.49808	0.49758	0.49708
0.70	0.49659	0.49609	0.49559	0.49510	0.49460	0.49411	0.49361	0.49312	0.49263	0.49214
0.71	0.49164	0.49115	0.49066	0.49017	0.48968	0.48919	0.48870	0.48821	0.48773	0.48724
0.72	0.48675	0.48627	0.48578	0.48529	0.48481	0.48432	0.48384	0.48336	0.48287	0.48239
0.73	0.48191	0.48143	0.48095	0.48047	0.47999	0.47951	0.47903	0.47855	0.47807	0.47759
0.74	0.47711	0.47664	0.47616	0.47568	0.47521	0.47473	0.47426	0.47379	0.47331	0.47284
0.75	0.47237	0.47189	0.47142	0.47095	0.47048	0.47001	0.46954	0.46907	0.46860	0.46813
0.76	0.46767	0.46720	0.46673	0.46627	0.46580	0.46533	0.46487	0.46440	0.46394	0.46348
0.77	0.46301	0.46255	0.46209	0.46163	0.46116	0.46070	0.46024	0.45978	0.45932	0.45886
0.78	0.45841	0.45795	0.45749	0.45703	0.45658	0.45612	0.45566	0.45521	0.45475	0.45430
0.79	0.45384	0.45339	0.45294	0.45249	0.45203	0.45158	0.45113	0.45068	0.45023	0.44978
0.80	0.44933	0.44888	0.44843	0.44798	0.44754	0.44709	0.44664	0.44619	0.44575	0.44530
0.81	0.44486	0.44441	0.44397	0.44353	0.44308	0.44264	0.44220	0.44175	0.44131	0.44087
0.82	0.44043	0.43999	0.43955	0.43911	0.43867	0.43823	0.43780	0.43736	0.43692	0.43649
0.83	0.43605	0.43561	0.43518	0.43474	0.43431	0.43387	0.43344	0.43301	0.43257	0.43214
0.84	0.43171	0.43128	0.43085	0.43042	0.42999	0.42956	0.42913	0.42870	0.42827	0.42784
0.85	0.42741	0.42699	0.42656	0.42613	0.42571	0.42528	0.42486	0.42443	0.42401	0.42359
0.86	0.42316	0.42274	0.42232	0.42189	0.42147	0.42105	0.42063	0.42021	0.41979	0.41937
0.87	0.41895	0.41853	0.41811	0.41770	0.41728	0.41686	0.41645	0.41603	0.41561	0.41520
0.88	0.41478	0.41437	0.41395	0.41354	0.41313	0.41271	0.41230	0.41189	0.41148	0.41107
0.89	0.41066	0.41025	0.40984	0.40943	0.40902	0.40861	0.40820	0.40779	0.40738	0.40698
0.90	0.40657	0.40616	0.40576	0.40535	0.40495	0.40454	0.40414	0.40373	0.40333	0.40293
0.91	0.40252	0.40212	0.40172	0.40132	0.40092	0.40052	0.40012	0.39972	0.39932	0.39892
0.92	0.39851	0.39812	0.39772	0.39733	0.39693	0.39653	0.39614	0.39574	0.39534	0.39495
0.93	0.39455	0.39416	0.39377	0.39337	0.39298	0.39259	0.39219	0.39180	0.39141	0.39102
0.94	0.39063	0.39024	0.38985	0.38946	0.38907	0.38868	0.38829	0.38790	0.38752	0.38713
0.95	0.38674	0.38635	0.38597	0.38558	0.38520	0.38481	0.38443	0.38404	0.38366	0.38328
0.96	0.38289	0.38251	0.38213	0.38175	0.38136	0.38098	0.38060	0.38022	0.37984	0.37946
0.97	0.37908	0.37870	0.37833	0.37795	0.37757	0.37719	0.37682	0.37644	0.37606	0.37569
0.98	0.37531	0.37494	0.37456	0.37419	0.37381	0.37344	0.37307	0.37269	0.37232	0.37195
0.99	0.37158	0.37121	0.37083	0.37046	0.37009	0.36972	0.36935	0.36898	0.36862	0.36825
1.00	0.36788	0.36751	0.36714	0.36678	0.36641	0.36604	0.36568	0.36531	0.36495	0.36458
1.01	0.36422	0.36385	0.36349	0.36313	0.36277	0.36240	0.36204	0.36168	0.36132	0.36096
1.02	0.36059	0.36023	0.35987	0.35951	0.35916	0.35880	0.35844	0.35808	0.35772	0.35736
1.03	0.35701	0.35665	0.35629	0.35594	0.35558	0.35523	0.35487	0.35452	0.35416	0.35381
1.04	0.35345	0.35310	0.35275	0.35240	0.35204	0.35169	0.35134	0.35099	0.35064	0.35029
1.05	0.34994	0.34959	0.34924	0.34889	0.34854	0.34819	0.34784	0.34750	0.34715	0.34680
1.06	0.34646	0.34611	0.34576	0.34542	0.34507	0.34473	0.34438	0.34404	0.34370	0.34335
1.07	0.34301	0.34267	0.34232	0.34198	0.34164	0.34130	0.34096	0.34062	0.34028	0.33994
1.08	0.33960	0.33926	0.33892	0.33858	0.33824	0.33790	0.33756	0.33723	0.33689	0.33655
1.09	0.33622	0.33588	0.33554	0.33521	0.33487	0.33454	0.33421	0.33387	0.33354	0.33320

*Column headings provide third decimal place for value of F.

F	0.00*	0.01	0.02	0.03	0.04	0.05	0.06	0.07	0.08	0.09
1.1	0.33287	0.32956	0.32628	0.32303	0.31982	0.31664	0.31349	0.31037	0.30728	0.30422
1.2	0.30119	0.29820	0.29523	0.29229	0.28938	0.28650	0.28365	0.28083	0.27804	0.27527
1.3	0.27253	0.26982	0.26714	0.26448	0.26185	0.25924	0.25666	0.25411	0.25158	0.24908
1.4	0.24660	0.24414	0.24171	0.23931	0.23693	0.23457	0.23224	0.22992	0.22764	0.22537
1.5	0.22313	0.22090	0.21871	0.21654	0.21438	0.21225	0.21014	0.20805	0.20598	0.20393
1.6	0.20190	0.19989	0.19790	0.19593	0.19398	0.19205	0.19014	0.18825	0.18637	0.18452
1.7	0.18268	0.18087	0.17907	0.17728	0.17552	0.17377	0.17204	0.17033	0.16864	0.16696
1.8	0.16530	0.16365	0.16203	0.16041	0.15882	0.15724	0.15567	0.15412	0.15259	0.15107
1.9	0.14957	0.14808	0.14661	0.14515	0.14370	0.14227	0.14086	0.13946	0.13807	0.13669
2.0	0.13533	0.13399	0.13266	0.13134	0.13003	0.12873	0.12745	0.12619	0.12493	0.12369
2.1	0.12246	0.12124	0.12003	0.11884	0.11765	0.11648	0.11533	0.11418	0.11304	0.11192
2.2	0.11080	0.10970	0.10861	0.10753	0.10645	0.10540	0.10435	0.10331	0.10228	0.10127
2.3	0.10030	0.09926	0.09827	0.09730	0.09633	0.09537	0.09442	0.09348	0.09255	0.09163
2.4	0.09072	0.08982	0.08892	0.08804	0.08716	0.08629	0.08543	0.08458	0.08374	0.08291
2.5	0.08209	0.08127	0.08046	0.07966	0.07887	0.07808	0.07731	0.07654	0.07577	0.07502
2.6	0.07427	0.07353	0.07280	0.07208	0.07136	0.07065	0.06995	0.06925	0.06856	0.06788
2.7	0.06721	0.06654	0.06587	0.06522	0.06457	0.06393	0.06329	0.06266	0.06204	0.06142
2.8	0.06081	0.06020	0.05960	0.05901	0.05843	0.05784	0.05727	0.05670	0.05614	0.05558
2.9	0.05502	0.05448	0.05393	0.05340	0.05287	0.05234	0.05182	0.05130	0.05079	0.05029
3.0	0.04979	0.04929	0.04880	0.04832	0.04784	0.04736	0.04689	0.04642	0.04596	0.04550
3.1	0.04505	0.04460	0.04416	0.04372	0.04328	0.04285	0.04243	0.04200	0.04159	0.04117
3.2	0.04076	0.04036	0.03996	0.03956	0.03916	0.03877	0.03839	0.03800	0.03763	0.03725
3.3	0.03688	0.03652	0.03615	0.03579	0.03544	0.03508	0.03473	0.03439	0.03405	0.03371
3.4	0.03337	0.03304	0.03271	0.03239	0.03206	0.03175	0.03143	0.03112	0.03081	0.03050
3.5	0.03020	0.02990	0.02960	0.02930	0.02901	0.02872	0.02844	0.02816	0.02788	0.02760
3.6	0.02732	0.02705	0.02678	0.02652	0.02625	0.02599	0.02573	0.02548	0.02522	0.02498
3.7	0.02472	0.02448	0.02423	0.02399	0.02375	0.02352	0.02328	0.02305	0.02282	0.02260
3.8	0.02237	0.02215	0.02193	0.02171	0.02149	0.02128	0.02107	0.02086	0.02065	0.02045
3.9	0.02024	0.02004	0.01984	0.01964	0.01945	0.01925	0.01906	0.01887	0.01869	0.01850
4.0	0.01832	0.01813	0.01795	0.01777	0.01760	0.01742	0.01725	0.01708	0.01691	0.01674
4.1	0.01657	0.01641	0.01624	0.01608	0.01592	0.01576	0.01561	0.01545	0.01530	0.01515
4.2	0.01500	0.01485	0.01470	0.01455	0.01441	0.01426	0.01412	0.01398	0.01384	0.01370
4.3	0.01357	0.01343	0.01330	0.01317	0.01304	0.01291	0.01278	0.01265	0.01253	0.01240
4.4	0.01228	0.01215	0.01203	0.01191	0.01180	0.01168	0.01156	0.01145	0.01133	0.01122
4.5	0.01111	0.01100	0.01089	0.01078	0.01067	0.01057	0.01046	0.01036	0.01025	0.01015
4.6	0.01005	0.00995	0.00985	0.00975	0.00966	0.00956	0.00947	0.00937	0.00928	0.00919
4.7	0.00909	0.00900	0.00891	0.00883	0.00874	0.00865	0.00857	0.00848	0.00840	0.00831
4.8	0.00823	0.00815	0.00807	0.00799	0.00791	0.00783	0.00775	0.00767	0.00760	0.00752
4.9	0.00745	0.00737	0.00730	0.00723	0.00715	0.00708	0.00701	0.00694	0.00687	0.00681
5.0	0.00674	0.00667	0.00660	0.00654	0.00647	0.00641	0.00635	0.00628	0.00622	0.00616

F	0.0†	0.1	0.2	0.3	0.4	0.5	0.6	0.7	0.8	0.9
6.0	0.00248	0.00224	0.00203	0.00184	0.00166	0.00150	0.00136	0.00123	0.00111	0.00101
7.0	0.00091	0.00083	0.00075	0.00068	0.00061	0.00055	0.00050	0.00045	0.00041	0.00037
8.0	0.00034	0.00030	0.00027	0.00025	0.00022	0.00020	0.00018	0.00016	0.00015	0.00014
9.0	0.00012	0.00011	0.00010	0.00009	0.00008	0.00007	0.00007	0.00006	0.00006	0.00005

*Column headings provide second decimal place for value of F.
†Column headings provide first decimal place for value of F.

TABLE B.2   Areas of the normal curve between $-Z$ **and** $+Z$ **standard deviations from the mean**

Z	0.00	0.01	0.02	0.03	0.04	0.05	0.06	0.07	0.08	0.09
0.0	0.0000	0.0080	0.0160	0.0239	0.0319	0.0399	0.0478	0.0558	0.0638	0.0717
0.1	0.0797	0.0876	0.0955	0.1034	0.1113	0.1192	0.1271	0.1350	0.1428	0.1507
0.2	0.1585	0.1663	0.1741	0.1819	0.1897	0.1974	0.2051	0.2128	0.2205	0.2282
0.3	0.2358	0.2434	0.2510	0.2586	0.2661	0.2737	0.2812	0.2886	0.2961	0.3035
0.4	0.3108	0.3182	0.3255	0.3328	0.3401	0.3473	0.3545	0.3616	0.3759	0.3829
0.5	0.3829	0.3899	0.3969	0.4039	0.4108	0.4177	0.4245	0.4313	0.4381	0.4448
0.6	0.4515	0.4581	0.4647	0.4713	0.4768	0.4843	0.4907	0.4971	0.5035	0.5098
0.7	0.5161	0.5223	0.5285	0.5346	0.5407	0.5467	0.5527	0.5587	0.5646	0.5705
0.8	0.5763	0.5821	0.5878	0.5935	0.5991	0.6047	0.6102	0.6157	0.6211	0.6265
0.9	0.6319	0.6372	0.6424	0.6476	0.6528	0.6579	0.6629	0.6680	0.6729	0.6778
1.0	0.6827	0.6875	0.6923	0.6970	0.7017	0.7063	0.7109	0.7154	0.7199	0.7243
1.1	0.7287	0.7330	0.7373	0.7415	0.7457	0.7499	0.7540	0.7580	0.7620	0.7660
1.2	0.7699	0.7737	0.7775	0.7813	0.7850	0.7887	0.7923	0.7959	0.7995	0.8029
1.3	0.8064	0.8098	0.8132	0.8165	0.8198	0.8230	0.8262	0.8293	0.8324	0.8355
1.4	0.8385	0.8415	0.8444	0.8473	0.8501	0.8529	0.8557	0.8584	0.8611	0.8638
1.5	0.8664	0.8690	0.8715	0.8740	0.8764	0.8789	0.8812	0.8836	0.8859	0.8882
1.6	0.8904	0.8926	0.8948	0.8969	0.8990	0.9011	0.9031	0.9051	0.9070	0.9090
1.7	0.9109	0.9127	0.9146	0.9164	0.9181	0.9199	0.9216	0.9233	0.9249	0.9265
1.8	0.9281	0.9297	0.9312	0.9328	0.9342	0.9357	0.9371	0.9385	0.9399	0.9412
1.9	0.9426	0.9439	0.9451	0.9464	0.9476	0.9488	0.9500	0.9512	0.9523	0.9534
2.0	0.9545	0.9556	0.9566	0.9576	0.9586	0.9596	0.9606	0.9615	0.9625	0.9634
2.1	0.9643	0.9652	0.9660	0.9668	0.9676	0.9684	0.9692	0.9700	0.9707	0.9715
2.2	0.9722	0.9729	0.9736	0.9743	0.9749	0.9756	0.9762	0.9768	0.9774	0.9780
2.3	0.9786	0.9791	0.9797	0.9802	0.9807	0.9812	0.9817	0.9822	0.9827	0.9832
2.4	0.9836	0.9840	0.9845	0.9849	0.9853	0.9857	0.9861	0.9865	0.9869	0.9872
2.5	0.9876	0.9879	0.9883	0.9886	0.9889	0.9892	0.9895	0.9898	0.9901	0.9904
2.6	0.9907	0.9909	0.9912	0.9915	0.9917	0.9920	0.9922	0.9924	0.9926	0.9929
2.7	0.9931	0.9933	0.9935	0.9937	0.9939	0.9940	0.9942	0.9944	0.9946	0.9947
2.8	0.9949	0.9950	0.9952	0.9953	0.9955	0.9956	0.9958	0.9959	0.9960	0.9961
2.9	0.9963	0.9964	0.9965	0.9966	0.9967	0.9968	0.9969	0.9970	0.9971	0.9972
3.0	0.99731	0.99739	0.99748	0.99756	0.99764	0.99772	0.99779	0.99786	0.99794	0.99800
3.1	0.99807	0.99813	0.99820	0.99826	0.99831	0.99837	0.99842	0.99847	0.99852	0.99857
3.2	0.99863	0.99868	0.99872	0.99876	0.99880	0.99884	0.99888	0.99892	0.99896	0.99900
3.3	0.99903	0.99907	0.99911	0.99914	0.99917	0.99920	0.99923	0.99925	0.99928	0.99930
3.4	0.99933	0.99935	0.99938	0.99940	0.99942	0.99944	0.99946	0.99948	0.99950	0.99952
3.5	0.99953	0.99955	0.99957	0.99959	0.99960	0.99962	0.99963	0.99964	0.99966	0.99967
3.6	0.99968	0.99969	0.99970	0.99971	0.99972	0.99973	0.99974	0.99975	0.99976	0.99977
3.7	0.99978	0.99979	0.99980	0.99981	0.99982	0.99982	0.99983	0.99984	0.99984	0.99985
3.8	0.99986	0.99986	0.99987	0.99987	0.99988	0.99988	0.99989	0.99989	0.99990	0.99990
3.9	0.99990	0.99991	0.99991	0.99991	0.99992	0.99992	0.99992	0.99993	0.99993	0.99994

TABLE B.3   Values of the t distribution

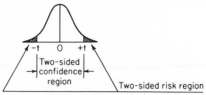

Two-sided confidence region
Two-sided risk region
Values of t – two sided

a risk...	0.90	0.80	0.70	0.60	0.50	0.40	0.30	0.20	0.10	0.05	0.02	0.01
Confidence level (prob.)	0.10	0.20	0.30	0.40	0.50	0.60	0.70	0.80	0.90	0.95	0.98	0.99
Degrees of freedom												
1	0.158	0.325	0.510	0.727	1.000	1.376	1.963	3.078	6.314	12.706	31.821	63.657
2	0.142	0.289	0.445	0.617	0.816	1.061	1.386	1.886	2.920	4.303	6.965	9.925
3	0.137	0.277	0.424	0.584	0.765	0.978	1.250	1.638	2.353	3.182	4.541	5.841
4	0.134	0.271	0.414	0.569	0.741	0.941	1.190	1.533	2.132	2.776	3.747	4.604
5	0.132	0.267	0.408	0.559	0.727	0.920	1.156	1.476	2.015	2.571	3.365	4.032
6	0.131	0.265	0.404	0.553	0.718	0.906	1.134	1.440	1.943	2.447	3.143	3.707
7	0.130	0.263	0.402	0.549	0.711	0.896	1.119	1.415	1.895	2.365	2.998	3.499
8	0.130	0.262	0.399	0.546	0.706	0.889	1.108	1.397	1.860	2.306	2.896	3.355
9	0.129	0.261	0.398	0.543	0.703	0.883	1.100	1.383	1.833	2.262	2.821	3.250
10	0.129	0.260	0.397	0.542	0.700	0.879	1.093	1.372	1.812	2.228	2.764	3.169
11	0.129	0.260	0.396	0.540	0.697	0.876	1.088	1.363	1.796	2.201	2.718	3.106
12	0.128	0.259	0.395	0.539	0.695	0.873	1.083	1.356	1.782	2.179	2.681	3.055
13	0.128	0.259	0.394	0.538	0.694	0.870	1.079	1.350	1.771	2.160	2.650	3.012
14	0.128	0.258	0.393	0.537	0.692	0.868	1.076	1.345	1.761	2.145	2.624	2.977
15	0.128	0.258	0.393	0.536	0.691	0.866	1.074	1.341	1.753	2.131	2.602	2.947
16	0.128	0.258	0.392	0.535	0.690	0.865	1.071	1.337	1.746	2.120	2.583	2.921
17	0.128	0.257	0.392	0.534	0.689	0.863	1.069	1.333	1.740	2.110	2.567	2.898
18	0.127	0.257	0.392	0.534	0.688	0.862	1.067	1.330	1.734	2.101	2.552	2.878
19	0.127	0.257	0.391	0.533	0.688	0.861	1.066	1.328	1.729	2.093	2.539	2.861
20	0.127	0.257	0.391	0.533	0.687	0.860	1.064	1.325	1.725	2.086	2.528	2.845
21	0.127	0.257	0.391	0.532	0.686	0.859	1.063	1.323	1.721	2.080	2.518	2.831
22	0.127	0.256	0.390	0.532	0.686	0.858	1.061	1.321	1.717	2.074	2.508	2.819
23	0.127	0.256	0.390	0.532	0.685	0.858	1.060	1.319	1.714	2.069	2.500	2.807
24	0.127	0.256	0.390	0.531	0.685	0.857	1.059	1.318	1.711	2.064	2.492	2.797
25	0.127	0.256	0.390	0.531	0.684	0.856	1.058	1.316	1.708	2.060	2.485	2.787
26	0.127	0.256	0.390	0.531	0.684	0.856	1.058	1.315	1.706	2.056	2.479	2.779
27	0.127	0.256	0.389	0.531	0.684	0.855	1.057	1.314	1.703	2.052	2.473	2.771
28	0.127	0.256	0.389	0.530	0.683	0.855	1.056	1.313	1.701	2.048	2.467	2.763
29	0.127	0.256	0.389	0.530	0.683	0.854	1.055	1.311	1.699	2.045	2.462	2.756
30	0.127	0.256	0.389	0.530	0.683	0.854	1.055	1.310	1.697	2.042	2.457	2.750
40	0.127	0.255	0.388	0.529	0.681	0.851	1.051	1.303	1.684	2.021	2.423	2.704
60	0.127	0.254	0.387	0.527	0.679	0.848	1.046	1.296	1.671	2.000	2.390	2.660
120	0.126	0.254	0.386	0.526	0.677	0.845	1.041	1.289	1.658	1.980	2.358	2.617
∞	0.126	0.253	0.385	0.524	0.674	0.842	1.036	1.282	1.645	1.960	2.326	2.576
Confidence level (prob.)	0.55	0.60	0.65	0.70	0.75	0.80	0.85	0.90	0.95	0.975	0.99	0.995
a risk....	0.45	0.40	0.35	0.30	0.25	0.20	0.15	0.10	0.05	0.025	0.01	0.005

Values of t—two-sided

Values of t—one-sided

TABLE B.4   Percentiles of the χ^2 distribution

(Column headings are percentages *greater than* the table values)

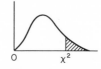

Degrees of freedom	99.5	99.0	98.0	95.0	90.0	80.0	70.0
1	0.000039	0.00016	0.00063	0.00393	0.0158	0.0642	0.148
2	0.0100	0.0201	0.0404	0.1026	0.211	0.446	0.713
3	0.0717	0.115	0.185	0.352	0.584	1.005	1.424
4	0.207	0.297	0.429	0.711	1.064	1.649	2.195
5	0.412	0.554	0.752	1.145	1.610	2.343	3.000
6	0.676	0.872	1.134	1.635	2.204	3.070	3.828
7	0.989	1.239	1.564	2.167	2.833	3.822	4.671
8	1.344	1.646	2.032	2.733	3.490	4.594	5.527
9	1.735	2.088	2.532	3.325	4.168	5.380	6.393
10	2.156	2.558	3.059	3.940	4.865	6.179	7.267
11	2.603	3.053	3.609	4.575	5.578	6.989	8.148
12	3.074	3.571	4.178	5.226	6.304	7.807	9.034
13	3.565	4.107	4.765	5.892	7.042	8.634	9.926
14	4.075	4.660	5.368	6.571	7.790	9.467	10.821
15	4.601	5.229	5.985	7.261	8.547	10.307	11.721
16	5.142	5.812	6.614	7.962	9.312	11.152	12.624
17	5.697	6.408	7.255	8.672	10.085	12.002	13.531
18	6.265	7.015	7.906	9.390	10.865	12.857	14.440
19	6.844	7.633	8.567	10.117	11.651	13.716	15.352
20	7.434	8.260	9.237	10.851	12.443	14.578	16.266
21	8.034	8.897	9.915	11.591	13.240	15.445	17.182
22	8.643	9.542	10.600	12.338	14.042	16.314	18.101
23	9.260	10.196	11.293	13.091	14.848	17.187	19.021
24	9.886	10.856	11.992	13.848	15.659	18.062	19.943
25	10.520	11.524	12.697	14.611	16.473	18.940	20.867
26	11.160	12.198	13.409	15.379	17.292	19.820	21.792
27	11.808	12.879	14.125	16.151	18.114	20.703	22.719
28	12.461	13.565	14.847	16.928	18.939	21.588	23.647
29	13.121	14.257	15.574	17.708	19.768	22.475	24.577
30	13.787	14.954	16.306	18.493	20.599	23.364	25.508
40	20.707	22.164	23.824	26.509	29.051	32.352	34.876
60	35.534	37.485	39.689	43.188	46.459	50.647	53.815
80	51.172	53.540	56.204	60.391	64.278	69.213	72.920
100	67.328	70.065	73.134	77.929	82.358	87.950	92.133
200	152.241	156.432	161.099	168.279	174.835	183.006	189.052

Values in this table are reproduced from "Handbook of Statistical Tables" by L. B. Owen and from "Statistical Methods for Research Workers" by R. A. Fisher with the permission of the authors and their publishers, Addison-Wesley Publishing Company, Inc., Reading, Mass., and Oliver & Boyd Ltd., Edinburgh, respectively. Values not given in these references were obtained from the formula

$$\chi_a^2 = \nu \left(1 - \frac{2}{9\nu} + z_a \sqrt{\frac{2}{9\nu}} \right)^3$$

50.0	30.0	20.0	10.0	5.0	2.0	1.0	0.5
0.455	1.074	1.642	2.706	3.841	5.412	6.635	7.879
1.386	2.408	3.219	4.605	5.991	7.824	9.210	10.597
2.366	3.665	4.642	6.251	7.815	9.837	11.341	12.838
3.557	4.878	5.989	7.779	9.488	11.668	13.277	14.860
4.351	6.064	7.289	9.236	11.070	13.388	15.086	16.750
5.348	7.231	8.558	10.645	12.592	15.033	16.812	18.548
6.346	8.383	9.803	12.017	14.067	16.622	18.475	20.278
7.344	9.524	11.030	13.362	15.507	18.168	20.090	21.955
8.343	10.656	12.242	14.684	16.919	19.697	21.666	23.589
9.342	11.781	13.442	15.987	18.307	21.161	23.209	25.188
10.341	12.899	14.631	17.275	19.675	22.618	24.725	26.757
11.340	14.011	15.812	18.549	21.026	24.054	26.217	28.299
12.340	15.119	16.985	19.812	22.362	25.472	27.688	29.819
13.339	16.222	18.151	21.064	23.685	26.873	29.141	31.319
14.339	17.322	19.311	22.307	24.996	28.259	30.578	32.801
15.338	18.418	20.465	23.542	26.296	29.633	32.000	34.267
16.338	19.511	21.615	24.769	27.587	30.995	33.409	35.718
17.338	20.601	22.760	25.989	28.869	32.346	34.805	37.156
18.338	21.689	23.900	27.204	30.144	33.687	36.191	38.582
19.337	22.775	25.038	28.412	31.410	35.020	37.566	39.997
20.337	23.858	26.171	29.615	32.671	36.343	38.932	41.401
21.337	24.939	27.301	30.813	33.924	37.659	40.289	42.796
22.337	26.018	28.429	32.007	35.172	38.968	41.638	44.181
23.337	27.096	29.553	33.196	36.415	40.270	42.980	45.559
24.337	28.172	30.675	34.382	37.652	41.566	44.314	46.928
25.336	29.246	31.795	35.563	38.885	42.856	45.642	48.290
26.336	30.319	32.912	36.741	40.113	44.140	46.963	49.645
27.336	31.391	34.027	37.916	41.337	45.419	48.278	50.993
28.336	32.461	35.139	39.087	42.557	46.693	49.588	52.336
29.336	33.530	36.250	40.256	43.773	47.962	50.892	53.672
39.335	44.163	47.263	51.805	55.758	60.796	63.691	66.766
59.335	65.225	68.969	74.397	79.082	84.588	88.379	91.952
79.334	86.122	90.403	96.578	101.879	108.082	112.329	116.321
99.334	106.908	111.667	118.498	123.342	131.154	135.807	140.169
199.333	209.997	216.618	226.021	233.994	243.198	249.445	255.264

TABLE B.5 Percentiles of the χ^2/d.f. distribution

(Column headings are percentages *greater than* the table values)

Degrees of freedom	99.5	99.0	98.0	95.0	90.0	80.0	70.0
1	0.000039	0.00016	0.00063	0.00393	0.0158	0.0642	0.14
2	0.0050	0.0101	0.0202	0.0513	0.105	0.223	0.35
3	0.0239	0.0383	0.0617	0.117	0.195	0.335	0.47
4	0.0517	0.074	0.107	0.178	0.266	0.412	0.54
5	0.0823	0.111	0.150	0.229	0.322	0.469	0.60
6	0.113	0.145	0.189	0.273	0.367	0.512	0.63
7	0.141	0.177	0.223	0.310	0.405	0.546	0.66
8	0.168	0.206	0.254	0.342	0.436	0.574	0.69
9	0.193	0.232	0.281	0.369	0.463	0.598	0.71
10	0.216	0.256	0.306	0.394	0.487	0.618	0.72
11	0.237	0.278	0.328	0.416	0.507	0.635	0.74
12	0.256	0.298	0.348	0.436	0.525	0.651	0.75
13	0.274	0.316	0.367	0.453	0.542	0.664	0.76
14	0.291	0.333	0.383	0.469	0.556	0.676	0.77
15	0.307	0.349	0.399	0.484	0.570	0.687	0.78
16	0.321	0.363	0.413	0.498	0.582	0.697	0.78
17	0.335	0.377	0.427	0.510	0.593	0.706	0.79
18	0.348	0.390	0.439	0.522	0.604	0.714	0.80
19	0.360	0.402	0.451	0.532	0.613	0.722	0.80
20	0.372	0.413	0.462	0.543	0.622	0.729	0.81
21	0.383	0.424	0.472	0.552	0.630	0.735	0.81
22	0.393	0.434	0.482	0.561	0.638	0.742	0.82
23	0.403	0.443	0.491	0.569	0.645	0.748	0.82
24	0.412	0.452	0.500	0.577	0.652	0.753	0.83
25	0.421	0.461	0.508	0.584	0.659	0.758	0.83
26	0.429	0.469	0.516	0.592	0.665	0.762	0.83
27	0.437	0.477	0.523	0.598	0.671	0.767	0.84
28	0.445	0.484	0.530	0.605	0.676	0.771	0.84
29	0.452	0.491	0.537	0.610	0.682	0.775	0.84
30	0.460	0.498	0.544	0.616	0.687	0.779	0.85
40	0.518	0.554	0.596	0.663	0.726	0.809	0.87
60	0.592	0.625	0.661	0.720	0.774	0.844	0.89
80	0.640	0.669	0.703	0.755	0.803	0.865	0.91
100	0.673	0.701	0.731	0.779	0.824	0.879	0.92
200	0.761	0.782	0.805	0.841	0.874	0.915	0.94

Values in this table were obtained by dividing values in Table B.4 by the number of degrees of freedom.

50.0	30.0	20.0	10.0	5.0	2.0	1.0	0.5
0.455	1.074	1.642	2.706	3.841	5.412	6.635	7.879
0.693	1.204	1.610	2.302	2.996	3.912	4.605	5.298
0.789	1.222	1.547	2.084	2.605	3.279	3.780	4.279
0.839	1.219	1.497	1.945	2.372	2.917	3.319	3.715
0.870	1.213	1.458	1.847	2.214	2.678	3.017	3.350
0.891	1.205	1.426	1.774	2.099	2.505	2.802	3.091
0.907	1.198	1.400	1.717	2.010	2.375	2.639	2.897
0.918	1.191	1.379	1.670	1.938	2.271	2.511	2.744
0.927	1.184	1.360	1.632	1.880	2.189	2.407	2.621
0.934	1.178	1.344	1.599	1.831	2.116	2.321	2.519
0.940	1.173	1.330	1.570	1.789	2.056	2.248	2.432
0.945	1.168	1.318	1.546	1.752	2.005	2.185	2.358
0.949	1.163	1.307	1.524	1.720	1.959	2.130	2.294
0.953	1.159	1.297	1.505	1.692	1.919	2.082	2.237
0.956	1.155	1.287	1.487	1.666	1.884	2.039	2.187
0.959	1.151	1.279	1.471	1.644	1.852	2.000	2.142
0.961	1.148	1.271	1.457	1.623	1.823	1.965	2.101
0.963	1.145	1.264	1.444	1.604	1.797	1.934	2.064
0.965	1.142	1.258	1.432	1.587	1.773	1.905	2.031
0.967	1.139	1.252	1.421	1.571	1.751	1.878	2.000
0.968	1.136	1.246	1.410	1.556	1.731	1.854	1.971
0.970	1.134	1.241	1.401	1.542	1.712	1.831	1.945
0.971	1.131	1.236	1.392	1.529	1.694	1.810	1.921
0.972	1.129	1.231	1.383	1.517	1.678	1.791	1.898
0.973	1.127	1.227	1.375	1.506	1.663	1.773	1.877
0.974	1.125	1.223	1.368	1.496	1.648	1.755	1.857
0.975	1.123	1.219	1.361	1.486	1.635	1.739	1.839
0.976	1.121	1.215	1.354	1.476	1.622	1.724	1.821
0.977	1.119	1.212	1.348	1.467	1.610	1.710	1.805
0.978	1.118	1.208	1.342	1.459	1.599	1.696	1.789
0.983	1.104	1.182	1.295	1.394	1.520	1.592	1.669
0.989	1.087	1.149	1.240	1.318	1.410	1.473	1.533
0.992	1.077	1.130	1.207	1.273	1.351	1.404	1.454
0.993	1.069	1.117	1.185	1.243	1.312	1.358	1.402
0.997	1.050	1.083	1.130	1.170	1.216	1.247	1.276

TABLE B.6   Values of the F distribution

<table>
<tr><td rowspan="2"></td><td rowspan="2">Prob.
γ*</td><td colspan="9">Degrees of freedom</td></tr>
<tr><td>1</td><td>2</td><td>3</td><td>4</td><td>5</td><td>6</td><td>7</td><td>8</td><td>9</td></tr>

<tr><td rowspan="10">1</td><td>0.005</td><td>$0.0^4 62$</td><td>0.0051</td><td>0.018</td><td>0.032</td><td>0.044</td><td>0.054</td><td>0.062</td><td>0.068</td><td>0.073</td></tr>
<tr><td>0.01</td><td>$0.0^3 25$</td><td>0.010</td><td>0.029</td><td>0.047</td><td>0.062</td><td>0.073</td><td>0.082</td><td>0.089</td><td>0.095</td></tr>
<tr><td>0.05</td><td>0.0062</td><td>0.054</td><td>0.099</td><td>0.130</td><td>0.151</td><td>0.167</td><td>0.179</td><td>0.188</td><td>0.195</td></tr>
<tr><td>0.10</td><td>0.025</td><td>0.117</td><td>0.181</td><td>0.220</td><td>0.246</td><td>0.265</td><td>0.279</td><td>0.289</td><td>0.298</td></tr>
<tr><td>0.25</td><td>0.172</td><td>0.389</td><td>0.494</td><td>0.553</td><td>0.591</td><td>0.617</td><td>0.637</td><td>0.650</td><td>0.661</td></tr>
<tr><td>0.75</td><td>5.83</td><td>7.50</td><td>8.20</td><td>8.58</td><td>8.82</td><td>8.98</td><td>9.10</td><td>9.19</td><td>9.26</td></tr>
<tr><td>0.90</td><td>39.86</td><td>49.50</td><td>53.59</td><td>55.83</td><td>57.24</td><td>58.20</td><td>58.91</td><td>59.44</td><td>59.86</td></tr>
<tr><td>0.95</td><td>161.5</td><td>199.5</td><td>215.7</td><td>224.6</td><td>230.2</td><td>234.0</td><td>236.8</td><td>238.9</td><td>240.5</td></tr>
<tr><td>0.99</td><td>4052.</td><td>5000.</td><td>5403.</td><td>5625.</td><td>5764.</td><td>5859.</td><td>5928.</td><td>5981.</td><td>6023.</td></tr>
<tr><td>0.995</td><td>16211</td><td>20000</td><td>21615</td><td>22500</td><td>23056</td><td>23437</td><td>23715</td><td>23925</td><td>24091</td></tr>

<tr><td rowspan="10">2</td><td>0.005</td><td>$0.0^4 50$</td><td>0.0050</td><td>0.020</td><td>0.038</td><td>0.055</td><td>0.069</td><td>0.081</td><td>0.091</td><td>0.099</td></tr>
<tr><td>0.01</td><td>$0.0^3 20$</td><td>0.010</td><td>0.032</td><td>0.056</td><td>0.075</td><td>0.092</td><td>0.105</td><td>0.116</td><td>0.125</td></tr>
<tr><td>0.05</td><td>0.0050</td><td>0.053</td><td>0.105</td><td>0.144</td><td>0.173</td><td>0.194</td><td>0.211</td><td>0.224</td><td>0.235</td></tr>
<tr><td>0.10</td><td>0.020</td><td>0.111</td><td>0.183</td><td>0.231</td><td>0.265</td><td>0.289</td><td>0.307</td><td>0.321</td><td>0.333</td></tr>
<tr><td>0.25</td><td>0.133</td><td>0.333</td><td>0.439</td><td>0.500</td><td>0.540</td><td>0.568</td><td>0.588</td><td>0.604</td><td>0.616</td></tr>
<tr><td>0.75</td><td>2.57</td><td>3.00</td><td>3.15</td><td>3.23</td><td>3.28</td><td>3.31</td><td>3.34</td><td>3.35</td><td>3.37</td></tr>
<tr><td>0.90</td><td>8.53</td><td>9.00</td><td>9.16</td><td>9.24</td><td>9.29</td><td>9.33</td><td>9.35</td><td>9.37</td><td>9.38</td></tr>
<tr><td>0.95</td><td>18.51</td><td>19.00</td><td>19.16</td><td>19.25</td><td>19.30</td><td>19.33</td><td>19.36</td><td>19.37</td><td>19.39</td></tr>
<tr><td>0.99</td><td>98.50</td><td>99.00</td><td>99.17</td><td>99.25</td><td>99.30</td><td>99.33</td><td>99.36</td><td>99.37</td><td>99.39</td></tr>
<tr><td>0.995</td><td>198.5</td><td>199.0</td><td>199.2</td><td>199.2</td><td>199.3</td><td>199.3</td><td>199.4</td><td>199.4</td><td>199.4</td></tr>

<tr><td rowspan="10">3</td><td>0.005</td><td>$0.0^4 46$</td><td>0.0050</td><td>0.021</td><td>0.041</td><td>0.060</td><td>0.077</td><td>0.092</td><td>0.104</td><td>0.115</td></tr>
<tr><td>0.01</td><td>$0.0^3 19$</td><td>0.010</td><td>0.034</td><td>0.060</td><td>0.083</td><td>0.102</td><td>0.118</td><td>0.132</td><td>0.143</td></tr>
<tr><td>0.05</td><td>0.0046</td><td>0.052</td><td>0.108</td><td>0.152</td><td>0.185</td><td>0.210</td><td>0.230</td><td>0.246</td><td>0.259</td></tr>
<tr><td>0.10</td><td>0.019</td><td>0.109</td><td>0.185</td><td>0.239</td><td>0.276</td><td>0.304</td><td>0.325</td><td>0.342</td><td>0.356</td></tr>
<tr><td>0.25</td><td>0.122</td><td>0.317</td><td>0.424</td><td>0.489</td><td>0.531</td><td>0.561</td><td>0.582</td><td>0.600</td><td>0.613</td></tr>
<tr><td>0.75</td><td>2.02</td><td>2.28</td><td>2.36</td><td>2.39</td><td>2.41</td><td>2.42</td><td>2.43</td><td>2.44</td><td>2.44</td></tr>
<tr><td>0.90</td><td>5.54</td><td>5.46</td><td>5.39</td><td>5.34</td><td>5.31</td><td>5.28</td><td>5.27</td><td>5.25</td><td>5.24</td></tr>
<tr><td>0.95</td><td>10.13</td><td>9.55</td><td>9.28</td><td>9.12</td><td>9.01</td><td>8.94</td><td>8.89</td><td>8.85</td><td>8.81</td></tr>
<tr><td>0.99</td><td>34.12</td><td>30.82</td><td>29.46</td><td>28.71</td><td>28.24</td><td>27.91</td><td>27.67</td><td>27.50</td><td>27.35</td></tr>
<tr><td>0.995</td><td>55.55</td><td>49.80</td><td>47.47</td><td>46.20</td><td>45.39</td><td>44.84</td><td>44.43</td><td>44.13</td><td>43.88</td></tr>

<tr><td rowspan="10">4</td><td>0.005</td><td>$0.0^4 44$</td><td>0.0050</td><td>0.022</td><td>0.043</td><td>0.064</td><td>0.083</td><td>0.100</td><td>0.114</td><td>0.126</td></tr>
<tr><td>0.01</td><td>$0.0^3 18$</td><td>0.010</td><td>0.035</td><td>0.063</td><td>0.088</td><td>0.109</td><td>0.127</td><td>0.143</td><td>0.156</td></tr>
<tr><td>0.05</td><td>0.0044</td><td>0.052</td><td>0.110</td><td>0.157</td><td>0.193</td><td>0.221</td><td>0.243</td><td>0.261</td><td>0.275</td></tr>
<tr><td>0.10</td><td>0.018</td><td>0.108</td><td>0.187</td><td>0.243</td><td>0.284</td><td>0.314</td><td>0.338</td><td>0.356</td><td>0.371</td></tr>
<tr><td>0.25</td><td>0.117</td><td>0.309</td><td>0.418</td><td>0.484</td><td>0.528</td><td>0.560</td><td>0.583</td><td>0.601</td><td>0.615</td></tr>
<tr><td>0.75</td><td>1.81</td><td>2.00</td><td>2.05</td><td>2.06</td><td>2.07</td><td>2.08</td><td>2.08</td><td>2.08</td><td>2.08</td></tr>
<tr><td>0.90</td><td>4.54</td><td>4.32</td><td>4.19</td><td>4.11</td><td>4.05</td><td>4.01</td><td>3.98</td><td>3.95</td><td>3.94</td></tr>
<tr><td>0.95</td><td>7.71</td><td>6.94</td><td>6.59</td><td>6.39</td><td>6.26</td><td>6.16</td><td>6.09</td><td>6.04</td><td>6.00</td></tr>
<tr><td>0.99</td><td>21.20</td><td>18.00</td><td>16.69</td><td>15.98</td><td>15.52</td><td>15.21</td><td>14.98</td><td>14.80</td><td>14.66</td></tr>
<tr><td>0.995</td><td>31.33</td><td>26.28</td><td>24.26</td><td>23.16</td><td>22.46</td><td>21.98</td><td>21.62</td><td>21.35</td><td>21.14</td></tr>

<tr><td rowspan="10">5</td><td>0.005</td><td>$0.0^4 43$</td><td>0.0050</td><td>0.022</td><td>0.045</td><td>0.067</td><td>0.087</td><td>0.105</td><td>0.120</td><td>0.134</td></tr>
<tr><td>0.01</td><td>$0.0^3 17$</td><td>0.010</td><td>0.035</td><td>0.064</td><td>0.091</td><td>0.114</td><td>0.134</td><td>0.151</td><td>0.165</td></tr>
<tr><td>0.05</td><td>0.0043</td><td>0.052</td><td>0.111</td><td>0.160</td><td>0.198</td><td>0.228</td><td>0.252</td><td>0.271</td><td>0.287</td></tr>
<tr><td>0.10</td><td>0.017</td><td>0.108</td><td>0.188</td><td>0.247</td><td>0.290</td><td>0.322</td><td>0.347</td><td>0.367</td><td>0.383</td></tr>
<tr><td>0.25</td><td>0.113</td><td>0.305</td><td>0.415</td><td>0.483</td><td>0.528</td><td>0.560</td><td>0.584</td><td>0.604</td><td>0.618</td></tr>
<tr><td>0.75</td><td>1.69</td><td>1.85</td><td>1.88</td><td>1.89</td><td>1.89</td><td>1.89</td><td>1.89</td><td>1.89</td><td>1.89</td></tr>
<tr><td>0.90</td><td>4.06</td><td>3.78</td><td>3.62</td><td>3.52</td><td>3.45</td><td>3.40</td><td>3.37</td><td>3.34</td><td>3.32</td></tr>
<tr><td>0.95</td><td>6.61</td><td>5.79</td><td>5.41</td><td>5.19</td><td>5.05</td><td>4.95</td><td>4.88</td><td>4.82</td><td>4.77</td></tr>
<tr><td>0.99</td><td>16.26</td><td>13.27</td><td>12.06</td><td>11.39</td><td>10.97</td><td>10.67</td><td>10.46</td><td>10.29</td><td>10.16</td></tr>
<tr><td>0.995</td><td>22.78</td><td>18.31</td><td>16.53</td><td>15.56</td><td>14.94</td><td>14.51</td><td>14.20</td><td>13.96</td><td>13.77</td></tr>
</table>

Degrees of freedom for denominator

*Actual variance ratios will, by chance, be less than the table value with the stated probability γ. The probability a of exceeding the table value is equal to $1 - \gamma$. F values for $\nu_1 \nu_2$ not given in the references were found from the formula

$$F_{\gamma,\,\nu_2 \nu_1} = \frac{1}{F_{1-\gamma,\,\nu_1,\,\nu_2}} = \frac{1}{F_{a,\,\nu_1,\,\nu_2}}$$

10	15	20	30	40	50	60	100	∞
0.078	0.093	0.101	0.109	0.113	0.116	0.118	0.121	0.127
0.100	0.115	0.124	0.132	0.137	0.139	0.141	0.145	0.151
0.201	0.220	0.230	0.240	0.245	0.248	0.250	0.254	0.261
0.304	0.325	0.336	0.347	0.353	0.356	0.358	0.362	0.370
0.670	0.698	0.712	0.727	0.734	0.738	0.741	0.747	0.756
9.32	9.49	9.58	9.67	9.71	9.74	9.76	9.78	9.85
60.20	61.22	61.74	62.26	62.53	62.68	62.79	62.96	63.33
241.9	246.0	248.0	250.1	251.1	251.8	252.2	252.9	254.3
6056.	6157.	6209.	6261.	6287.	6302.	6313.	6330.	6366.
24224	24630	24836	25044	25148	25210	25253	25325	25465
0.106	0.130	0.143	0.157	0.165	0.169	0.173	0.179	0.189
0.132	0.157	0.171	0.186	0.193	0.198	0.201	0.207	0.217
0.244	0.272	0.286	0.302	0.309	0.314	0.317	0.324	0.334
0.342	0.371	0.386	0.402	0.410	0.415	0.418	0.424	0.434
0.626	0.657	0.672	0.689	0.697	0.702	0.705	0.711	0.721
3.38	3.41	3.43	3.44	3.45	3.45	3.46	3.47	3.48
9.39	9.42	9.44	9.46	9.47	9.47	9.47	9.48	9.49
19.40	19.43	19.45	19.46	19.47	19.48	19.48	19.49	19.50
99.40	99.43	99.45	99.47	99.47	99.48	99.48	99.49	99.50
199.4	199.4	199.4	199.5	199.5	199.5	199.5	199.5	199.5
0.124	0.154	0.172	0.191	0.201	0.207	0.211	0.220	0.234
0.153	0.185	0.203	0.222	0.232	0.238	0.242	0.251	0.264
0.270	0.304	0.323	0.342	0.352	0.358	0.363	0.370	0.384
0.367	0.402	0.420	0.439	0.449	0.455	0.459	0.467	0.480
0.624	0.658	0.675	0.693	0.702	0.708	0.711	0.719	0.730
2.44	2.46	2.46	2.47	2.47	2.47	2.47	2.47	2.47
5.23	5.20	5.18	5.17	5.16	5.15	5.15	5.14	5.13
8.79	8.70	8.66	8.62	8.59	8.58	8.57	8.55	8.53
27.23	26.87	26.69	26.50	26.41	26.36	26.32	26.24	26.12
43.69	43.08	42.78	42.47	42.31	42.22	42.15	42.03	41.83
0.137	0.172	0.193	0.216	0.229	0.237	0.242	0.253	0.269
0.167	0.204	0.226	0.249	0.261	0.269	0.274	0.285	0.301
0.288	0.327	0.349	0.372	0.384	0.391	0.396	0.407	0.422
0.384	0.424	0.445	0.467	0.478	0.485	0.490	0.500	0.514
0.627	0.664	0.683	0.702	0.712	0.718	0.722	0.731	0.743
2.08	2.08	2.08	2.08	2.08	2.08	2.08	2.08	2.08
3.92	3.87	3.84	3.82	3.80	3.80	3.79	3.78	3.76
5.96	5.86	5.80	5.75	5.72	5.70	5.69	5.66	5.63
14.55	14.20	14.20	13.84	13.74	13.69	13.65	13.59	13.46
20.97	20.44	20.17	19.89	19.75	19.67	19.61	19.51	19.32
0.146	0.186	0.210	0.237	0.251	0.260	0.266	0.279	0.299
0.177	0.219	0.244	0.270	0.285	0.293	0.299	0.312	0.331
0.301	0.345	0.369	0.395	0.408	0.417	0.422	0.432	0.452
0.397	0.440	0.463	0.488	0.501	0.508	0.514	0.524	0.541
0.631	0.669	0.690	0.711	0.722	0.728	0.732	0.741	0.755
1.89	1.89	1.88	1.88	1.88	1.88	1.87	1.87	1.87
3.30	3.24	3.21	3.17	3.16	3.15	3.14	3.13	3.10
4.74	4.62	4.56	4.50	4.46	4.44	4.43	4.41	4.36
10.05	9.72	9.55	9.38	9.29	9.24	9.20	9.13	9.02
13.62	13.15	12.90	12.66	12.53	12.46	12.40	12.31	12.14

Values given in this table are reproduced by permission from two sources: Tables of Percentage Points of the Inverted Beta (F) Distribution, by Prof. E. S. Pearson from M. Merrington and C. M. Thomson, *Biometrika,* vol. 33, p. 73 (1943); and "Handbook of Statistical Tables," D. B. Owen, Sandia Corporation, Addison-Wesley Publishing Company, Inc., Reading, Mass., p. 63, 1962.

TABLE B.6   Values of the F distribution (Continued)

Degrees of freedom for denominator

Prob. γ		1	2	3	4	5	6	7	8	9
										Degrees of freedom
6	0.005	$0.0^4 43$	0.0050	0.022	0.045	0.069	0.090	0.109	0.126	0.140
	0.01	$0.0^3 17$	0.010	0.036	0.066	0.094	0.118	0.139	0.157	0.172
	0.05	0.0043	0.052	0.112	0.162	0.202	0.233	0.259	0.279	0.296
	0.10	0.017	0.107	0.189	0.249	0.294	0.327	0.354	0.375	0.392
	0.25	0.111	0.302	0.413	0.481	0.524	0.561	0.586	0.606	0.622
	0.75	1.62	1.76	1.78	1.79	1.79	1.78	1.78	1.78	1.77
	0.90	3.78	3.46	3.29	3.18	3.11	3.05	3.01	2.98	2.96
	0.95	5.99	5.14	4.76	4.53	4.39	4.28	4.21	4.15	4.10
	0.99	13.74	10.92	9.78	9.15	8.75	8.47	8.26	8.10	7.98
	0.995	18.64	14.54	12.92	12.03	11.46	11.07	10.79	10.57	10.39
7	0.005	$0.0^4 42$	0.0050	0.023	0.046	0.070	0.093	0.113	0.130	0.145
	0.01	$0.0^3 17$	0.010	0.036	0.067	0.096	0.121	0.143	0.162	0.178
	0.05	0.0042	0.052	0.113	0.164	0.205	0.238	0.264	0.286	0.304
	0.10	0.017	0.107	0.190	0.251	0.297	0.332	0.359	0.381	0.399
	0.25	0.110	0.300	0.412	0.481	0.528	0.562	0.588	0.608	0.624
	0.75	1.57	1.70	1.72	1.72	1.71	1.71	1.70	1.70	1.69
	0.90	3.59	3.26	3.07	2.96	2.88	2.83	2.78	2.75	2.72
	0.95	5.59	4.74	4.35	4.12	3.97	3.87	3.79	3.73	3.68
	0.99	12.25	9.55	8.45	7.85	7.46	7.19	6.99	6.84	6.72
	0.995	16.24	12.40	10.88	10.05	9.52	9.16	8.89	8.68	8.51
8	0.005	$0.0^4 42$	0.0050	0.027	0.047	0.072	0.095	0.115	0.133	0.149
	0.01	$0.0^3 17$	0.010	0.036	0.068	0.097	0.123	0.146	0.166	0.183
	0.05	0.0042	0.052	0.113	0.166	0.208	0.241	0.268	0.291	0.310
	0.10	0.017	0.107	0.190	0.253	0.299	0.335	0.363	0.386	0.405
	0.25	0.109	0.298	0.411	0.481	0.529	0.563	0.589	0.610	0.627
	0.75	1.54	1.66	1.67	1.66	1.66	1.65	1.64	1.64	1.64
	0.90	3.46	3.11	2.92	2.81	2.73	2.67	2.62	2.59	2.56
	0.95	5.32	4.46	4.07	3.84	3.69	3.58	3.50	3.44	3.39
	0.99	11.26	8.65	7.59	7.01	6.63	6.37	6.18	6.03	5.91
	0.995	14.69	11.04	9.60	8.81	8.30	7.95	7.69	7.50	7.34
9	0.005	$0.0^4 42$	0.0050	0.023	0.047	0.073	0.096	0.117	0.136	0.153
	0.01	$0.0^3 17$	0.010	0.037	0.068	0.098	0.125	0.149	0.169	0.187
	0.05	0.0040	0.052	0.113	0.167	0.210	0.244	0.272	0.296	0.315
	0.10	0.017	0.107	0.191	0.254	0.302	0.338	0.367	0.390	0.410
	0.25	0.108	0.297	0.410	0.480	0.529	0.564	0.591	0.612	0.629
	0.75	1.51	1.62	1.63	1.63	1.62	1.61	1.60	1.60	1.59
	0.90	3.36	3.01	2.81	2.69	2.61	2.55	2.51	2.47	2.44
	0.95	5.12	4.26	3.86	3.63	3.48	3.37	3.29	3.23	3.18
	0.99	10.56	8.02	6.99	6.42	6.06	5.80	5.61	5.47	5.35
	0.995	13.61	10.11	8.72	7.96	7.47	7.13	6.88	6.69	6.54
10	0.005	$0.0^4 41$	0.0050	0.023	0.048	0.073	0.098	0.119	0.139	0.156
	0.01	$0.0^3 17$	0.010	0.037	0.069	0.100	0.127	0.151	0.172	0.190
	0.05	0.0041	0.052	0.114	0.168	0.211	0.246	0.275	0.299	0.319
	0.10	0.017	0.106	0.191	0.255	0.303	0.340	0.370	0.394	0.414
	0.25	0.107	0.296	0.409	0.480	0.529	0.565	0.592	0.613	0.631
	0.75	1.49	1.60	1.60	1.59	1.59	1.58	1.57	1.56	1.56
	0.90	3.28	2.92	2.73	2.61	2.52	2.46	2.41	2.38	2.35
	0.95	4.96	4.10	3.71	3.48	3.33	3.22	3.14	3.07	3.02
	0.99	10.04	7.56	6.55	5.99	5.64	5.39	5.20	5.06	4.94
	0.995	12.83	9.43	8.08	7.34	6.87	6.54	6.30	6.12	5.97

for numerator

10	15	20	30	40	50	60	100	∞
0.153	0.197	0.224	0.253	0.269	0.279	0.286	0.301	0.324
0.186	0.232	0.258	0.288	0.304	0.313	0.321	0.334	0.357
0.311	0.358	0.385	0.413	0.428	0.437	0.444	0.457	0.476
0.406	0.453	0.478	0.505	0.519	0.526	0.533	0.546	0.564
0.635	0.675	0.696	0.718	0.729	0.736	0.741	0.751	0.765
1.77	1.76	1.76	1.75	1.75	1.75	1.74	1.74	1.74
2.94	2.87	2.84	2.80	2.78	2.77	2.76	2.75	2.72
4.06	3.94	3.87	3.81	3.77	3.75	3.74	3.71	3.67
7.87	7.56	7.40	7.23	7.14	7.09	7.06	6.99	6.88
10.25	9.81	9.59	9.36	9.24	9.17	9.12	9.03	8.88
0.159	0.206	0.235	0.267	0.285	0.296	0.304	0.319	0.345
0.192	0.241	0.270	0.303	0.320	0.331	0.339	0.355	0.379
0.319	0.369	0.398	0.428	0.445	0.455	0.461	0.476	0.498
0.414	0.463	0.491	0.519	0.534	0.543	0.550	0.562	0.582
0.637	0.679	0.702	0.725	0.737	0.745	0.749	0.760	0.775
1.69	1.68	1.67	1.66	1.66	1.66	1.65	1.65	1.65
2.70	2.63	2.59	2.56	2.54	2.52	2.51	2.50	2.47
3.64	3.51	3.44	3.38	3.34	3.32	3.30	3.27	3.23
6.62	6.31	6.16	5.99	5.91	5.86	5.82	5.75	5.65
8.38	7.97	7.75	7.53	7.42	7.35	7.31	7.22	7.08
0.164	0.214	0.244	0.279	0.299	0.311	0.319	0.337	0.364
0.198	0.250	0.281	0.315	0.334	0.346	0.354	0.372	0.398
0.326	0.379	0.409	0.441	0.459	0.469	0.477	0.493	0.516
0.421	0.472	0.500	0.531	0.547	0.556	0.562	0.578	0.599
0.640	0.684	0.707	0.730	0.743	0.751	0.756	0.767	0.783
1.63	1.62	1.61	1.60	1.59	1.59	1.59	1.58	1.58
2.54	2.46	2.42	2.38	2.36	2.35	2.34	2.32	2.29
3.35	3.22	3.15	3.08	3.04	3.02	3.01	2.97	2.93
5.81	5.52	5.36	5.20	5.12	5.07	5.03	4.96	4.86
7.21	6.81	6.61	6.40	6.29	6.22	6.18	6.09	5.95
0.168	0.220	0.253	0.290	0.310	0.324	0.332	0.351	0.382
0.202	0.257	0.289	0.326	0.346	0.358	0.368	0.386	0.415
0.331	0.386	0.418	0.452	0.471	0.483	0.490	0.508	0.532
0.426	0.479	0.509	0.541	0.558	0.568	0.575	0.588	0.613
0.643	0.687	0.711	0.736	0.749	0.757	0.762	0.773	0.791
1.59	1.57	1.56	1.55	1.55	1.54	1.54	1.53	1.53
2.42	2.34	2.30	2.25	2.23	2.22	2.21	2.19	2.16
3.14	3.01	2.94	2.86	2.83	2.80	2.79	2.76	2.71
5.26	4.96	4.81	4.65	4.57	4.52	4.48	4.42	4.31
6.42	6.03	5.83	5.62	5.52	5.45	5.41	5.32	5.19
0.171	0.226	0.260	0.299	0.321	0.334	0.344	0.365	0.397
0.206	0.263	0.297	0.336	0.357	0.370	0.380	0.400	0.431
0.336	0.393	0.426	0.462	0.481	0.493	0.502	0.518	0.546
0.430	0.486	0.516	0.549	0.567	0.578	0.586	0.602	0.625
0.645	0.691	0.714	0.740	0.754	0.762	0.767	0.779	0.797
1.55	1.53	1.52	1.51	1.51	1.50	1.50	1.49	1.48
2.32	2.24	2.20	2.16	2.13	2.12	2.11	2.09	2.06
2.98	2.85	2.77	2.70	2.66	2.64	2.62	2.59	2.54
4.85	4.56	4.41	4.25	4.17	4.12	4.08	4.01	3.91
5.85	5.47	5.27	5.07	4.97	4.90	4.86	4.77	4.64

TABLE B.6   Values of the F distribution (Continued)

<table>
<tr><th rowspan="2"></th><th rowspan="2">Prob.
γ</th><th colspan="9">Degrees of freedom</th></tr>
<tr><th>1</th><th>2</th><th>3</th><th>4</th><th>5</th><th>6</th><th>7</th><th>8</th><th>9</th></tr>

<tr><td rowspan="10">12</td><td>0.005</td><td>$0.0^4 39$</td><td>0.0050</td><td>0.023</td><td>0.048</td><td>0.075</td><td>0.100</td><td>0.122</td><td>0.143</td><td>0.16</td></tr>
<tr><td>0.01</td><td>$0.0^3 16$</td><td>0.010</td><td>0.037</td><td>0.070</td><td>0.101</td><td>0.130</td><td>0.155</td><td>0.176</td><td>0.19</td></tr>
<tr><td>0.05</td><td>0.0041</td><td>0.052</td><td>0.114</td><td>0.169</td><td>0.214</td><td>0.250</td><td>0.280</td><td>0.305</td><td>0.32</td></tr>
<tr><td>0.10</td><td>0.016</td><td>0.106</td><td>0.192</td><td>0.257</td><td>0.306</td><td>0.344</td><td>0.375</td><td>0.400</td><td>0.42</td></tr>
<tr><td>0.25</td><td>0.106</td><td>0.295</td><td>0.408</td><td>0.480</td><td>0.530</td><td>0.566</td><td>0.594</td><td>0.616</td><td>0.63</td></tr>
<tr><td>0.75</td><td>1.46</td><td>1.56</td><td>1.56</td><td>1.55</td><td>1.54</td><td>1.53</td><td>1.52</td><td>1.51</td><td>1.5</td></tr>
<tr><td>0.90</td><td>3.18</td><td>2.81</td><td>2.61</td><td>2.48</td><td>2.39</td><td>2.33</td><td>2.28</td><td>2.24</td><td>2.2</td></tr>
<tr><td>0.95</td><td>4.75</td><td>3.89</td><td>3.49</td><td>3.26</td><td>3.11</td><td>3.00</td><td>2.91</td><td>2.85</td><td>2.8</td></tr>
<tr><td>0.99</td><td>9.33</td><td>6.93</td><td>5.95</td><td>5.41</td><td>5.06</td><td>4.82</td><td>4.64</td><td>4.50</td><td>4.3</td></tr>
<tr><td>0.995</td><td>11.75</td><td>8.51</td><td>7.23</td><td>6.52</td><td>6.07</td><td>5.76</td><td>5.52</td><td>5.35</td><td>5.2</td></tr>

<tr><td rowspan="10">15</td><td>0.005</td><td>$0.0^4 39$</td><td>0.0050</td><td>0.023</td><td>0.049</td><td>0.076</td><td>0.102</td><td>0.125</td><td>0.147</td><td>0.16</td></tr>
<tr><td>0.01</td><td>$0.0^3 16$</td><td>0.010</td><td>0.037</td><td>0.070</td><td>0.103</td><td>0.132</td><td>0.158</td><td>0.181</td><td>0.20</td></tr>
<tr><td>0.05</td><td>0.0041</td><td>0.051</td><td>0.115</td><td>0.170</td><td>0.216</td><td>0.254</td><td>0.285</td><td>0.311</td><td>0.33</td></tr>
<tr><td>0.10</td><td>0.016</td><td>0.106</td><td>0.192</td><td>0.258</td><td>0.309</td><td>0.348</td><td>0.380</td><td>0.406</td><td>0.42</td></tr>
<tr><td>0.25</td><td>0.105</td><td>0.293</td><td>0.407</td><td>0.480</td><td>0.531</td><td>0.568</td><td>0.596</td><td>0.618</td><td>0.63</td></tr>
<tr><td>0.75</td><td>1.43</td><td>1.52</td><td>1.52</td><td>1.51</td><td>1.49</td><td>1.48</td><td>1.47</td><td>1.46</td><td>1.4</td></tr>
<tr><td>0.90</td><td>3.07</td><td>2.70</td><td>2.49</td><td>2.36</td><td>2.27</td><td>2.21</td><td>2.16</td><td>2.12</td><td>2.0</td></tr>
<tr><td>0.95</td><td>4.54</td><td>3.68</td><td>3.29</td><td>3.06</td><td>2.90</td><td>2.79</td><td>2.71</td><td>2.64</td><td>2.5</td></tr>
<tr><td>0.99</td><td>8.68</td><td>6.36</td><td>5.42</td><td>4.89</td><td>4.56</td><td>4.32</td><td>4.14</td><td>4.00</td><td>3.8</td></tr>
<tr><td>0.995</td><td>10.80</td><td>7.70</td><td>6.48</td><td>5.80</td><td>5.37</td><td>5.07</td><td>4.85</td><td>4.67</td><td>4.5</td></tr>

<tr><td rowspan="10">20</td><td>0.005</td><td>$0.0^4 39$</td><td>0.0050</td><td>0.023</td><td>0.050</td><td>0.077</td><td>0.104</td><td>0.129</td><td>0.151</td><td>0.17</td></tr>
<tr><td>0.01</td><td>$0.0^3 16$</td><td>0.010</td><td>0.037</td><td>0.071</td><td>0.105</td><td>0.135</td><td>0.162</td><td>0.187</td><td>0.20</td></tr>
<tr><td>0.05</td><td>0.0040</td><td>0.051</td><td>0.115</td><td>0.172</td><td>0.219</td><td>0.258</td><td>0.290</td><td>0.318</td><td>0.34</td></tr>
<tr><td>0.10</td><td>0.016</td><td>0.106</td><td>0.193</td><td>0.260</td><td>0.312</td><td>0.353</td><td>0.385</td><td>0.412</td><td>0.43</td></tr>
<tr><td>0.25</td><td>0.104</td><td>0.292</td><td>0.407</td><td>0.480</td><td>0.531</td><td>0.569</td><td>0.598</td><td>0.622</td><td>0.64</td></tr>
<tr><td>0.75</td><td>1.40</td><td>1.49</td><td>1.48</td><td>1.47</td><td>1.45</td><td>1.44</td><td>1.43</td><td>1.42</td><td>1.4</td></tr>
<tr><td>0.90</td><td>2.97</td><td>2.59</td><td>2.38</td><td>2.25</td><td>2.16</td><td>2.09</td><td>2.04</td><td>2.00</td><td>1.9</td></tr>
<tr><td>0.95</td><td>4.35</td><td>3.49</td><td>3.10</td><td>2.87</td><td>2.71</td><td>2.60</td><td>2.51</td><td>2.45</td><td>2.3</td></tr>
<tr><td>0.99</td><td>8.10</td><td>5.85</td><td>4.94</td><td>4.43</td><td>4.10</td><td>3.87</td><td>3.70</td><td>3.56</td><td>3.4</td></tr>
<tr><td>0.995</td><td>9.94</td><td>6.99</td><td>5.82</td><td>5.17</td><td>4.76</td><td>4.47</td><td>4.26</td><td>4.09</td><td>3.9</td></tr>

<tr><td rowspan="10">25</td><td>0.005</td><td>$0.0^4 40$</td><td>0.0050</td><td>0.023</td><td>0.050</td><td>0.078</td><td>0.106</td><td>0.131</td><td>0.154</td><td>0.17</td></tr>
<tr><td>0.01</td><td>$0.0^3 16$</td><td>0.010</td><td>0.038</td><td>0.072</td><td>0.106</td><td>0.137</td><td>0.165</td><td>0.190</td><td>0.21</td></tr>
<tr><td>0.05</td><td>0.0040</td><td>0.051</td><td>0.116</td><td>0.173</td><td>0.221</td><td>0.260</td><td>0.293</td><td>0.321</td><td>0.34</td></tr>
<tr><td>0.10</td><td>0.016</td><td>0.106</td><td>0.193</td><td>0.261</td><td>0.313</td><td>0.355</td><td>0.388</td><td>0.416</td><td>0.43</td></tr>
<tr><td>0.25</td><td>0.104</td><td>0.291</td><td>0.406</td><td>0.480</td><td>0.532</td><td>0.570</td><td>0.600</td><td>0.624</td><td>0.64</td></tr>
<tr><td>0.75</td><td>1.39</td><td>1.47</td><td>1.46</td><td>1.44</td><td>1.42</td><td>1.41</td><td>1.40</td><td>1.39</td><td>1.3</td></tr>
<tr><td>0.90</td><td>2.92</td><td>2.53</td><td>2.32</td><td>2.18</td><td>2.09</td><td>2.02</td><td>1.97</td><td>1.93</td><td>1.8</td></tr>
<tr><td>0.95</td><td>4.24</td><td>3.39</td><td>2.99</td><td>2.76</td><td>2.60</td><td>2.49</td><td>2.40</td><td>2.34</td><td>2.2</td></tr>
<tr><td>0.99</td><td>7.77</td><td>5.57</td><td>4.68</td><td>4.18</td><td>3.86</td><td>3.63</td><td>3.46</td><td>3.32</td><td>3.2</td></tr>
<tr><td>0.995</td><td>9.48</td><td>6.60</td><td>5.46</td><td>4.84</td><td>4.43</td><td>4.15</td><td>3.94</td><td>3.78</td><td>3.6</td></tr>

<tr><td rowspan="10">30</td><td>0.005</td><td>$0.0^4 40$</td><td>0.0050</td><td>0.024</td><td>0.050</td><td>0.079</td><td>0.107</td><td>0.133</td><td>0.156</td><td>0.17</td></tr>
<tr><td>0.01</td><td>$0.0^3 16$</td><td>0.010</td><td>0.038</td><td>0.072</td><td>0.107</td><td>0.138</td><td>0.167</td><td>0.192</td><td>0.21</td></tr>
<tr><td>0.05</td><td>0.0040</td><td>0.051</td><td>0.116</td><td>0.174</td><td>0.222</td><td>0.263</td><td>0.296</td><td>0.325</td><td>0.34</td></tr>
<tr><td>0.10</td><td>0.016</td><td>0.106</td><td>0.193</td><td>0.262</td><td>0.315</td><td>0.357</td><td>0.391</td><td>0.420</td><td>0.44</td></tr>
<tr><td>0.25</td><td>0.103</td><td>0.290</td><td>0.406</td><td>0.480</td><td>0.532</td><td>0.571</td><td>0.601</td><td>0.625</td><td>0.64</td></tr>
<tr><td>0.75</td><td>1.38</td><td>1.45</td><td>1.44</td><td>1.42</td><td>1.41</td><td>1.39</td><td>1.38</td><td>1.37</td><td>1.3</td></tr>
<tr><td>0.90</td><td>2.88</td><td>2.49</td><td>2.28</td><td>2.14</td><td>2.05</td><td>1.98</td><td>1.93</td><td>1.88</td><td>1.8</td></tr>
<tr><td>0.95</td><td>4.17</td><td>3.32</td><td>2.92</td><td>2.69</td><td>2.53</td><td>2.42</td><td>2.33</td><td>2.27</td><td>2.2</td></tr>
<tr><td>0.99</td><td>7.56</td><td>5.39</td><td>4.51</td><td>4.02</td><td>3.70</td><td>3.47</td><td>3.30</td><td>3.17</td><td>3.0</td></tr>
<tr><td>0.995</td><td>9.18</td><td>6.35</td><td>5.24</td><td>4.62</td><td>4.23</td><td>3.95</td><td>3.74</td><td>3.58</td><td>3.4</td></tr>
</table>

Degrees of freedom for denominator

for numerator

10	15	20	30	40	50	60	100	∞
0.177	0.235	0.272	0.315	0.339	0.355	0.365	0.388	0.424
0.212	0.273	0.310	0.352	0.375	0.391	0.401	0.422	0.458
0.343	0.404	0.439	0.478	0.499	0.513	0.522	0.541	0.571
0.438	0.496	0.528	0.564	0.583	0.595	0.604	0.621	0.647
0.649	0.695	0.721	0.748	0.762	0.771	0.777	0.789	0.808
1.50	1.48	1.47	1.45	1.45	1.44	1.44	1.43	1.42
2.19	2.10	2.06	2.01	1.99	1.97	1.96	1.94	1.90
2.75	2.62	2.54	2.47	2.43	2.40	2.38	2.35	2.30
4.30	4.01	3.86	3.70	3.62	3.57	3.54	3.47	3.36
5.09	4.72	4.53	4.33	4.23	4.17	4.12	4.04	3.90
0.183	0.246	0.286	0.333	0.360	0.377	0.389	0.415	0.457
0.219	0.284	0.324	0.370	0.397	0.413	0.425	0.450	0.490
0.351	0.416	0.454	0.496	0.519	0.535	0.545	0.565	0.600
0.446	0.507	0.542	0.581	0.602	0.614	0.624	0.641	0.672
0.652	0.701	0.728	0.757	0.772	0.782	0.788	0.802	0.822
1.45	1.43	1.41	1.40	1.39	1.38	1.38	1.38	1.36
2.06	1.97	1.92	1.87	1.85	1.83	1.82	1.79	1.76
2.54	2.40	2.33	2.25	2.20	2.18	2.16	2.12	2.07
3.80	3.52	3.37	3.21	3.13	3.08	3.05	2.98	2.87
4.42	4.07	3.88	3.69	3.59	3.52	3.48	3.39	3.26
0.190	0.258	0.301	0.354	0.385	0.405	0.419	0.448	0.500
0.227	0.297	0.340	0.392	0.422	0.441	0.455	0.483	0.532
0.360	0.430	0.471	0.518	0.544	0.562	0.572	0.595	0.637
0.454	0.520	0.557	0.600	0.623	0.637	0.648	0.671	0.704
0.656	0.708	0.736	0.767	0.784	0.794	0.801	0.816	0.840
1.40	1.37	1.36	1.34	1.33	1.32	1.32	1.31	1.29
1.94	1.84	1.79	1.74	1.71	1.69	1.68	1.65	1.61
2.35	2.20	2.12	2.04	1.99	1.97	1.95	1.91	1.84
3.37	3.09	2.94	2.78	2.69	2.64	2.61	2.54	2.42
3.85	3.50	3.32	3.12	3.02	2.96	2.92	2.83	2.69
0.194	0.265	0.312	0.369	0.403	0.425	0.440	0.474	0.533
0.231	0.305	0.352	0.407	0.440	0.462	0.476	0.509	0.564
0.366	0.438	0.482	0.532	0.561	0.577	0.591	0.617	0.663
0.460	0.528	0.568	0.613	0.637	0.654	0.665	0.688	0.727
0.659	0.713	0.742	0.774	0.792	0.803	0.811	0.827	0.852
1.37	1.34	1.22	1.31	1.29	1.29	1.28	1.27	1.25
1.87	1.77	1.72	1.66	1.63	1.61	1.59	1.57	1.52
2.24	2.09	2.01	1.92	1.87	1.84	1.82	1.78	1.71
3.13	2.85	2.70	2.54	2.45	2.41	2.36	2.30	2.17
3.54	3.20	3.01	2.82	2.72	2.66	2.61	2.53	2.38
0.197	0.271	0.320	0.381	0.416	0.441	0.457	0.495	0.559
0.235	0.311	0.260	0.419	0.454	0.476	0.493	0.529	0.590
0.370	0.445	0.490	0.543	0.573	0.592	0.606	0.637	0.685
0.464	0.534	0.575	0.623	0.649	0.667	0.678	0.704	0.746
0.661	0.716	0.746	0.780	0.798	0.810	0.818	0.835	0.862
1.35	1.32	1.30	1.28	1.27	1.26	1.26	1.25	1.23
1.82	1.72	1.67	1.61	1.57	1.55	1.54	1.51	1.46
2.16	2.01	1.93	1.84	1.79	1.76	1.74	1.70	1.62
2.98	2.70	2.55	2.39	2.30	2.25	2.21	2.13	2.01
3.34	3.01	2.82	2.63	2.52	2.46	2.42	2.32	2.18

TABLE B.6   Values of the F distribution (Continued)

Degrees of freedom for denominator	Prob. γ	1	2	3	4	5	6	7	8	9
						Degrees of freedom				
40	0.005	$0.0^4 40$	0.0050	0.024	0.051	0.080	0.108	0.135	0.159	0.18
	0.01	$0.0^3 16$	0.010	0.038	0.073	0.108	0.140	0.169	0.195	0.21
	0.05	0.0040	0.051	0.116	0.175	0.224	0.265	0.299	0.329	0.35
	0.10	0.016	0.106	0.194	0.263	0.317	0.360	0.394	0.424	0.44
	0.25	0.103	0.290	0.405	0.480	0.533	0.572	0.603	0.627	0.64
	0.75	1.36	1.44	1.42	1.40	1.39	1.37	1.36	1.35	1.3
	0.90	2.84	2.44	2.23	2.09	2.00	1.93	1.87	1.83	1.7
	0.95	4.08	3.23	2.84	2.61	2.45	2.34	2.25	2.18	2.1
	0.99	7.31	5.18	4.31	3.83	3.51	3.29	3.12	2.99	2.8
	0.995	8.83	6.07	4.98	4.37	3.99	3.71	3.51	3.35	3.2
60	0.005	$0.0^4 40$	0.0050	0.024	0.051	0.081	0.110	0.137	0.162	0.18
	0.01	$0.0^3 16$	0.010	0.038	0.073	0.109	0.142	0.172	0.199	0.22
	0.05	0.0040	0.051	0.116	0.176	0.226	0.267	0.303	0.333	0.35
	0.10	0.016	0.106	0.194	0.264	0.318	0.362	0.398	0.428	0.45
	0.25	0.102	0.289	0.405	0.480	0.534	0.573	0.604	0.629	0.65
	0.75	1.35	1.42	1.41	1.38	1.37	1.35	1.33	1.32	1.3
	0.90	2.79	2.39	2.18	2.04	1.95	1.87	1.82	1.77	1.7
	0.95	4.00	3.15	2.76	2.53	2.37	2.25	2.17	2.10	2.0
	0.99	7.08	4.98	4.13	3.65	3.34	3.12	2.95	2.82	2.7
	0.995	8.49	5.80	4.73	4.14	3.76	3.49	3.29	3.13	3.0
80	0.005	$0.0^4 40$	0.0050	0.024	0.051	0.081	0.110	0.138	0.163	0.18
	0.01	$0.0^3 16$	0.010	0.038	0.074	0.109	0.142	0.173	0.200	0.22
	0.05	0.0040	0.051	0.117	0.176	0.226	0.269	0.304	0.334	0.36
	0.10	0.016	0.106	0.194	0.264	0.319	0.364	0.400	0.429	0.45
	0.25	0.102	0.289	0.405	0.481	0.534	0.574	0.604	0.630	0.65
	0.75	1.34	1.41	1.40	1.38	1.36	1.34	1.33	1.31	1.3
	0.90	2.77	2.37	2.15	2.02	1.92	1.85	1.79	1.75	1.7
	0.95	3.96	3.11	2.72	2.49	2.33	2.21	2.13	2.06	2.0
	0.99	6.96	4.88	4.04	3.56	3.26	3.04	2.87	2.74	2.6
	0.995	8.34	5.67	4.61	4.03	3.65	3.39	3.19	3.03	2.9
120	0.005	$0.0^4 39$	0.0050	0.024	0.051	0.081	0.111	0.139	0.165	0.18
	0.01	$0.0^3 16$	0.010	0.038	0.038	0.074	0.110	0.143	0.174	0.202	0.22
	0.05	0.0039	0.051	0.117	0.177	0.227	0.270	0.306	0.337	0.36
	0.10	0.016	0.105	0.194	0.265	0.320	0.365	0.401	0.432	0.45
	0.25	0.102	0.288	0.405	0.481	0.534	0.574	0.606	0.631	0.64
	0.75	1.34	1.40	1.39	1.37	1.35	1.33	1.31	1.30	1.2
	0.90	2.75	2.35	2.13	1.99	1.90	1.82	1.77	1.72	1.6
	0.95	3.92	3.07	2.68	2.45	2.29	2.18	2.09	2.02	1.9
	0.99	6.85	4.79	3.95	3.48	3.17	2.96	2.79	2.66	2.5
	0.995	8.18	5.54	4.50	3.92	3.55	3.28	3.09	2.93	2.8
∞	0.005	$0.0^4 39$	0.0050	0.024	0.052	0.082	0.113	0.141	0.168	0.19
	0.01	$0.0^3 16$	0.010	0.038	0.074	0.111	0.145	0.177	0.206	0.23
	0.05	0.0039	0.051	0.117	0.178	0.229	0.273	0.310	0.342	0.36
	0.10	0.016	0.105	0.195	0.266	0.322	0.367	0.405	0.436	0.46
	0.25	0.102	0.288	0.404	0.481	0.535	0.576	0.608	0.634	0.65
	0.75	1.32	1.39	1.37	1.35	1.33	1.31	1.29	1.28	1.2
	0.90	2.71	2.30	2.08	1.94	1.85	1.77	1.72	1.67	1.6
	0.95	3.84	3.00	2.60	2.37	2.21	2.10	2.01	1.94	1.8
	0.99	6.63	4.61	3.78	3.32	3.02	2.80	2.64	2.51	2.4
	0.995	7.88	5.30	4.28	3.72	3.35	3.09	2.90	2.74	2.6

for numerator

10	15	20	30	40	50	60	100	∞
0.201	0.279	0.331	0.396	0.436	0.463	0.481	0.524	0.599
0.240	0.319	0.371	0.435	0.473	0.498	0.516	0.556	0.628
0.376	0.454	0.502	0.558	0.591	0.613	0.627	0.658	0.717
0.469	0.542	0.585	0.636	0.664	0.683	0.696	0.724	0.772
0.664	0.720	0.752	0.787	0.806	0.819	0.828	0.846	0.877
1.33	1.30	1.28	1.25	1.24	1.23	1.22	1.21	1.19
1.76	1.66	1.61	1.54	1.51	1.48	1.47	1.43	1.38
2.08	1.92	1.84	1.74	1.69	1.66	1.65	1.59	1.51
2.80	2.52	2.37	2.20	2.11	2.06	2.02	1.94	1.80
3.12	2.78	2.60	2.40	2.30	2.23	2.18	2.09	1.93
0.206	0.287	0.343	0.414	0.458	0.488	0.510	0.559	0.652
0.245	0.328	0.383	0.453	0.495	0.524	0.545	0.592	0.679
0.382	0.463	0.514	0.575	0.611	0.633	0.652	0.690	0.759
0.475	0.550	0.596	0.650	0.682	0.702	0.717	0.750	0.806
0.667	0.725	0.758	0.796	0.816	0.830	0.840	0.860	0.896
1.30	1.27	1.25	1.22	1.21	1.20	1.19	1.17	1.15
1.71	1.60	1.54	1.48	1.44	1.41	1.40	1.36	1.29
1.99	1.84	1.75	1.65	1.59	1.56	1.53	1.48	1.39
2.63	2.35	2.20	2.03	1.94	1.88	1.84	1.75	1.60
2.90	2.57	2.39	2.19	2.08	2.01	1.96	1.86	1.69
0.208	0.291	0.350	0.424	0.470	0.502	0.526	0.579	0.685
0.247	0.333	0.390	0.463	0.507	0.537	0.559	0.609	0.709
0.384	0.467	0.521	0.582	0.621	0.645	0.667	0.705	0.782
0.477	0.554	0.601	0.656	0.690	0.713	0.727	0.762	0.825
0.668	0.727	0.761	0.800	0.821	0.835	0.845	0.870	0.906
1.29	1.26	1.23	1.21	1.19	1.18	1.17	1.16	1.12
1.68	1.57	1.51	1.44	1.40	1.38	1.36	1.32	1.24
1.95	1.79	1.70	1.60	1.54	1.51	1.48	1.43	1.32
2.55	2.27	2.12	1.94	1.85	1.79	1.75	1.67	1.49
2.80	2.47	2.29	2.09	1.97	1.90	1.85	1.77	1.56
0.211	0.297	0.356	0.434	0.484	0.520	0.545	0.605	0.733
0.250	0.338	0.397	0.474	0.522	0.556	0.579	0.636	0.755
0.388	0.473	0.527	0.594	0.634	0.661	0.682	0.727	0.819
0.480	0.560	0.609	0.667	0.702	0.726	0.742	0.781	0.855
0.670	0.730	0.765	0.805	0.828	0.843	0.853	0.877	0.923
1.28	1.24	1.22	1.19	1.18	1.17	1.16	1.14	1.12
1.65	1.55	1.48	1.41	1.37	1.34	1.32	1.27	1.19
1.91	1.75	1.66	1.55	1.50	1.46	1.43	1.37	1.25
2.47	2.19	2.03	1.86	1.76	1.70	1.66	1.56	1.38
2.71	2.37	2.19	1.98	1.87	1.80	1.75	1.64	1.43
0.216	0.307	0.372	0.460	0.518	0.559	0.592	0.671	0.843*
0.256	0.349	0.413	0.499	0.554	0.595	0.625	0.699	0.858*
0.394	0.484	0.543	0.617	0.663	0.694	0.720	0.781	0.896*
0.487	0.570	0.622	0.687	0.726	0.752	0.774	0.826	0.919*
0.674	0.736	0.773	0.816	0.842	0.860	0.872	0.901	0.957*
1.25	1.22	1.19	1.16	1.14	1.13	1.12	1.09	1.04*
1.60	1.49	1.42	1.34	1.30	1.26	1.24	1.18	1.08*
1.83	1.67	1.57	1.46	1.39	1.35	1.32	1.24	1.11*
2.32	2.04	1.88	1.70	1.59	1.52	1.47	1.36	1.15*
2.52	2.19	2.00	1.79	1.67	1.59	1.53	1.40	1.17*

*These values are for v_1 = 500; for v_1 and v_2 = ∞, all values are 1.00.

TABLE B.7   Squares and square roots

n	n^2	$\sqrt{n}$	n	n^2	$\sqrt{n}$	n	n^2	$\sqrt{n}$	n	n^2	$\sqrt{n}$
1	1	1.0000	51	2,601	7.1414	101	10,201	10.0499	151	22,801	12.2882
2	4	1.4142	52	2,704	7.2111	102	10,404	10.0995	152	23,104	12.3288
3	9	1.7321	53	2,809	7.2801	103	10,609	10.1489	153	23,409	12.3693
4	16	2.0000	54	2,916	7.3485	104	10,816	10,1980	154	23,716	12.4097
5	25	2.2361	55	3,025	7.4162	105	11,025	10.2470	155	24,025	12.4499
6	36	2,4495	56	3,136	7.4833	106	11,236	10.2956	156	24,336	12.4900
7	49	2.6458	57	3,249	7.5498	107	11,449	10.3441	157	24,649	12.5300
8	64	2.8284	58	3,364	7.6158	108	11,664	10.3923	158	24,964	12.5698
9	81	3.0000	59	3,481	7.6811	109	11,881	10.4403	159	25,281	12.6095
10	100	3.1623	60	3,600	7.7460	110	12,100	10.4881	160	25,600	12.6491
11	121	3.3166	61	3,721	7.8102	111	12,321	10.5357	161	25,921	12.6886
12	144	3.4641	62	3,844	7.8740	112	12,544	10.5830	162	26,244	12.7279
13	169	3.6056	63	3,969	7.9373	113	12,769	10.6301	163	26,569	12.7671
14	196	3.7417	64	4,096	8.0000	114	12,996	10.6771	164	26,896	12.8062
15	225	3.8730	65	4,225	8.0623	115	13,225	10.7238	165	27,225	12.8452
16	256	4.0000	66	4,356	8.1240	116	13,456	10.7703	166	27,556	12.8841
17	289	4.1231	67	4,489	8.1854	117	13,689	10.8167	167	27,889	12.9228
18	324	4.2426	68	4,624	8.2462	118	13,924	10.8628	168	28,224	12.9615
19	361	4.3589	69	4,761	8.3066	119	14,161	10.9087	169	28,561	13.0000
20	400	4.4721	70	4,900	8.3666	120	14,400	10.9545	170	28,900	13.0384
21	441	4.5826	71	5,041	8.4261	121	14,641	11.0000	171	29,241	13.0767
22	484	4.6904	72	5,184	8.4853	122	14,884	11.0454	172	29,584	13.1149
23	529	4.7958	73	5,329	8.5440	123	15,129	11.0905	173	29,929	13.1529
24	576	4.8990	74	5,476	8.6023	124	15,376	11.1355	174	30,276	13.1909
25	625	5.0000	75	5,625	8.6603	125	15,625	11.1803	175	30,625	13.2288
26	676	5.0990	76	5,776	8.7178	126	15,876	11.2250	176	30,976	13.2665
27	729	5.1962	77	5,929	8.7750	127	16,129	11.2694	177	31,329	13.3041
28	784	5.2915	78	6,084	8.8318	128	16,384	11.3137	178	31,684	13.3417
29	841	5.3852	79	6,241	8.8882	129	16,641	11.3578	179	32,041	13.3791
30	900	5.4772	80	6,400	8.9443	130	16,900	11.4018	180	32,400	13.4164
31	961	5.5678	81	6,561	9.0000	131	17,161	11.4455	181	32,761	13.4536
32	1,024	5.6569	82	6,724	9.0554	132	17,424	11.8491	182	33,124	13.4907
33	1,089	5.7446	83	6,889	9.1104	133	17,689	11.5326	183	33,489	13.5277
34	1,156	5.8310	84	7,056	9.1652	134	17,956	11.5758	184	33,856	13.5647
35	1,225	5.9161	85	7,225	9.2195	135	18,225	11.6190	185	34,225	13.6015
36	1,296	6.0000	86	7,396	9.2736	136	18,496	11.6619	186	34,596	13.6382
37	1,369	6.0828	87	7,569	9.3274	137	18,769	11.7047	187	34,969	13.6748
38	1,444	6.1644	88	7,744	9.3808	138	19,044	11.7473	188	35,344	13.7113
39	1,521	6.2450	89	7,921	9.4340	139	19,321	11.7898	189	35,721	13.7477
40	1,600	6.3246	90	8,100	9.4868	140	19,600	11.8322	190	36,100	13.7840
41	1,681	6.4031	91	8,281	9.5394	141	19,881	11.8743	191	36,481	13.8203
42	1,764	6.4807	92	8,464	9.5917	142	20,164	11.9164	192	36,864	13.8564
43	1,849	6.5574	93	8,649	9.6437	143	20,449	11.9583	193	37,249	13.8924
44	1,936	6.6332	94	8,836	9.6954	144	20,736	12.0000	194	37,636	13.9284
45	2,025	6.7082	95	9,025	9.7468	145	21,025	12.0416	195	38,025	13.9642
46	2,116	6.7823	96	9,216	9.7980	146	21,316	12.0830	196	38,416	14.0000
47	2,209	6.8557	97	9,409	9.8489	147	21,609	12.1244	197	38,809	14.0357
48	2,304	6.9282	98	9,604	9.8995	148	21,904	12.1655	198	39,204	14.0712
49	2,401	7.0000	99	9,801	9.9499	149	22,201	12.2066	199	39,601	14.1067
50	2,500	7.0711	100	10,000	10.0000	150	22,500	12.2474	200	40,000	14.1421

n	n^2	$\sqrt{n}$	n	n^2	$\sqrt{n}$	n	n^2	$\sqrt{n}$	n	n^2	$\sqrt{n}$
201	40,401	14.1774	251	63,001	15.8430	301	90,601	17.3494	351	123,201	18.7350
202	40,804	14.2127	252	63,504	15.8745	302	91,204	17.3781	352	123,904	18.7617
203	41,209	14.2478	253	64,009	15.9060	303	91,809	17.4069	353	124,609	18.7883
204	41,616	14.2829	254	64,516	15.9374	304	92,416	17.4356	354	125,316	18.8149
205	42,025	14.3178	255	65,025	15.9687	305	93,025	17.4642	355	126,025	18.8414
206	42,436	14.3527	256	65,536	16.0000	306	93,636	17.4929	356	126,736	18.8680
207	42,849	14.3875	257	66,049	16.0312	307	94,249	17.5214	357	127,449	18.8944
208	43,264	14.4222	258	66,564	16.0624	308	94,864	17.5499	358	128,164	18.9209
209	43,681	14.4568	259	67,081	16.0935	309	95,481	17.5784	359	128,881	18.9473
210	44,100	14.4914	260	67,600	16.1245	310	96,100	17.6068	360	129,600	18.9737
211	44,521	14.5258	261	68,121	16.1555	311	96,721	17.6352	361	130,321	19.0000
212	44,944	14.5602	262	68,644	16.1864	312	97,344	17.6635	362	131,044	19.0263
213	45,369	14.5945	263	69,169	16.2173	313	97,969	17.6918	363	131,769	19.0526
214	45,796	14.6287	264	69,696	16.2481	314	98,596	17.7200	364	132,496	19.0788
215	46,225	14.6629	265	70,225	16.2788	315	99,225	17.7482	365	133,225	19.1050
216	46,656	14.6969	266	70,756	16.3095	316	99,856	17.7764	366	133,956	19.1311
217	47,089	14.7309	267	71,289	16.3401	317	100,489	17.8045	367	134,689	19.1572
218	47,524	14.7648	268	71,824	16.3707	318	101,124	17.8326	368	135,424	19.1833
219	47,961	14.7986	269	72,361	16.4012	319	101,761	17.8606	369	136,161	19.2094
220	48,400	14.8324	270	72,900	16.4317	320	102,400	17.8885	370	136,900	19.2354
221	48,841	14.8661	271	73,441	16.4621	321	103,041	17.9165	371	137,641	19.2614
222	49,284	14.8997	272	73,984	16.4924	322	103,684	17.9444	372	138,384	19.2873
223	49,729	14.9332	273	74,529	16.5227	323	104,329	17.9722	373	139,129	19.3132
224	50,176	14.9666	274	75,076	16.5529	324	104,976	18.0000	374	139,876	19.3391
225	50,625	15.0000	275	75,625	16.5831	325	105,625	18.0278	375	140,625	19.3649
226	51,076	15.0333	276	76,176	16.6132	326	106,276	18.0555	376	141,376	19.3907
227	51,529	15.0665	277	76,729	16.6433	327	106,929	18.0831	377	142,129	19.4165
228	51,984	15.0997	278	77,284	16.6733	328	107,584	18.1108	378	142,884	19.4422
229	52,441	15.1327	279	77,841	16.7033	329	108,241	18.1384	379	143,641	19.4679
230	52,900	15.1658	280	78,400	16.7332	330	108,900	18.1659	380	144,400	19.4936
231	53,361	15.1987	281	78,961	16.7631	331	109,561	18.1934	381	145,161	19.5192
232	53,824	15.2315	282	79,524	16.7929	332	110,224	18.2209	382	145,924	19.5448
233	54,289	15.2643	282	80,089	16.8226	333	110,889	18.2483	383	146,689	19.5704
234	54,756	15.2971	284	80,656	16.8523	334	111,556	18.2757	384	147,456	19.5959
235	55,225	15.3297	285	81,225	16.8819	335	112,225	18.3030	385	148,225	19.6214
236	55,696	15.3623	286	81,796	16.9115	336	112,896	18.3303	386	148,996	19.6469
237	56,169	15.3948	287	82,369	16.9411	337	113,569	18.3576	387	149,769	19.6723
238	56,644	15.4272	288	82,944	16.9706	338	114,244	18.3848	388	150,544	19.6977
239	57,121	15.4596	289	83,521	17.0000	339	114,921	18.4120	389	151,321	19.7231
240	57,600	15.4919	290	84,100	17.0294	340	115,600	18.4391	390	152,100	19.7484
241	58,081	15.5242	291	84,681	17.0587	341	116,281	18.4662	391	152,881	19.7737
242	58,564	15.5563	292	85,264	17.0880	342	116,964	18.4932	392	153,664	19.7990
243	59,049	15.5885	293	85,849	17.1172	343	117,649	18.5203	393	154,449	19.8242
244	59,536	15.6205	294	86,436	17.1464	344	118,336	18.5472	394	155,236	19.8494
245	60,025	15.6525	295	87,025	17.1756	345	119,025	18.5742	395	156,025	19.8746
246	60,516	15.6844	296	87,616	17.2047	346	119,716	18.6011	396	156,816	19.8997
247	61,009	15.7162	297	88,209	17.2337	347	120,409	18.6279	397	157,609	19.9249
248	61,504	15.7480	298	88,804	17.2627	348	121,104	18.6548	398	158,404	19.9499
249	62,001	15.7797	299	89,401	17.2916	349	121,801	18.6815	399	159,201	19.9750
250	62,500	15.8114	300	90,000	17.3205	350	122,500	18.7083	400	160,000	20.0000

TABLE B.7   Squares and square roots (Continued)

n	n^2	$\sqrt{n}$	n	n^2	$\sqrt{n}$	n	n^2	$\sqrt{n}$	n	n^2	$\sqrt{n}$
401	160,801	20.0250	451	203,401	21.2368	501	251,001	22.3830	551	303,601	23.4734
402	161,604	20.0499	452	204,304	21.2603	502	252,004	22.4054	552	304,704	23.4947
403	162,409	20.0749	453	205,209	21.2838	503	253,009	22.4277	553	305,809	23.5160
404	163,216	20.0998	454	206,116	21.3073	504	254,016	22.4499	554	306,916	23.5372
405	164,025	20.1246	455	207,025	21.3307	505	255,025	22.4722	555	308,025	23.5584
406	164,836	20.1494	456	207,936	21.3542	506	256,036	22.4944	556	309,136	23.5797
407	165,649	20.1742	457	208,849	21.3776	507	257,049	22.5167	557	310,249	23.6008
408	166,464	20.1990	458	209,764	21.4009	508	258,064	22.5389	558	311,364	23.6220
409	167,281	20.2237	459	210,681	21.4243	509	259,081	22.5610	559	312,481	23.6432
410	168,100	20.2485	460	211,600	21.4476	510	260,100	22.5832	560	313,600	23.6643
411	168,921	20.2731	461	212,521	21.4709	511	261,121	22.6053	561	314,721	23.6854
412	169,744	20.2978	462	213,444	21.4942	512	262,144	22.6274	562	315,844	23.7065
413	170,569	20.3224	463	214,369	21.5174	513	263,169	22.6495	563	316,969	23.7276
414	171,396	20.3470	464	215,296	21.5407	514	264,196	22.6716	564	318,096	23.7487
415	172,225	20.3715	465	216,225	21.5639	515	265,225	22.6936	565	319,225	23.7698
416	173,056	20.3961	466	217,156	21.5870	516	266,256	22.7156	566	320,356	23.7908
417	173,889	20.4206	467	218,089	21.6102	517	267,289	22.7376	567	321,489	23.8118
418	174,724	20.4450	468	219,024	21.6333	518	268,324	22.7596	568	322,624	23.8328
419	175,561	20.4695	469	219,964	21.6564	519	269,361	22.7816	569	323,761	23.8537
420	176,400	20.4939	470	220,900	21.6795	520	270,400	22.8035	570	324,900	23.8747
421	177,241	20.5183	471	221,841	21.7025	521	271,441	22.8254	571	326,041	23.8956
422	178,084	20.5426	472	222,784	21.7256	522	272,484	22.8473	572	327,184	23.9165
423	178,929	20.5670	473	223,729	21.7486	523	273,529	22.8692	573	328,329	23.9374
424	179,776	20.5913	474	224,676	21.7715	524	274,576	22.8910	574	329,476	23.9583
425	180,625	20.6155	475	225,625	21.7945	525	275,625	22.9129	575	330,625	23.9792
426	181,476	20.6398	476	226,576	21.8174	526	276,676	22.9347	576	331,776	24.0000
427	182,329	20.6640	477	227,529	21.8403	527	277,729	22.9565	577	332,929	24.0208
428	183,184	20.6882	478	228,484	21.8632	528	278,784	22.9783	578	334,084	24.0416
429	184,081	20.7123	479	229,441	21.8861	529	279,841	23.0000	579	335,241	24.0624
430	184,900	20.7364	480	230,400	21.9089	530	280,900	23.0217	580	336,400	24.0832
431	185,761	20.7605	481	231,361	21.9317	531	281,961	23.0434	581	337,561	24.1039
432	186,624	20.7846	482	232,324	21.9545	532	283,024	23.0651	582	338,724	24.1247
433	187,489	20.8087	483	233,289	21.9773	533	284.089	23.0868	583	339,889	24.1454
434	188,356	20.8327	484	234,256	22.0000	534	285,156	23.1084	584	341,056	24.1661
435	189,225	20.8567	485	235,225	22.0227	535	286,225	23.1301	585	342,225	24.1868
436	190,096	20.8806	486	236,196	22.0454	536	287,296	23.1517	586	343,396	24.2074
437	190,969	20.9045	487	237,169	22.0681	537	288,369	23.1733	587	344,569	24.2281
438	191,844	30.9284	488	238,144	22.0907	538	289,444	23.1948	588	345,744	24.2487
439	192,721	20.9523	489	239,121	22,1133	539	290,521	23.2164	589	346,921	24.2693
440	193,600	20.9762	490	240,100	22.1359	540	291,600	23.2379	590	348,100	24.2899
441	194,481	21.0000	491	241,081	22.1585	541	292,681	23.2594	591	349,281	24.3105
442	195,364	21.0238	492	242,064	22.1811	542	293,764	23.2809	592	350,464	24.3311
443	196,249	21.0476	493	243,049	22.2036	543	294,849	23.3024	593	351,649	24.3516
444	197,136	21.0713	494	244,036	22.2261	544	295,936	23.3238	594	352,836	24.3721
445	198,025	21.0950	495	245,025	22.2486	545	297,025	23.3452	595	354,025	24.3926
446	198,916	21.1187	496	246,016	22.2711	546	298,116	23.3666	596	355,216	24.4131
447	199,809	21.1424	497	247,009	22.2935	547	299,209	23.3880	597	356,409	24.4336
448	200,704	21.1660	498	248,004	22.3159	548	300,304	23.4094	598	357,604	24.4540
449	201,601	21.1896	499	249,001	22.3383	549	301,401	23.4307	599	358,801	24.4745
450	202,500	21.2132	500	250,000	22.3607	550	302,500	23.4521	600	360,000	24.4949

n	n^2	$\sqrt{n}$	n	n^2	$\sqrt{n}$	n	n^2	$\sqrt{n}$	n	n^2	$\sqrt{n}$
601	361,201	24.5153	651	423,801	25.5147	701	491,401	26.4764	751	564,001	27.4044
602	362,404	24.5357	652	425,104	25.5343	702	492,804	26.4953	752	565,504	27.4226
603	363,609	24.5561	653	426,409	25.5539	703	494,209	26.5141	753	567,009	27.4408
604	364,816	24.5764	654	427,716	25.5734	704	495,616	26.5330	754	568,516	27.4591
605	366,025	24.5967	655	429,025	25.5930	705	497,025	26.5518	755	570,025	27.4773
606	367,236	24.6171	656	430,336	25.6125	706	498,436	26.5707	756	571,536	27.4955
607	368,449	24.6374	657	431,649	25.6320	707	499,849	26.5895	757	573,049	27.5136
608	369,664	24.5477	658	432,964	25.6515	708	501,264	26.6083	758	574,564	27.5318
609	370,881	24.6779	659	434,281	25.6710	709	502,681	26.6271	759	576,081	27.5500
610	372,100	24.6982	660	435,600	25.6905	710	504,100	26.6458	760	577,600	27.5681
611	373,321	24.7184	661	436,921	26.7099	711	505,521	26.6646	761	579,121	27.5862
612	374,544	24.7386	662	438,244	25.7294	712	506,944	26.6833	762	580,644	27.6043
613	375,769	24.7588	663	439,569	25.7488	713	508,369	26.7021	763	582,169	27.6225
614	376,996	24.7790	664	440,896	25.7682	714	509,796	26.7208	764	583,696	27.6405
615	378,225	24.7992	665	442,225	25.7876	715	511,225	26.7395	765	585,225	27.6586
616	379,456	24.8193	666	443,556	25.8070	716	512,656	26.7582	766	586,756	27.6767
617	380,689	24.8395	667	444,889	25.8263	717	514,089	26.7769	767	588,289	27.6948
618	381,924	24.8596	668	446,224	25.8457	718	515,524	26.7955	768	589,824	27.7128
619	383,161	24.8797	669	447,561	25.8650	719	516,961	26.8142	769	591,261	27.7308
620	384,400	24.8998	670	448,900	25.8844	720	518,400	26.8328	770	592,900	27.7489
621	385,641	24.9199	671	450,241	25.9037	721	519,841	26.8514	771	594,441	27.7669
622	386,884	24.9399	672	451,584	25.9230	722	521,284	26.8701	772	595,984	27.7849
623	388,129	24.9600	673	452,929	25.9422	723	522,729	26.8887	773	597,529	27.8029
624	389,376	24.9800	674	454,276	25.9615	724	524,176	26.9072	774	599,076	27.8209
625	390,625	25.0000	675	455,625	25.9808	725	525,625	26.9258	775	600,625	27.8388
626	391,876	25.0200	676	456,976	26.0000	726	527,076	26.9444	776	602,176	27.8568
627	393,129	25.0400	677	458,329	26.0192	727	528,529	26.9629	777	603,729	27.8747
628	394,384	25.0599	678	459,684	26.0384	728	529,984	26.9815	778	605,284	27.8927
629	395,641	25.0799	679	461,041	26.0576	729	531,441	27.0000	779	606,841	27.9106
630	396,900	25.0998	680	462,400	26.0768	730	532,900	27.0185	780	608,400	27.9285
631	398,161	25.1197	681	463,761	26.0960	731	534,361	27.0370	781	609,961	27.9464
632	399,424	25.1396	682	465,124	26.1151	732	535,824	27.0555	782	611,524	27.9643
633	400,689	25.1595	683	466,489	26.1343	733	537,289	27.0740	783	613,089	27.9821
634	401,956	25.1794	684	467,856	26.1534	734	538,756	27.0924	784	614,656	28.0000
635	403,225	25.1992	685	469,225	26.1725	735	540,225	27.1109	785	616,225	28.0179
636	404,496	25.2190	686	470,596	26.1916	736	541,696	27.1293	786	617,796	28.0357
637	405,769	25.2389	687	471,969	26.2107	737	543,169	27.1477	787	619,369	28.0535
638	407,044	25.2587	688	473,344	26.2298	738	544,644	27.1662	788	620,944	28.0713
639	408,321	25.2784	689	474,721	26.2488	739	546,121	27.1846	789	622,521	28.0891
640	409,600	25.2982	690	476,100	26.2679	740	547,600	27.2029	790	624,100	28.1069
641	410,881	25.3180	691	477,481	26.2869	741	549,081	27.2213	791	625,681	28.1247
642	412,164	25.3377	692	478,864	26.3059	742	550,564	27.2397	792	627,264	28.1425
643	413,449	25.3574	693	480,249	26.3249	743	552,049	27.2580	793	628,849	28.1603
644	414,736	25.3772	694	481,636	26.3439	744	553,536	27.2764	794	630,436	28.1780
645	416,025	25.3969	695	483,025	26.3629	745	555,025	27.2947	795	632,025	28.1957
646	417,316	25.4165	696	484,416	26.3818	746	556,516	27.3130	796	633,616	28.2135
647	418,609	26.4362	697	485,809	26.4008	747	558,009	27.3313	797	635,209	28.2312
648	419,904	25.4558	698	487,204	26.4197	748	559,504	27.3496	798	636,804	28.2489
649	421,201	25.4755	699	488,601	26.4386	749	561,001	27.3679	799	638,401	28.2666
650	422,500	25.4951	700	490,000	26.4575	750	562,500	27.3861	800	640,000	28.2843

TABLE B.7 Squares and square roots (Continued)

n	n^2	$\sqrt{n}$	n	n^2	$\sqrt{n}$	n	n^2	$\sqrt{n}$	n	n^2	$\sqrt{n}$
801	641,601	28.3019	851	724,201	29.1719	901	811,801	30.0167	951	904,401	30.8383
802	643,204	28.3196	852	725,904	29.1890	902	813,604	30.0333	952	906,304	30.8545
803	644,809	28.3373	853	727,609	29.2062	903	815,409	30.0500	953	908,209	30.8707
804	646,416	28.3549	854	729,316	29.2233	904	817,216	30.0666	954	910,116	30.8869
805	648,025	28.3725	855	731,025	29.2404	905	819,025	30.0832	955	912,025	30.9031
806	649,636	28.3901	856	732,736	29.2575	906	820,836	30.0998	956	913,936	30.9192
807	651,249	28.4077	857	734,449	29.2746	907	822,649	30.1164	957	915,849	30.9354
808	652,864	28.4253	858	736,164	29.2916	908	824,464	30.1330	958	917,764	30.9516
809	654,481	28.4429	859	737,881	29.3087	909	826,281	30.1496	959	919,681	30.9677
810	656,100	28.4605	860	739,600	29.3258	910	828,100	30.1662	960	921,600	30.9839
811	657,721	28.4781	861	741,321	29.3428	911	829,921	30.1828	961	923,521	31.0000
812	659,344	28.4956	862	743,044	29.3598	912	831,744	30.1993	962	925,444	31.0161
813	660,969	28.5132	863	744,769	29.3769	913	833,569	30.2159	963	927,369	31.0322
814	662,596	28.5307	864	746,496	29.3939	914	835,396	30.2324	964	929,296	31.0483
815	664,225	28.5482	865	748,225	29.4109	915	837,225	30.2490	965	931,255	31.0644
816	665,856	28.5657	866	749,956	29.4279	916	839,056	30.2655	966	933,156	31.0805
817	667,489	28.5832	867	751,689	29.4449	917	840,889	30.2820	967	935,089	31.0966
818	669,124	28.6007	868	753,424	29.4618	918	842,724	30.2985	969	937,024	31.1127
819	670,761	28.6182	869	755,161	29.4788	919	844,561	30.3150	969	938,961	31.1288
820	672,400	28.6356	870	756,900	29.4958	920	846,400	30.3315	970	940,900	31.1448
821	674,041	28.6531	871	758,541	29.5127	921	848,241	30.3480	971	942,841	31.1609
822	675,684	28.6705	872	760,384	29.5296	922	850,084	30.3645	972	944,784	31.1769
823	677,329	28.6880	873	762,129	29.5466	923	851,929	30.3809	973	946,729	31.1929
824	678,976	28.7054	874	763,876	29.5635	924	853,776	30.3974	974	948,676	31.2090
825	680,625	28.7228	875	765,625	29.5804	925	855,625	30.4138	975	950,625	31.2250
826	682,276	28.7402	876	767,376	29.5973	926	857,476	30.4302	976	952,576	31.2410
827	683,929	28.7576	877	769,129	29.6142	927	859,329	30.4468	977	954,529	31.2570
828	685,584	28.7750	878	770,884	29.6311	928	861,184	30.4631	978	956,484	31.2730
829	687,241	28.7924	879	772,641	29.6479	929	863,041	30.4795	979	958,441	31.2890
830	688,900	28.8097	880	774,400	29.6648	930	864,900	30.4959	980	960,400	31.3050
831	690,561	28.8271	881	776,161	29.6816	931	866,761	30.5123	981	962,361	31.3209
832	692,224	28.8444	882	777,924	29.6985	932	868,624	30.5287	982	964,324	31.3369
833	693,889	28.8617	883	779,689	29.7153	933	870,489	30.5450	983	966,289	31.3528
834	695,556	28.8791	884	781,456	29.7321	934	872,356	30.5614	984	968,256	31.3688
835	697,225	28.8964	885	783,225	29.7489	935	874,225	30.5778	985	970,225	31.3847
836	698,896	28.9137	886	784,996	29.7658	936	876,096	30.5941	986	972,196	31.4006
837	700,569	28.9310	887	786,769	29.7825	937	877,969	30.6105	987	974,169	31.4166
838	702,244	28.9482	888	788,544	29.7993	938	879,844	30.6268	988	976,144	31.4325
839	703,921	28.9655	889	790,321	29.8161	939	881,721	30.6431	989	978,121	31.4484
840	705,600	28.9828	890	792,100	29.8329	940	883,600	30.6594	990	980,100	31.4643
841	707,281	29.0000	891	793,881	29.8496	941	885,481	30.6757	991	982,081	31.4802
842	708,964	29.0172	892	795,664	29.8664	942	887,364	30.6920	992	984,064	31.4960
843	710,649	29.0345	893	797,449	29.8831	943	889,249	30.7083	993	986,049	31.5119
844	712,336	29.0517	894	799,236	29.8998	944	891,136	30.7246	994	988,036	31.5278
845	714,025	29.0689	895	801,025	29.9166	945	893,025	30.7409	995	990,025	31.5436
846	715,716	29.0861	896	802,816	29.9333	946	894,916	30.7571	996	992,016	31.5595
847	717,409	29.1033	897	804,609	29.9500	947	896,809	30.7734	997	994,009	31.5753
848	719,104	29.1204	898	806,404	29.9666	948	898,704	30.7896	998	996,004	31.5911
849	720,801	29.1376	899	808,201	29.9833	949	900,601	30.8058	999	998,001	31.6070
850	722,500	29.1548	900	810,000	30.0000	950	902,500	30.8221	1000	1,000,000	31.6228

Appendix *C*

RELIABILITY TABLES

TABLE C.1   System reliability as a function of the number of active parallel components, reliability of each component, and number of failures permitted*

Permitted number of failures	R_i								
	0.50	0.60	0.75	0.90	0.95	0.975	0.99	0.995	0.999
	2 Components in Parallel								
0*	0.2500	0.3600	0.5625	0.8100	0.9025	0.9506	0.9801	0.99002	0.99800
1	0.7500	0.8400	0.9375	0.9900	0.9975	0.9994	0.9999	0.99998	1.00000
	3 Components in Parallel								
0	0.1250	0.2160	0.4219	0.7290	0.8574	0.9269	0.9703	0.98507	0.99700
1	0.5000	0.6480	0.8438	0.9720	0.9928	0.9982	0.9997	0.99993	1.00000
2	0.8750	0.9360	0.9844	0.9990	0.9999	1.0000	1.0000	1.00000	1.00000
	4 Components in Parallel								
0	0.0625	0.1296	0.3164	0.6561	0.8145	0.9037	0.9606	0.98015	0.99601
1	0.3125	0.4752	0.7383	0.9477	0.9860	0.9964	0.9994	0.99985	0.99999
2	0.6875	0.8208	0.9492	0.9963	0.9995	0.9999	1.0000	1.00000	1.00000
3	0.9375	0.9744	0.9961	0.9999	1.0000	1.0000	1.0000	1.00000	1.00000
	5 Components in Parallel								
0	0.0312	0.0778	0.2373	0.5905	0.7738	0.8811	0.9510	0.97525	0.99501
1	0.1875	0.3370	0.6328	0.9185	0.9774	0.9941	0.9990	0.99975	0.99999
2	0.5000	0.6826	0.8965	0.9914	0.9988	0.9998	1.0000	1.00000	1.00000
3	0.8125	0.9130	0.9844	0.9995	1.0000	1.0000	1.0000	1.00000	1.00000
4	0.9688	0.9898	0.9990	1.0000	1.0000	1.0000	1.0000	1.00000	1.00000
	6 Components in Parallel								
0	0.0156	0.0467	0.1780	0.5314	0.7351	0.8591	0.9415	0.97037	0.99401
1	0.1094	0.2333	0.5339	0.8857	0.9672	0.9912	0.9986	0.99963	0.99999
2	0.3438	0.5443	0.8306	0.9841	0.9978	0.9997	1.0000	1.00000	1.00000
3	0.6562	0.8208	0.9624	0.9987	0.9999	1.0000	1.0000	1.00000	1.00000
4	0.8906	0.9590	0.9954	0.9999	1.0000	1.0000	1.0000	1.00000	1.00000
5	0.9844	0.9959	0.9998	1.0000	1.0000	1.0000	1.0000	1.00000	1.00000
	7 Components in Parallel								
0	0.0078	0.0280	0.1335	0.4783	0.6983	0.8376	0.9321	0.96552	0.99302
1	0.0625	0.1586	0.4449	0.8503	0.9556	0.9879	0.9980	0.99948	0.99998
2	0.2266	0.4199	0.7564	0.9743	0.9962	0.9995	1.0000	1.00000	1.00000
3	0.5000	0.7102	0.9294	0.9973	0.9998	1.0000	1.0000	1.00000	1.00000
4	0.7734	0.9037	0.9871	0.9998	1.0000	1.0000	1.0000	1.00000	1.00000
5	0.9375	0.9812	0.9987	1.0000	1.0000	1.0000	1.0000	1.00000	1.00000
6	0.9922	0.9984	0.9999	1.0000	1.0000	1.0000	1.0000	1.00000	1.00000

TABLE C.1 System reliability as a function of the number of active parallel components, reliability of each component, and number of failures permitted* (Continued)

Permitted number of failures	R_i								
	0.50	0.60	0.75	0.90	0.95	0.975	0.99	0.995	0.999
	8 Components in Parallel								
0	0.0039	0.0168	0.1001	0.4305	0.6634	0.8167	0.9227	0.96069	0.9926
1	0.0352	0.1064	0.3671	0.8131	0.9428	0.9842	0.9973	0.99931	0.9999
2	0.1445	0.3154	0.6785	0.9619	0.9942	0.9992	0.9999	0.99999	1.0000
3	0.3633	0.5941	0.8862	0.9950	0.9996	1.0000	1.0000	1.00000	1.0000
4	0.6367	0.8263	0.9727	0.9996	1.0000	1.0000	1.0000	1.00000	1.0000
5	0.8555	0.9502	0.9958	1.0000	1.0000	1.0000	1.0000	1.00000	1.0000
6	0.9648	0.9915	0.9996	1.0000	1.0000	1.0000	1.0000	1.00000	1.0000
7	0.9961	0.9993	1.0000	1.0000	1.0000	1.0000	1.0000	1.00000	1.0000
	9 Components in Parallel								
0	0.0020	0.0101	0.0751	0.3874	0.6302	0.7962	0.9135	0.95589	0.9910
1	0.0195	0.0705	0.3003	0.7748	0.9288	0.9800	0.9966	0.99912	0.9999
2	0.0898	0.2318	0.6007	0.9470	0.9916	0.9988	0.9999	0.99999	1.0000
3	0.2539	0.4826	0.8343	0.9917	0.9994	1.0000	1.0000	1.00000	1.0000
4	0.5000	0.7334	0.9511	0.9991	1.0000	1.0000	1.0000	1.00000	1.0000
5	0.7461	0.9006	0.9900	0,9999	1.0000	1.0000	1.0000	1.00000	1.0000
6	0.9102	0.9750	0.9987	1.0000	1.0000	1.0000	1.0000	1.00000	1.0000
7	0.9805	0.9962	0.9999	1.0000	1.0000	1.0000	1.0000	1.00000	1.0000
	10 Components in Parallel								
0	0.0010	0.0060	0.0563	0.3487	0.5987	0.7763	0.9044	0.95111	0.9900
1	0.0107	0.0464	0.2440	0.7361	0.9139	0.9754	0.9957	0.99890	0.9999
2	0.0547	0.1673	0.5256	0.9298	0.9885	0.9984	0.9999	0.99999	1.0000
3	0.1719	0.3823	0.7759	0.9872	0.9990	0.9999	1.0000	1.00000	1.0000
4	0.3770	0.6331	0.9219	0.9984	0.9999	1.0000	1.0000	1.00000	1.0000
5	0.6230	0.8338	0.9803	0.9999	1.0000	1.0000	1.0000	1.00000	1.0000
6	0.8281	0.9452	0.9965	1.0000	1.0000	1.0000	1.0000	1.00000	1.0000
7	0.9453	0.9877	0.9996	1.0000	1.0000	1.0000	1.0000	1.00000	1.0000
	15 Components in Parallel								
0	0.0000	0.0005	0.0134	0.2059	0.4633	0.6840	0.8601	0.92757	0.9851
1	0.0005	0.0052	0.0802	0.5490	0.8290	0.9471	0.9904	0.99749	0.9999
2	0.0037	0.0271	0.2361	0.8159	0.9638	0.9943	0.9996	0.99995	1.0000
3	0.0176	0.0905	0.4613	0.9444	0.9945	0.9996	1.0000	1.00000	1.0000
4	0.0592	0.2173	0.6865	0.9873	0.9994	1.0000	1.0000	1.00000	1.0000

Permitted number of failures	R_i								
	0.50	0.60	0.75	0.90	0.95	0.975	0.99	0.995	0.999
	15 Components in Parallel (Continued)								
5	0.1509	0.4032	0.8516	0.9978	0.9999	1.0000	1.0000	1.00000	1.00000
6	0.3036	0.6098	0.9434	0.9997	1.0000	1.0000	1.0000	1.00000	1.00000
7	0.5000	0.7869	0.9827	1.0000	1.0000	1.0000	1.0000	1.00000	1.00000
8	0.6964	0.9050	0.9958	1.0000	1.0000	1.0000	1.0000	1.00000	1.00000
9	0.8491	0.9662	0.9992	1.0000	1.0000	1.0000	1.0000	1.00000	1.00000
	20 Components in Parallel								
0	0.0000	0.0000	0.0032	0.1216	0.3585	0.6027	0.8179	0.90461	0.98019
1	0.0000	0.0005	0.0243	0.3918	0.7358	0.9118	0.9831	0.99553	0.99981
2	0.0002	0.0036	0.0913	0.6769	0.9245	0.9870	0.9990	0.99987	1.00000
3	0.0013	0.0160	0.2252	0.8670	0.9841	0.9986	1.0000	1.00000	1.00000
4	0.0059	0.0510	0.4148	0.9568	0.9974	0.9999	1.0000	1.00000	1.00000
5	0.0207	0.1256	0.6172	0.9888	0.9997	1.0000	1.0000	1.00000	1.00000
6	0.0577	0.2500	0.7858	0.9976	1.0000	1.0000	1.0000	1.00000	1.00000
7	0.1316	0.4159	0.8982	0.9996	1.0000	1.0000	1.0000	1.00000	1.00000
8	0.2517	0.5956	0.9591	0.9999	1.0000	1.0000	1.0000	1.00000	1.00000
9	0.4119	0.7553	0.9861	1.0000	1.0000	1.0000	1.0000	1.00000	1.00000
10	0.5881	0.8725	0.9961	1.0000	1.0000	1.0000	1.0000	1.00000	1.00000
11	0.7483	0.9435	0.9991	1.0000	1.0000	1.0000	1.0000	1.00000	1.00000

*No failures permitted is the equivalent of a series system.

TABLE C.2a Reliability determined from number of tests and number of failures at 50 percent confidence

Number of failures	Number of tests									
	1	2	3	4	5	6	7	8	9	10
0	0.500	0.7071	0.7937	0.8409	0.8706	0.8909	0.9057	0.9170	0.9259	0.9330
1	.....	0.2929	0.5000	0.6143	0.6862	0.7356	0.7715	0.7989	0.8204	0.8377
2	.....	.....	0.2063	0.3857	0.5000	0.5786	0.6359	0.6795	0.7138	0.7414
3	.....	.....	.....	0.1591	0.3138	0.4214	0.5000	0.5598	0.6069	0.6449
4	.....	.....	.....	.....	0.1294	0.2644	0.3641	0.4402	0.5000	0.5483
5	.....	.....	.....	.....	.....	0.1091	0.2285	0.3205	0.3931	0.4517

	15	20	25	30	35	40	45	50	60	70
0	0.9548	0.9659	0.9727	0.9772	0.9804	0.9828	0.9847	0.9862	0.9885	0.9901
1	0.8906	0.9175	0.9338	0.9447	0.9525	0.9584	0.9630	0.9667	0.9722	0.9761
2	0.8257	0.8685	0.8945	0.9119	0.9243	0.9337	0.9410	0.9469	0.9557	0.9620
3	0.7606	0.8194	0.8551	0.8790	0.8961	0.9090	0.9190	0.9270	0.9391	0.9478
4	0.6955	0.7703	0.8156	0.8460	0.8678	0.8842	0.8970	0.9072	0.9226	0.9336
5	0.6303	0.7212	0.7762	0.8131	0.8395	0.8594	0.8749	0.8874	0.9060	0.9194
6	0.5652	0.6720	0.7368	0.7801	0.8113	0.8346	0.8529	0.8675	0.8895	0.9052
7	0.5000	0.6229	0.6973	0.7472	0.7830	0.8099	0.8308	0.8476	0.8729	0.8910
8	0.4348	0.5737	0.6578	0.7142	0.7547	0.7851	0.8088	0.8278	0.8563	0.8767
9	0.3697	0.5246	0.6184	0.6813	0.7264	0.7603	0.7867	0.8079	0.8398	0.8625
10	0.3045	0.4754	0.5789	0.6483	0.6981	0.7355	0.7647	0.7880	0.8232	0.8483
11	0.2394	0.4263	0.5395	0.6154	0.6698	0.7107	0.7426	0.7682	0.8066	0.8341
12	0.1743	0.3771	0.5000	0.5824	0.6415	0.6859	0.7206	0.7483	0.7900	0.8199
13	0.1094	0.3280	0.4605	0.5494	0.6132	0.6611	0.6985	0.7285	0.7735	0.8057
14	0.0452	0.2788	0.4211	0.5165	0.5849	0.6363	0.6765	0.7086	0.7569	0.7915
15	.....	0.2297	0.3816	0.4835	0.5566	0.6116	0.6544	0.6887	0.7403	0.7772

	80	90	100	150	200	300	400	500	750	1,000
0	0.9914	0.9923	0.9931	0.9954	0.9965	0.9977	0.9983	0.9986	0.9991	0.9993
1	0.9791	0.9814	0.9833	0.9888	0.9916	0.9944	0.9958	0.9966	0.9978	0.9983
2	0.9667	0.9704	0.9733	0.9822	0.9867	0.9911	0.9933	0.9947	0.9964	0.9973
3	0.9543	0.9594	0.9634	0.9756	0.9817	0.9878	0.9908	0.9927	0.9951	0.9963
4	0.9419	0.9483	0.9534	0.9689	0.9767	0.9844	0.9883	0.9907	0.9938	0.9953
5	0.9294	0.9372	0.9435	0.9623	0.9717	0.9811	0.9858	0.9887	0.9924	0.9943
6	0.9170	0.9262	0.9335	0.9556	0.9667	0.9778	0.9833	0.9867	0.9911	0.9933
7	0.9045	0.9151	0.9236	0.9490	0.9617	0.9745	0.9808	0.9847	0.9898	0.9923
8	0.8921	0.9040	0.9136	0.9423	0.9567	0.9711	0.9783	0.9827	0.9884	0.9913
9	0.8796	0.8930	0.9036	0.9357	0.9517	0.9678	0.9758	0.9807	0.9871	0.9903
10	0.8672	0.8819	0.8937	0.9290	0.9467	0.9645	0.9734	0.9787	0.9858	0.9893
11	0.8548	0.8708	0.8837	0.9224	0.9418	0.9611	0.9709	0.9767	0.9844	0.9883
12	0.8423	0.8598	0.8737	0.9157	0.9368	0.9578	0.9684	0.9747	0.9831	0.9873
13	0.8299	0.8487	0.8638	0.9091	0.9318	0.9545	0.9659	0.9727	0.9818	0.9863
14	0.8174	0.8376	0.8538	0.9024	0.9268	0.9512	0.9634	0.9707	0.9805	0.9853
15	0.8050	0.8266	0.8438	0.8958	0.9218	0.9478	0.9609	0.9687	0.9791	0.9843
16	0.7925	0.8155	0.8339	0.8891	0.9168	0.9445	0.9584	0.9667	0.9778	0.9833
17	0.7801	0.8044	0.8239	0.8825	0.9118	0.9412	0.9559	0.9647	0.9765	0.9823
18	0.7676	0.7933	0.8139	0.8758	0.9068	0.9378	0.9534	0.9627	0.9751	0.9813
19	0.7552	0.7823	0.8040	0.8692	0.9018	0.9345	0.9509	0.9607	0.9738	0.9803
20	0.7427	0.7712	0.7940	0.8625	0.8968	0.9312	0.9484	0.9587	0.9725	0.9793
21	0.7303	0.7601	0.7840	0.8559	0.8918	0.9279	0.9459	0.9567	0.9711	0.9783
22	0.7178	0.7491	0.7741	0.8492	0.8869	0.9245	0.9434	0.9547	0.9698	0.9773
23	0.7054	0.7380	0.7641	0.8426	0.8819	0.9212	0.9409	0.9527	0.9685	0.9763
24	0.6929	0.7269	0.7541	0.8359	0.8769	0.9179	0.9384	0.9507	0.9671	0.9753
25	0.6805	0.7159	0.7442	0.8293	0.8719	0.9145	0.9359	0.9487	0.9658	0.9743

ABLE C.2b   Reliability determined from number of tests and number
f failures at 75 percent confidence

umber of ailures	Number of tests									
	1	2	3	4	5	6	7	8	9	10
0	0.2500	0.5000	0.6300	0.7071	0.7579	0.7937	0.8203	0.8409	0.8572	0.8706
1	.....	0.1340	0.3264	0.4563	0.5458	0.6105	0.6593	0.6973	0.7277	0.7526
2	.....	.....	0.0914	0.2430	0.3594	0.4468	0.5139	0.5668	0.6095	0.6446
3	.....	.....	.....	0.0694	0.1938	0.2969	0.3788	0.4445	0.4980	0.5423
4	.....	.....	.....	.....	0.559	0.1612	0.2531	0.3291	0.3920	0.4445
5	.....	.....	.....	.....	.....	0.0468	0.1380	0.2206	0.2910	0.3507

	15	20	25	30	35	40	45	50	60	70
0	0.9117	0.9330	0.9461	0.9548	0.9612	0.9659	0.9697	0.9727	0.9772	0.9804
1	0.8304	0.8710	0.8959	0.9127	0.9249	0.9341	0.9413	0.9471	0.9558	0.9620
2	0.7551	0.8133	0.8491	0.8735	0.8910	0.9043	0.9147	0.9231	0.9357	0.9448
3	0.6831	0.7579	0.8042	0.8356	0.8584	0.8756	0.8891	0.9000	0.9163	0.9281
4	0.6135	0.7041	0.7604	0.7988	0.8266	0.8476	0.8641	0.8774	0.8974	0.9118
5	0.5459	0.6516	0.7176	0.7626	0.7953	0.8201	0.8395	0.8552	0.8788	0.8958
6	0.4796	0.6000	0.6755	0.7270	0.7645	0.7929	0.8153	0.8332	0.8604	0.8800
7	0.4150	0.5494	0.6340	0.6919	0.7341	0.7661	0.7913	0.8116	0.8422	0.8643
8	0.3518	0.4994	0.5929	0.6572	0.7040	0.7396	0.7675	0.7901	0.8242	0.8488
9	0.2902	0.4502	0.5524	0.6229	0.6742	0.7133	0.7440	0.7688	0.8063	0.8334
10	0.2301	0.4018	0.5123	0.5888	0.6447	0.6872	0.7206	0.7477	0.7886	0.8181
11	0.1720	0.3541	0.4727	0.5551	0.6153	0.6613	0.6974	0.7267	0.7709	0.8029
12	0.1163	0.3071	0.4335	0.5217	0.5862	0.6255	0.6744	0.7058	0.7534	0.7877
13	0.0642	0.2610	0.3947	0.4884	0.5573	0.6099	0.6515	0.6850	0.7359	0.7726
14	0.0190	0.2157	0.3564	0.4556	0.5286	0.5845	0.6287	0.6644	0.7185	0.7576
15	.....	0.1714	0.3184	0.4229	0.5001	0.5593	0.6060	0.6438	0.7012	0.7427

	80	90	100	150	200	300	400	500	750	1,000
0	0.9828	0.9847	0.9862	0.9908	0.9931	0.9954	0.9965	0.9972	0.9982	0.9986
1	0.9670	0.9704	0.9733	0.9822	0.9866	0.9910	0.9933	0.9946	0.9964	0.9973
2	0.9516	0.9569	0.9612	0.9740	0.9805	0.9870	0.9902	0.9922	0.9948	0.9961
3	0.9370	0.9439	0.9494	0.9662	0.9746	0.9830	0.9873	0.9898	0.9932	0.9949
4	0.9227	0.9317	0.9380	0.9585	0.9688	0.9792	0.9844	0.9875	0.9916	0.9937
5	0.9086	0.9186	0.9267	0.9509	0.9631	0.9754	0.9815	0.9852	0.9901	0.9926
6	0.8947	0.9063	0.9155	0.9434	0.9575	0.9716	0.9787	0.9829	0.9886	0.9915
7	0.8810	0.8940	0.9045	0.9360	0.9519	0.9679	0.9759	0.9807	0.9871	0.9903
8	0.8674	0.8819	0.8935	0.9287	0.9464	0.9642	0.9731	0.9785	0.9856	0.9892
9	0.8538	0.8698	0.8826	0.9214	0.9409	0.9605	0.9703	0.9762	0.9841	0.9881
10	0.8404	0.8580	0.8718	0.9141	0.9354	0.9568	0.9676	0.9740	0.9827	0.9870
11	0.8270	0.8459	0.8610	0.9069	0.9300	0.9532	0.9648	0.9718	0.9812	0.9859
12	0.8137	0.8340	0.8503	0.8997	0.9245	0.9496	0.9621	0.9697	0.9798	0.9848
13	0.8004	0.8222	0.8396	0.8925	0.9191	0.9459	0.9594	0.9675	0.9783	0.9837
14	0.7872	0.8104	0.8290	0.8853	0.9137	0.9423	0.9567	0.9653	0.9769	0.9826
15	0.7741	0.7987	0.8184	0.8782	0.9084	0.9387	0.9540	0.9632	0.9754	0.9815
16	0.7610	0.7870	0.8079	0.8711	0.9030	0.9352	0.9513	0.9610	0.9740	0.9805
17	0.7479	0.7753	0.7974	0.8640	0.8977	0.9316	0.9486	0.9588	0.9725	0.9794
18	0.7349	0.7637	0.7869	0.8570	0.8924	0.9280	0.9459	0.9567	0.9711	0.9783
19	0.7220	0.7521	0.7764	0.8499	0.8871	0.9245	0.9433	0.9546	0.9697	0.9772
20	0.7090	0.7406	0.7660	0.8429	0.8818	0.9209	0.9406	0.9524	0.9682	0.9762
21	0.6961	0.7291	0.7556	0.8359	0.8765	0.9174	0.9379	0.9503	0.9668	0.9751
22	0.6833	0.7176	0.7452	0.8289	0.8712	0.9138	0.9353	0.9482	0.9654	0.9740
23	0.6705	0.7061	0.7348	0.8219	0.8659	0.9103	0.9326	0.9460	0.9640	0.9730
24	0.6577	0.6947	0.7245	0.8149	0.8607	0.9068	0.9300	0.9439	0.9625	0.9719
25	0.6449	0.6833	0.7142	0.8079	0.8554	0.9033	0.9273	0.9418	0.9611	0.9708

TABLE C.2c Reliability determined from number of tests and number of failures at 90 percent confidence

Number of failures	Number of tests									
	1	2	3	4	5	6	7	8	9	10
0	0.1000	0.3162	0.4642	0.5623	0.6310	0.6813	0.7197	0.7499	0.7743	0.7943
1	.....	0.0513	0.1958	0.3205	0.4161	0.4897	0.5474	0.5938	0.6316	0.6631
2	.....	.....	0.0345	0.1426	0.2466	0.3332	0.4038	0.4618	0.5099	0.5504
3	.....	.....	.....	0.0260	0.1122	0.2009	0.2786	0.3446	0.4006	0.4483
4	.....	.....	.....	.....	0.0209	0.0926	0.1696	0.2397	0.3010	0.3542
5	.....	.....	.....	.....	.....	0.0174	0.0788	0.1469	0.2104	0.2673

	15	20	25	30	35	40	45	50	60	70
0	0.8577	0.8912	0.9120	0.9261	0.9362	0.9441	0.9501	0.9550	0.9624	0.9676
1	0.7644	0.8190	0.8531	0.8764	0.8933	0.9062	0.9163	0.9244	0.9367	0.9456
2	0.6827	0.7552	0.8009	0.8321	0.8550	0.8723	0.8860	0.8970	0.9137	0.9257
3	0.6072	0.6958	0.7520	0.7907	0.8190	0.8405	0.8575	0.8712	0.8920	0.9070
4	0.5360	0.6393	0.7053	0.7510	0.7844	0.8100	0.8302	0.8464	0.8711	0.8890
5	0.4683	0.5851	0.6603	0.7126	0.7510	0.7804	0.8036	0.8224	0.8509	0.8715
6	0.4035	0.5327	0.6167	0.6753	0.7185	0.7515	0.7777	0.7989	0.8311	0.8543
7	0.3415	0.4820	0.5742	0.6389	0.6866	0.7233	0.7523	0.7758	0.8116	0.8376
8	0.2822	0.4327	0.5326	0.6032	0.6553	0.6955	0.7273	0.7531	0.7925	0.8290
9	0.2256	0.3847	0.4920	0.5681	0.6246	0.6682	0.7027	0.7308	0.7735	0.8046
10	0.1720	0.3381	0.4522	0.5337	0.5944	0.6412	0.6784	0.7087	0.7549	0.7885
11	0.1218	0.2929	0.4133	0.4999	0.5646	0.6146	0.6545	0.6869	0.7364	0.7724
12	0.0758	0.2491	0.3751	0.4666	0.5352	0.5884	0.6308	0.6653	0.7181	0.7566
13	0.0360	0.2067	0.3367	0.4338	0.5062	0.5624	0.6073	0.6440	0.7000	0.7409
14	0.0070	0.1659	0.3011	0.4015	0.4775	0.5368	0.5841	0.6228	0.6820	0.7253
15	.....	0.1269	0.2653	0.3697	0.4492	0.5114	0.5611	0.6018	0.6642	0.7098

	80	90	100	150	200	300	400	500	750	1,000
0	0.9716	0.9747	0.9772	0.9848	0.9886	0.9924	0.9943	0.9954	0.9969	0.9977
1	0.9522	0.9575	0.9617	0.9743	0.9807	0.9871	0.9903	0.9922	0.9948	0.9961
2	0.9348	0.9419	0.9476	0.9649	0.9736	0.9824	0.9867	0.9894	0.9929	0.9947
3	0.9183	0.9273	0.9344	0.9560	0.9669	0.9779	0.9834	0.9867	0.9911	0.9933
4	0.9025	0.9131	0.9216	0.9474	0.9604	0.9735	0.9801	0.9841	0.9894	0.9920
5	0.8871	0.8994	0.9092	0.9390	0.9541	0.9693	0.9769	0.9815	0.9877	0.9907
6	0.8720	0.8859	0.8971	0.9308	0.9479	0.9652	0.9738	0.9790	0.9860	0.9895
7	0.8572	0.8727	0.8851	0.9228	0.9418	0.9611	0.9707	0.9766	0.9844	0.9883
8	0.8426	0.8596	0.8733	0.9148	0.9358	0.9570	0.9677	0.9741	0.9827	0.9870
9	0.8282	0.8467	0.8616	0.9069	0.9299	0.9531	0.9647	0.9717	0.9811	0.9858
10	0.8140	0.8340	0.8501	0.8991	0.9240	0.9491	0.9617	0.9694	0.9795	0.9846
11	0.7998	0.8214	0.8387	0.8914	0.9182	0.9452	0.9588	0.9670	0.9780	0.9834
12	0.7858	0.8088	0.8274	0.8837	0.9124	0.9413	0.9559	0.9646	0.9764	0.9823
13	0.7720	0.7964	0.8161	0.8761	0.9066	0.9374	0.9530	0.9623	0.9748	0.9811
14	0.7582	0.7841	0.8050	0.8686	0.9009	0.9336	0.9501	0.9600	0.9733	0.9799
15	0.7445	0.7718	0.7939	0.8610	0.8952	0.9298	0.9472	0.9577	0.9717	0.9788
16	0.7309	0.7596	0.7828	0.8536	0.8896	0.9260	0.9443	0.9554	0.9702	0.9776
17	0.7174	0.7475	0.7719	0.8461	0.8839	0.9222	0.9415	0.9531	0.9687	0.9765
18	0.7040	0.7355	0.7610	0.8387	0.8783	0.9184	0.9386	0.9508	0.9671	0.9753
19	0.6906	0.7235	0.7501	0.8313	0.8727	0.9147	0.9358	0.9486	0.9656	0.9742
20	0.6773	0.7116	0.7393	0.8240	0.8672	0.9110	0.9330	0.9463	0.9641	0.9730
21	0.6641	0.6997	0.7286	0.8167	0.8616	0.9072	0.9302	0.9440	0.9626	0.9719
22	0.6509	0.6879	0.7178	0.8094	0.8561	0.9035	0.9274	0.9418	0.9611	0.9708
23	0.6378	0.6761	0.7072	0.8021	0.8506	0.8998	0.9246	0.9396	0.9596	0.9697
24	0.6248	0.6644	0.6966	0.7949	0.8451	0.8961	0.9218	0.9373	0.9581	0.9685
25	0.6118	0.6528	0.6860	0.7877	0.8397	0.8924	0.9190	0.9351	0.9566	0.9674

TABLE C.2d   Reliability determined from number of tests and number of failures at 95 percent confidence

Number of failures	Number of tests									
	1	2	3	4	5	6	7	8	9	10
0	0.0500	0.2236	0.3684	0.4729	0.5493	0.6070	0.6518	0.6877	0.7169	0.7411
1	.....	0.0253	0.1354	0.2486	0.3426	0.4182	0.4793	0.5293	0.5709	0.6058
2	.....	.....	0.0170	0.0976	0.1893	0.2713	0.3413	0.4003	0.4504	0.4931
3	.....	.....	.....	0.0127	0.0764	0.1532	0.2253	0.2892	0.3449	0.3934
4	.....	.....	.....	.....	0.0102	0.0628	0.1288	0.1929	0.2514	0.3035
5	.....	.....	.....	.....	.....	0.0085	0.0534	0.1111	0.1688	0.2224
	15	20	25	30	35	40	45	50	60	70
0	0.8190	0.8609	0.8871	0.9050	0.9180	0.9278	0.9356	0.9418	0.9513	0.9581
1	0.7206	0.7839	0.8239	0.8514	0.8715	0.8868	0.8989	0.9086	0.9234	0.9340
2	0.6366	0.7174	0.7690	0.8047	0.8308	0.8508	0.8666	0.8794	0.8988	0.9128
3	0.5603	0.6563	0.7183	0.7614	0.7931	0.8174	0.8366	0.8522	0.8758	0.8929
4	0.4892	0.5990	0.6704	0.7204	0.7573	0.7856	0.8080	0.8262	0.8539	0.8740
5	0.4226	0.5445	0.6246	0.6810	0.7228	0.7550	0.7805	0.8012	0.8327	0.8557
6	0.3596	0.4922	0.5805	0.6430	0.6894	0.7253	0.7537	0.7768	0.8121	0.8378
7	0.3000	0.4420	0.5378	0.6061	0.6570	0.6963	0.7276	0.7531	0.7920	0.8204
8	0.2437	0.3936	0.4964	0.5701	0.6253	0.6680	0.7020	0.7298	0.7723	0.8032
9	0.1909	0.3469	0.4561	0.5349	0.5942	0.6402	0.6770	0.7069	0.7528	0.7864
10	0.1417	0.3020	0.4168	0.5005	0.5638	0.6130	0.6523	0.6844	0.7337	0.7697
11	0.0967	0.2587	0.3786	0.4669	0.5339	0.5862	0.6280	0.6623	0.7148	0.7533
12	0.0568	0.2171	0.3414	0.4339	0.5045	0.5597	0.6041	0.6404	0.6962	0.7371
13	0.0242	0.1773	0.3051	0.4016	0.4756	0.5337	0.5805	0.6188	0.6778	0.7210
14	0.0034	0.1395	0.2699	0.3699	0.4472	0.5080	0.5571	0.5975	0.6595	0.7052
15	.....	0.1041	0.2356	0.3389	0.4192	0.4828	0.5340	0.5764	0.6415	0.6894
	80	90	100	150	200	300	400	500	750	1,000
0	0.9632	0.9673	0.9705	0.9802	0.9851	0.9901	0.9925	0.9940	0.9960	0.9970
1	0.9421	0.9484	0.9534	0.9688	0.9765	0.9843	0.9882	0.9905	0.9937	0.9953
2	0.9234	0.9317	0.9384	0.9586	0.9689	0.9792	0.9843	0.9875	0.9916	0.9937
3	0.9059	0.9161	0.9243	0.9491	0.9617	0.9744	0.9807	0.9846	0.9897	0.9923
4	0.8892	0.9012	0.9108	0.9400	0.9548	0.9698	0.9773	0.9818	0.9878	0.9909
5	0.8731	0.8867	0.8977	0.9312	0.9482	0.9653	0.9739	0.9791	0.9860	0.9895
6	0.8573	0.8727	0.8850	0.9226	0.9417	0.9609	0.9706	0.9765	0.9843	0.9882
7	0.8419	0.8589	0.8725	0.9141	0.9353	0.9566	0.9674	0.9739	0.9825	0.9869
8	0.8268	0.8453	0.8603	0.9058	0.9290	0.9524	0.9642	0.9713	0.9808	0.9856
9	0.8119	0.8320	0.8482	0.8976	0.9228	0.9482	0.9611	0.9688	0.9792	0.9843
10	0.7972	0.8188	0.8363	0.8895	0.9167	0.9441	0.9580	0.9663	0.9775	0.9831
11	0.7827	0.8058	0.8245	0.8815	0.9106	0.9400	0.9549	0.9638	0.9758	0.9819
12	0.7683	0.7929	0.8128	0.8736	0.9046	0.9360	0.9518	0.9614	0.9742	0.9806
13	0.7541	0.7802	0.8013	0.8657	0.8986	0.9320	0.9488	0.9590	0.9726	0.9794
14	0.7400	0.7676	0.7898	0.8579	0.8927	0.9280	0.9458	0.9566	0.9710	0.9782
15	0.7261	0.7550	0.7785	0.8502	0.8869	0.9241	0.9428	0.9542	0.9694	0.9770
16	0.7122	0.7426	0.7672	0.8425	0.8810	0.9201	0.9399	0.9518	0.9678	0.9758
17	0.6985	0.7303	0.7560	0.8349	0.8752	0.9162	0.9369	0.9494	0.9662	0.9746
18	0.6849	0.7180	0.7449	0.8273	0.8695	0.9123	0.9340	0.9471	0.9646	0.9734
19	0.6713	0.7058	0.7338	0.8197	0.8637	0.9085	0.9311	0.9447	0.9630	0.9722
20	0.6579	0.6937	0.7228	0.8122	0.8580	0.9046	0.9282	0.9424	0.9615	0.9711
21	0.6445	0.6817	0.7119	0.8047	0.8523	0.9008	0.9253	0.9401	0.9599	0.9699
22	0.6312	0.6697	0.7010	0.7972	0.8467	0.8969	0.9224	0.9378	0.9584	0.9687
23	0.6180	0.6578	0.6902	0.7898	0.8410	0.8931	0.9195	0.9355	0.9568	0.9676
24	0.6049	0.6460	0.6794	0.7824	0.8354	0.8893	0.9167	0.9332	0.9553	0.9664
25	0.5918	0.6342	0.6687	0.7751	0.8298	0.8856	0.9138	0.9309	0.9538	0.9653

TABLE C.2e Reliability determined from number of tests and number
of failures at 99 percent confidence

Number of failures	Number of tests									
	1	2	3	4	5	6	7	8	9	10
0	0.0100	0.1000	0.2154	0.3162	0.3981	0.4642	0.5179	0.5623	0.5995	0.6310
1	.....	0.0050	0.0589	0.1409	0.2221	0.2943	0.3566	0.4101	0.4560	0.4957
2	.....	.....	0.0033	0.0420	0.1056	0.1731	0.2363	0.2932	0.3437	0.3883
3	.....	.....	.....	0.0025	0.0327	0.0847	0.1423	0.1982	0.2500	0.2971
4	.....	.....	.....	.....	0.0020	0.0268	0.0708	0.1210	0.1710	0.2183
5	.....	.....	.....	.....	.....	0.0017	0.0227	0.0608	0.1053	0.1504
	15	20	25	30	35	40	45	50	60	70
0	0.7356	0.7943	0.8318	0.8577	0.8767	0.8913	0.9027	0.9120	0.9261	0.9363
1	0.6321	0.7112	0.7625	0.7984	0.8249	0.8453	0.8614	0.8745	0.8944	0.9089
2	0.5468	0.6417	0.7041	0.7481	0.7808	0.8060	0.8260	0.8423	0.8672	0.8853
3	0.4715	0.5793	0.6512	0.7024	0.7406	0.7701	0.7936	0.8128	0.8421	0.8636
4	0.4031	0.5218	0.6021	0.6597	0.7029	0.7364	0.7631	0.7850	0.8185	0.8430
5	0.3403	0.4680	0.5558	0.6192	0.6670	0.7043	0.7341	0.7584	0.7959	0.8233
6	0.2823	0.4171	0.5117	0.5805	0.6326	0.6734	0.7061	0.7329	0.7741	0.8043
7	0.2287	0.3691	0.4694	0.5433	0.5995	0.6436	0.6790	0.7080	0.7529	0.7858
8	0.1795	0.3234	0.4289	0.5074	0.5674	0.6146	0.6526	0.6839	0.7322	0.7678
9	0.1346	0.2801	0.3900	0.4726	0.5362	0.5864	0.6269	0.6603	0.7120	0.7501
10	0.0944	0.2390	0.3524	0.4388	0.5059	0.5589	0.6018	0.6372	0.6922	0.7328
11	0.0594	0.2001	0.3163	0.4061	0.4761	0.5320	0.5772	0.6146	0.6727	0.7157
12	0.0307	0.1634	0.2814	0.3742	0.4472	0.5056	0.5531	0.5924	0.6536	0.6989
13	0.0102	0.1292	0.2479	0.3433	0.4189	0.4797	0.5294	0.5706	0.6347	0.6824
14	0.0007	0.0975	0.2156	0.3132	0.3913	0.4544	0.5060	0.5491	0.6162	0.6661
15	.....	0.0688	0.1848	0.2839	0.3643	0.4295	0.4831	0.5279	0.5978	0.6499
	80	90	100	150	200	300	400	500	750	1,000
0	0.9441	0.9501	0.9550	0.9698	0.9772	0.9848	0.9886	0.9908	0.9939	0.9954
1	0.9199	0.9285	0.9355	0.9566	0.9673	0.9781	0.9835	0.9868	0.9912	0.9934
2	0.8990	0.9099	0.9186	0.9451	0.9586	0.9723	0.9792	0.9833	0.9888	0.9916
3	0.8799	0.8927	0.9030	0.9346	0.9507	0.9669	0.9751	0.9801	0.9867	0.9900
4	0.8617	0.8764	0.8883	0.9246	0.9431	0.9618	0.9713	0.9770	0.9846	0.9884
5	0.8443	0.8608	0.8741	0.9149	0.9358	0.9569	0.9676	0.9740	0.9826	0.9869
6	0.8274	0.8457	0.8604	0.9056	0.9287	0.9521	0.9640	0.9711	0.9807	0.9855
7	0.8111	0.8310	0.8471	0.8965	0.9218	0.9475	0.9604	0.9683	0.9788	0.9841
8	0.7951	0.8166	0.8341	0.8876	0.9150	0.9429	0.9570	0.9655	0.9769	0.9827
9	0.7794	0.8025	0.8213	0.8788	0.9084	0.9384	0.9536	0.9628	0.9751	0.9813
10	0.7640	0.7886	0.8087	0.8702	0.9018	0.9340	0.9503	0.9601	0.9733	0.9800
11	0.7488	0.7750	0.7963	0.8617	0.8954	0.9296	0.9470	0.9575	0.9715	0.9786
12	0.7339	0.7616	0.7841	0.8533	0.8890	0.9253	0.9437	0.9549	0.9698	0.9773
13	0.7191	0.7483	0.7720	0.8450	0.8827	0.9210	0.9405	0.9523	0.9681	0.9760
14	0.7046	0.7352	0.7600	0.8368	0.8764	0.9168	0.9373	0.9497	0.9663	0.9747
15	0.6902	0.7222	0.7482	0.8287	0.8702	0.9126	0.9341	0.9471	0.9646	0.9734
16	0.6760	0.7094	0.7365	0.8206	0.8641	0.9084	0.9310	0.9446	0.9629	0.9721
17	0.6619	0.6966	0.7250	0.8126	0.8580	0.9043	0.9279	0.9421	0.9612	0.9709
18	0.6479	0.6840	0.7135	0.8047	0.8519	0.9002	0.9248	0.9396	0.9596	0.9696
19	0.6341	0.6716	0.7021	0.7968	0.8459	0.8961	0.9217	0.9371	0.9579	0.9684
20	0.6204	0.6592	0.6908	0.7890	0.8399	0.8921	0.9186	0.9347	0.9562	0.9671
21	0.6069	0.6469	0.6796	0.7812	0.8340	0.8880	0.9156	0.9322	0.9546	0.9659
22	0.5934	0.6347	0.6685	0.7735	0.8281	0.8840	0.9125	0.9298	0.9530	0.9646
23	0.5801	0.6226	0.6574	0.7658	0.8222	0.8800	0.9095	0.9273	0.9513	0.9634
24	0.5668	0.6106	0.6464	0.7582	0.8164	0.8761	0.9065	0.9249	0.9497	0.9622
25	0.5537	0.5987	0.6355	0.7506	0.8105	0.8721	0.9035	0.9225	0.9481	0.9610

Tables C.3 show reliability determined from
equivalent missions and number of failures
at five different confidence levels.

TABLE C.3a   Reliability determined from equivalent missions and number of failures at 50 percent confidence

Number of failures*		Number of equivalent missions									
		0.5	1.0	1.5	2.0	2.5	3.0	3.5	4.0	4.5	5.0
A	B										
1	0	0.2500	0.5000	0.6300	0.7071	0.7579	0.7937	0.8204	0.8409	0.8573	0.8706
2	1	0.0348	0.1867	0.3266	0.4320	0.5110	0.5715	0.6190	0.6573	0.6887	0.7148
3	2	0.0048	0.0690	0.1682	0.2626	0.3431	0.4101	0.4658	0.5125	0.5520	0.5858
4	3	0.0006	0.0254	0.0865	0.1595	0.2302	0.2941	0.3502	0.3993	0.4422	0.4798
5	4	0.0001	0.0094	0.0444	0.0968	0.1544	0.2108	0.2633	0.3111	0.3542	0.3929
		5.5	6.0	6.5	7.0	7.5	8.0	8.5	9.0	9.5	10.5
1	0	0.8816	0.8909	0.8989	0.9057	0.9117	0.9170	0.9217	0.9259	0.9296	0.9330
2	1	0.7370	0.7560	0.7724	0.7868	0.7995	0.8107	0.8208	0.8299	0.8380	0.8455
3	2	0.6150	0.6404	0.6627	0.6825	0.7001	0.7159	0.7301	0.7430	0.7547	0.7654
4	3	0.5129	0.5423	0.5684	0.5918	0.6129	0.6319	0.6492	0.6650	0.6794	0.6927
5	4	0.4277	0.4591	0.4874	0.5131	0.5364	0.5577	0.5772	0.5951	0.6116	0.6268
6	5	0.3567	0.3887	0.4180	0.4449	0.4695	0.4923	0.5132	0.5326	0.5505	0.5672
7	6	0.2974	0.3290	0.3584	0.3857	0.4110	0.4344	0.4563	0.4766	0.4956	0.5133
8	7	0.2480	0.2785	0.3073	0.3343	0.3597	0.3834	0.4057	0.4265	0.4461	0.4645
9	8	0.2068	0.2358	0.2635	0.2898	0.3148	0.3384	0.3606	0.3817	0.4015	0.4203
10	9	0.1724	0.1996	0.2259	0.2513	0.2755	0.2986	0.3206	0.3415	0.3614	0.3803
		11.0	12.0	13.0	14.0	15.0	16.0	17.0	18.0	19.0	20.0
1	0	0.9389	0.9439	0.9481	0.9517	0.9549	0.9576	0.9601	0.9622	0.9642	0.9659
2	1	0.8585	0.8695	0.8789	0.8870	0.8941	0.9004	0.9060	0.9110	0.9154	0.9195
3	2	0.7842	0.8002	0.8141	0.8261	0.8367	0.8461	0.8545	0.8620	0.8687	0.8749
4	3	0.7162	0.7364	0.7539	0.7693	0.7829	0.7949	0.8057	0.8155	0.8243	0.8323
5	4	0.6540	0.6776	0.6982	0.7163	0.7324	0.7468	0.7598	0.7714	0.7820	0.7917
6	5	0.5972	0.6234	0.6465	0.6670	0.6852	0.7016	0.7164	0.7298	0.7420	0.7531
7	6	0.5454	0.5736	0.5987	0.6210	0.6411	0.6591	0.6755	0.6904	0.7040	0.7164
8	7	0.4980	0.5278	0.5544	0.5782	0.5997	0.6192	0.6369	0.6531	0.6679	0.6815
9	8	0.4547	0.4856	0.5133	0.5384	0.5611	0.5817	0.6005	0.6178	0.6336	0.6483
10	9	0.4152	0.4468	0.4753	0.5013	0.5249	0.5465	0.5662	0.5844	0.6012	0.6167
11	10	0.3791	0.4110	0.4401	0.4667	0.4910	0.5134	0.5339	0.5528	0.5704	0.5866
12	11	0.3462	0.3782	0.4076	0.4345	0.4594	0.4823	0.5034	0.5230	0.5411	0.5580
13	12	0.3161	0.3480	0.3774	0.4046	0.4298	0.4531	0.4747	0.4947	0.5134	0.5308
14	13	0.2886	0.3201	0.3495	0.3767	0.4020	0.4256	0.4475	0.4680	0.4871	0.5049
15	14	0.2636	0.2945	0.3236	0.3507	0.3761	0.3998	0.4220	0.4427	0.4621	0.4803
16	15	0.2407	0.2710	0.2996	0.3266	0.3519	0.3756	0.3979	0.4188	0.4384	0.4569
17	16	0.2198	0.2493	0.2774	0.3041	0.3292	0.3528	0.3751	0.3961	0.4159	0.4346
18	17	0.2007	0.2294	0.2569	0.2831	0.3079	0.3315	0.3537	0.3747	0.3946	0.4134
19	18	0.1832	0.2111	0.2379	0.2636	0.2881	0.3114	0.3335	0.3545	0.3744	0.3932
20	19	0.1673	0.1942	0.2203	0.2454	0.2695	0.2925	0.3145	0.3353	0.3552	0.3740

*If testing ended after a failure, use column A; if testing terminated after a given time, use column B.

Number of failures		Number of equivalent missions										
		30.0	40.0	50.0	60.0	80.0	100.0	200.0	400.0	600.0	800.0	1,000.0
A	*B*											
1	0	0.9772	0.9828	0.9862	0.9885	0.9914	0.9931	0.9965	0.9983	0.9988	0.9991	0.9993
2	1	0.9456	0.9589	0.9670	0.9724	0.9792	0.9834	0.9916	0.9958	0.9972	0.9979	0.9983
3	2	0.9147	0.9353	0.9479	0.9564	0.9671	0.9736	0.9867	0.9933	0.9956	0.9967	0.9973
4	3	0.8848	0.9123	0.9292	0.9406	0.9551	0.9639	0.9818	0.9909	0.9939	0.9954	0.9963
5	4	0.8558	0.8898	9.9108	0.9251	0.9433	0.9544	0.9769	0.9884	0.9922	0.9942	0.9953
6	5	0.8278	0.8678	0.8928	0.9098	0.9316	0.9449	0.9720	0.9859	0.9906	0.9929	0.9943
7	6	0.8007	0.8464	0.8751	0.8948	0.9200	0.9355	0.9672	0.9835	0.9889	0.9917	0.9934
8	7	0.7744	0.8255	0.8578	0.8800	0.9086	0.9262	0.9624	0.9810	0.9873	0.9905	0.9924
9	8	0.7490	0.8052	0.8408	0.8655	0.8973	0.9170	0.9576	0.9786	0.9857	0.9892	0.9914
10	9	0.7245	0.7853	0.8242	0.8512	0.8862	0.9078	0.9528	0.9761	0.9840	0.9880	0.9904
11	10	0.7007	0.7659	0.8079	0.8371	0.8752	0.8988	0.9481	0.9737	0.9824	0.9868	0.9894
12	11	0.6778	0.7470	0.7919	0.8233	0.8643	0.8899	0.9433	0.9713	0.9807	0.9855	0.9884
13	12	0.6556	0.7285	0.7762	0.8097	0.8536	0.8810	0.9386	0.9688	0.9791	0.9843	0.9874
14	13	0.6341	0.7106	0.7608	0.7963	0.8429	0.8722	0.9339	0.9664	0.9775	0.9831	0.9864
15	14	0.6133	0.6930	0.7458	0.7831	0.8325	0.8636	0.9293	0.9640	0.9758	0.9818	0.9854
16	15	0.5932	0.6759	0.7310	0.7702	0.8221	0.8550	0.9246	0.9616	0.9742	0.9806	0.9845
17	16	0.5737	0.6592	0.7165	0.7575	0.8119	0.8465	0.9200	0.9592	0.9726	0.9794	0.9835
18	17	0.5549	0.6430	0.7023	0.7449	0.8018	0.8381	0.9155	0.9568	0.9710	0.9782	0.9825
19	18	0.5367	0.6271	0.6884	0.7326	0.7919	0.8297	0.9109	0.9544	0.9694	0.9769	0.9815
20	19	0.5191	0.6116	0.6748	0.7205	0.7820	0.8215	0.9063	0.9520	0.9678	0.9757	0.9805
21	20	0.5021	0.5965	0.6614	0.7086	0.7723	0.8133	0.9018	0.9496	0.9661	0.9745	0.9795
22	21	0.4857	0.5818	0.6483	0.6969	0.7627	0.8052	0.8973	0.9473	0.9645	0.9733	0.9786
23	22	0.4697	0.5674	0.6355	0.6854	0.7533	0.7972	0.8929	0.9449	0.9629	0.9721	0.9776
24	23	0.4543	0.5534	0.6229	0.6741	0.7439	0.7893	0.8884	0.9425	0.9613	0.9708	0.9766
25	24	0.4394	0.5397	0.6106	0.6629	0.7347	0.7814	0.8840	0.9402	0.9597	0.9696	0.9756
30	29	0.3720	0.4763	0.5525	0.6099	0.6902	0.7433	0.8621	0.9285	0.9518	0.9636	0.9708
35	34	0.3149	0.4204	0.4999	0.5612	0.6483	0.7070	0.8409	0.9170	0.9439	0.9576	0.9659
40	39	0.2665	0.3710	0.4523	0.5163	0.6091	0.6726	0.8201	0.9056	0.9360	0.9516	0.9611
45	44	0.2256	0.3274	0.4093	0.4750	0.5722	0.6398	0.7999	0.8943	0.9283	0.9457	0.9563
50	49	0.1910	0.2889	0.3703	0.4370	0.5375	0.6086	0.7801	0.8832	0.9206	0.9398	0.9515
60	59	0.1369	0.2250	0.3032	0.3699	0.4743	0.5507	0.7421	0.8614	0.9053	0.9281	0.9421
70	69	0.981	0.1752	0.2483	0.3131	0.4186	0.4983	0.7059	0.8402	0.8904	0.9166	0.9327
80	79	0.703	0.1365	0.2033	0.2651	0.3694	0.4508	0.6715	0.8194	0.8757	0.9052	0.9234
90	89	0.503	0.1063	0.1664	0.2244	0.3260	0.4079	0.6387	0.7992	0.8612	0.8940	0.9142
100	99	0.361	0.0828	0.1362	0.1899	0.2877	0.3691	0.6076	0.7795	0.8470	0.8829	0.9051
120	119	0.0185	0.0502	0.0913	0.1361	0.2241	0.3022	0.5497	0.7414	0.8192	0.8611	0.8872
140	139	0.0095	0.0305	0.0612	0.0975	0.1745	0.2474	0.4974	0.7053	0.7923	0.8398	0.8697
160	159	0.0049	0.0185	0.0410	0.0699	0.1359	0.2026	0.4501	0.6709	0.7664	0.8191	0.8524
180	179	0.0025	0.0112	0.0275	0.0501	0.1058	0.1659	0.4073	0.6382	0.7412	0.7989	0.8356
200	199	0.0013	0.0068	0.0184	0.0359	0.0824	0.1358	0.3685	0.6070	0.7169	0.7791	0.8190

TABLE C.3b   Reliability determined from equivalent missions and number of failures at 75 percent confidence

Number of failures		Number of equivalent missions									
		0.5	1.0	1.5	2.0	2.5	3.0	3.5	4.0	4.5	5.0
A	B										
1	0	0.0625	0.2500	0.3968	0.5000	0.5743	0.6300	0.6729	0.7071	0.7348	0.7579
2	1	0.0046	0.0677	0.1661	0.2602	0.3406	0.4076	0.4633	0.5101	0.5497	0.5836
3	2	0.0004	0.0198	0.0733	0.1408	0.2084	0.2707	0.3262	0.3753	0.4184	0.4565
4	3	0.0000	0.0060	0.0332	0.0777	0.1295	0.1821	0.2323	0.2788	0.3213	0.3599
5	4	0.0000	0.0019	0.0153	0.0434	0.0813	0.1235	0.1665	0.2083	0.2480	0.2851
		5.5	6.0	6.5	7.0	7.5	8.0	8.5	9.0	9.5	10.0
1	0	0.7772	0.7937	0.8079	0.8204	0.8312	0.8409	0.8495	0.8572	0.8642	0.8705
2	1	0.6129	0.6384	0.6608	0.6807	0.6984	0.7142	0.7285	0.7414	0.7532	0.7640
3	2	0.4903	0.5203	0.5471	0.5712	0.5929	0.6126	0.6305	0.6469	0.6619	0.6757
4	3	0.3949	0.4267	0.4556	0.4819	0.5060	0.5280	0.5482	0.5668	0.5840	0.5999
5	4	0.3196	0.3514	0.3809	0.4081	0.4332	0.4564	0.4780	0.4980	0.5166	0.5340
6	5	0.2594	0.2902	0.3192	0.3463	0.3717	0.3954	0.4176	0.4384	0.4578	0.4760
7	6	0.2110	0.2402	0.2680	0.2945	0.3195	0.3431	0.3654	0.3864	0.4062	0.4249
8	7	0.1719	0.1991	0.2254	0.2507	0.2749	0.2980	0.3200	0.3409	0.3608	0.3797
9	8	0.1403	0.1652	0.1898	0.2137	0.2368	0.2592	0.2806	0.3011	0.3207	0.3395
10	9	0.1146	0.1373	0.1599	0.1823	0.2042	0.2255	0.2462	0.2661	0.2853	0.3038
		11.0	12.0	13.0	14.0	15.0	16.0	17.0	18.0	19.0	20.0
1	0	0.8816	0.8909	0.8988	0.9057	0.9117	0.9170	0.9217	0.9259	0.9296	0.9330
2	1	0.7829	0.7990	0.8129	0.8250	0.8357	0.8451	0.8535	0.8611	0.8679	0.8740
3	2	0.7002	0.7213	0.7397	0.7558	0.7700	0.7827	0.7940	0.8043	0.8136	0.8220
4	3	0.6284	0.6533	0.6750	0.6942	0.7113	0.7266	0.7404	0.7529	0.7642	0.7745
5	4	0.5653	0.5928	0.6171	0.6388	0.6582	0.6756	0.6914	0.7057	0.7188	0.7307
6	5	0.5093	0.5387	0.5650	0.5885	0.6097	0.6288	0.6462	0.6621	0.6766	0.6900
7	6	0.4593	0.4901	0.5177	0.5426	0.5652	0.5857	0.6044	0.6216	0.6373	0.6519
8	7	0.4146	0.4462	0.4748	0.5007	0.5243	0.5459	0.5657	0.5839	0.6007	0.6162
9	8	0.3745	0.4065	0.4356	0.4623	0.4867	0.5091	0.5297	0.5487	0.5663	0.5827
10	9	0.3385	0.3705	0.3999	0.4270	0.4519	0.4749	0.4962	0.5159	0.5342	0.5512
11	10	0.3062	0.3379	0.3673	0.3946	0.4198	0.4432	0.4649	0.4851	0.5040	0.5215
12	11	0.2770	0.3083	0.3375	0.3647	0.3901	0.4237	0.4358	0.4564	0.4756	0.4936
13	12	0.2507	0.2814	0.3102	0.3372	0.3626	0.3863	0.4085	0.4294	0.4489	0.4673
14	13	0.2270	0.2569	0.2852	0.3119	0.3371	0.3608	0.3831	0.4041	0.4238	0.4424
15	14	0.2056	0.2346	0.2622	0.2886	0.3135	0.3371	0.3593	0.3803	0.4002	0.4190
16	15	0.1863	0.2143	0.2412	0.2670	0.2916	0.3149	0.3371	0.3581	0.3780	0.3968
17	16	0.1688	0.1958	0.2219	0.2471	0.2713	0.2943	0.3163	0.3371	0.3570	0.3759
18	17	0.1530	0.1789	0.2042	0.2287	0.2524	0.2751	0.2968	0.3175	0.3372	0.3561
19	18	0.1387	0.1635	0.1879	0.2118	0.2349	0.2571	0.2785	0.2990	0.3186	0.3374
20	19	0.1258	0.1495	0.1730	0.1961	0.2186	0.2404	0.2614	0.2816	0.3011	0.3197

Number of failures		Number of equivalent missions										
		30.0	40.0	50.0	60.0	80.0	100.0	200.0	400.0	600.0	800.0	1,000.0
A	B											
1	0	0.9548	0.9659	0.9727	0.9772	0.9828	0.9862	0.9931	0.9965	0.9977	0.9983	0.9986
2	1	0.9142	0.9349	0.9476	0.9561	0.9669	0.9734	0.9866	0.9933	0.9955	0.9966	0.9973
3	2	0.8775	0.9066	0.9246	0.9367	0.9522	0.9616	0.9806	0.9902	0.9935	0.9951	0.9961
4	3	0.8434	0.8801	0.9029	0.9184	0.9381	0.9502	0.9748	0.9873	0.9915	0.9936	0.9949
5	4	0.8113	0.8548	0.8821	0.9007	0.9246	0.9392	0.9691	0.9844	0.9896	0.9922	0.9937
6	5	0.7808	0.8306	0.8620	0.8836	0.9114	0.9285	0.9636	0.9816	0.9877	0.9908	0.9926
7	6	0.7518	0.8074	0.8427	0.8671	0.8985	0.9180	0.9581	0.9788	0.9858	0.9894	0.9915
8	7	0.7241	0.7850	0.8239	0.8509	0.8860	0.9077	0.9527	0.9761	0.9840	0.9880	0.9904
9	8	0.6976	0.7633	0.8057	0.8352	0.8737	0.8976	0.9474	0.9734	0.9822	0.9866	0.9893
10	9	0.6722	0.7424	0.7880	0.8199	0.8616	0.8877	0.9422	0.9707	0.9803	0.9852	0.9882
11	10	0.6479	0.7222	0.7708	0.8049	0.8498	0.8779	0.9370	0.9680	0.9785	0.9839	0.9871
12	11	0.6246	0.7026	0.7540	0.7903	0.8382	0.8683	0.9318	0.9653	0.9767	0.9825	0.9860
13	12	0.6021	0.6836	0.7376	0.7760	0.8268	0.8588	0.9267	0.9627	0.9750	0.9812	0.9849
14	13	0.5806	0.6651	0.7217	0.7620	0.8156	0.8495	0.9217	0.9600	0.9732	0.9798	0.9838
15	14	0.5599	0.6473	0.7061	0.7483	0.8045	0.8403	0.9167	0.9574	0.9714	0.9785	0.9828
16	15	0.5400	0.6299	0.6909	0.7348	0.7937	0.8312	0.9117	0.9548	0.9697	0.9772	0.9817
17	16	0.5208	0.6131	0.6761	0.7217	0.7830	0.8223	0.9068	0.9523	0.9679	0.9758	0.9806
18	17	0.5024	0.5967	0.6616	0.7088	0.7725	0.8134	0.9019	0.9497	0.9662	0.9745	0.9796
19	18	0.4846	0.5808	0.6475	0.6962	0.7621	0.8047	0.8970	0.9471	0.9644	0.9732	0.9785
20	19	0.4675	0.5654	0.6337	0.6838	0.7519	0.7961	0.8922	0.9446	0.9627	0.9719	0.9775
21	20	0.4511	0.5504	0.6202	0.6716	0.7419	0.7875	0.8874	0.9420	0.9610	0.9706	0.9764
22	21	0.4352	0.5358	0.6071	0.6597	0.7320	0.7791	0.8827	0.9395	0.9593	0.9693	0.9754
23	22	0.4200	0.5217	0.5942	0.6480	0.7223	0.7708	0.8780	0.9370	0.9575	0.9680	0.9743
24	23	0.4052	0.5079	0.5816	0.6366	0.7127	0.7626	0.8733	0.9345	0.9558	0.9667	0.9733
25	24	0.3911	0.4945	0.5693	0.6253	0.7032	0.7545	0.8686	0.9320	0.9541	0.9654	0.9722
30	29	0.3275	0.4329	0.5118	0.5723	0.6579	0.7154	0.8458	0.9197	0.9457	0.9590	0.9671
35	34	0.2745	0.3792	0.4603	0.5239	0.6158	0.6785	0.8237	0.9076	0.9374	0.9527	0.9620
40	39	0.2302	0.3323	0.4142	0.4798	0.5765	0.6436	0.8023	0.8947	0.9292	0.9464	0.9569
45	44	0.1932	0.2914	0.3729	0.4395	0.5398	0.6106	0.7814	0.8840	0.9211	0.9402	0.9519
50	49	0.1622	0.2556	0.3357	0.4027	0.5055	0.5794	0.7612	0.8725	0.9131	0.9341	0.9469
60	59	0.1145	0.1968	0.2724	0.3383	0.4436	0.5219	0.7224	0.8500	0.8973	0.9219	0.9370
70	69	0.0809	0.1517	0.2211	0.2844	0.3894	0.4703	0.6858	0.8281	0.8818	0.9100	0.9273
80	79	0.0572	0.1170	0.1797	0.2392	0.3420	0.4239	0.6511	0.8069	0.8667	0.8983	0.9177
90	89	0.0405	0.0903	0.1460	0.2012	0.3004	0.3821	0.6182	0.7862	0.8519	0.8867	0.9083
100	99	0.0287	0.0697	0.1187	0.1693	0.2640	0.3446	0.5870	0.7661	0.8373	0.8753	0.8989
120	119	0.0144	0.0416	0.0786	0.1200	0.2039	0.2803	0.5294	0.7276	0.8090	0.8530	0.8806
140	139	0.0073	0.0249	0.0520	0.0852	0.1576	0.2281	0.4776	0.6911	0.7817	0.8313	0.8626
160	159	0.0037	0.0149	0.0345	0.0605	0.1219	0.1857	0.4310	0.6565	0.7554	0.8102	0.8451
180	179	0.0018	0.0089	0.0229	0.0430	0.0944	0.1513	0.3890	0.6237	0.7300	0.7897	0.8279
200	199	0.0009	0.0053	0.0152	0.0305	0.0730	0.1233	0.3511	0.5925	0.7055	0.7698	0.8111

TABLE C.3c Reliability determined from equivalent missions and number of failures at 90 percent confidence

Number of failures		Number of equivalent missions									
A	B	0.5	1.0	1.5	2.0	2.5	3.0	3.5	4.0	4.5	5.0
1	0	0.0100	0.1000	0.2155	0.3162	0.3981	0.4642	0.5180	0.5624	0.5995	0.6310
2	1	0.0004	0.0205	0.0748	0.1430	0.0210	0.2735	0.3291	0.3782	0.4213	0.4594
3	2	0.0000	0.0049	0.0288	0.0699	0.1190	0.1696	0.2186	0.2643	0.3064	0.3449
4	3	0.0000	0.0013	0.0116	0.0354	0.0691	0.1079	0.1482	0.1882	0.2266	0.2628
5	4	0.0000	0.0003	0.0048	0.0184	0.0409	0.0696	0.1019	0.1356	0.1693	0.2022
		5.5	6.0	6.5	7.0	7.5	8.0	8.5	9.0	9.5	10.0
1	0	0.6579	0.6813	0.7017	0.7197	0.7357	0.7499	0.7627	0.7743	0.7848	0.7943
2	1	0.4930	0.5230	0.5497	0.5737	0.5954	0.6150	0.6328	0.6491	0.6640	0.6778
3	2	0.3799	0.4119	0.4409	0.4675	0.4918	0.5141	0.5346	0.5536	0.5711	0.5873
4	3	0.2968	0.3284	0.3578	0.3850	0.4103	0.4338	0.4557	0.4760	0.4950	0.5127
5	4	0.2338	0.2639	0.2924	0.3192	0.3445	0.3682	0.3905	0.4114	0.4311	0.4496
6	5	0.1852	0.2132	0.2401	0.2658	0.2904	0.3137	0.3358	0.3568	0.3767	0.3956
7	6	0.1474	0.1728	0.1978	0.2221	0.2455	0.2681	0.2897	0.3103	0.3300	0.3488
8	7	0.1176	0.1406	0.1635	0.1861	0.2082	0.2296	0.2504	0.2704	0.2897	0.3082
9	8	0.0942	0.1147	0.1354	0.1562	0.1768	0.1970	0.2168	0.2360	0.2547	0.2727
10	9	0.0756	0.0937	0.1124	0.1314	0.1504	0.1694	0.1880	0.2063	0.2242	0.2416
		11.0	12.0	13.0	14.0	15.0	16.0	17.0	18.0	19.0	20.0
1	0	0.8111	0.8254	0.8377	0.8483	0.8577	0.8660	0.8733	0.8799	0.8859	0.8913
2	1	0.7022	0.7232	0.7414	0.7574	0.7716	0.7842	0.7955	0.8057	0.8149	0.8233
3	2	0.6164	0.6418	0.6640	0.6837	0.7013	0.7170	0.7312	0.7440	0.7557	0.7663
4	3	0.5448	0.5731	0.5981	0.6205	0.6406	0.6587	0.6750	0.6899	0.7035	0.7160
5	4	0.4835	0.5137	0.5407	0.5650	0.5869	0.6068	0.6249	0.6414	0.6566	0.6705
6	5	0.4304	0.4617	0.4900	0.5156	0.5389	0.5601	0.5795	0.5974	0.6138	0.6289
7	6	0.3839	0.4158	0.4448	0.4713	0.4955	0.5178	0.5382	0.5570	0.5745	0.5906
8	7	0.3430	0.3750	0.4044	0.4314	0.4562	0.4792	0.5004	0.5200	0.5382	0.5551
9	8	0.3069	0.3386	0.3680	0.3953	0.4205	0.4439	0.4656	0.4858	0.5046	0.5222
10	9	0.2749	0.3061	0.3353	0.3625	0.3879	0.4115	0.4336	0.4542	0.4735	0.4915
11	10	0.2465	0.2770	0.3057	0.3327	0.3580	0.3818	0.4040	0.4249	0.4445	0.4629
12	11	0.2212	0.2508	0.2789	0.3056	0.3307	0.3544	0.3767	0.3977	0.4175	0.4361
13	12	0.1986	0.2272	0.2547	0.2808	0.3056	0.3291	0.3514	0.3724	0.3922	0.4110
14	13	0.1784	0.2060	0.2326	0.2582	0.2826	0.3058	0.3279	0.3488	0.3687	0.3876
15	14	0.1604	0.1869	0.2126	0.2375	0.2614	0.2842	0.3061	0.3269	0.3467	0.3655
16	15	0.1443	0.1696	0.1944	0.2185	0.2418	0.2643	0.2858	0.3064	0.3261	0.3449
17	16	0.1299	0.1540	0.1778	0.2012	0.2239	0.2458	0.2670	0.2873	0.3068	0.3254
18	17	0.1170	0.1399	0.1627	0.1852	0.2073	0.2287	0.2494	0.2694	0.2887	0.3072
19	18	0.1053	0.1271	0.1489	0.1706	0.1920	0.2128	0.2331	0.2527	0.2717	0.2900
20	19	0.0949	0.1155	0.1364	0.1572	0.1778	0.1981	0.2179	0.2372	0.2558	0.2739

Number of failures		Number of equivalent missions										
A	B	30.0	40.0	50.0	60.0	80.0	100.0	200.0	400.0	600.0	800.0	1,000.0
1	0	0.9261	0.9441	0.9550	0.9624	0.9716	0.9772	0.9886	0.9943	0.9962	0.9971	0.9977
2	1	0.8784	0.9073	0.9252	0.9372	0.9525	0.9619	0.9807	0.9903	0.9935	0.9951	0.9961
3	2	0.8374	0.8754	0.8990	0.9151	0.9356	0.9482	0.9737	0.9868	0.9912	0.9934	0.9947
4	3	0.8004	0.8462	0.8749	0.8946	0.9199	0.9354	0.9671	0.9834	0.9889	0.9917	0.9933
5	4	0.7661	0.8189	0.8523	0.8753	0.9049	0.9232	0.9608	0.9802	0.9868	0.9901	0.9920
6	5	0.7341	0.7931	0.8307	0.8568	0.8905	0.9114	0.9547	0.9771	0.9847	0.9885	0.9908
7	6	0.7039	0.7685	0.8101	0.8390	0.8766	0.9000	0.9487	0.9740	0.9826	0.9869	0.9895
8	7	0.6755	0.7451	0.7902	0.8219	0.8632	0.8890	0.9428	0.9710	0.9806	0.9854	0.9883
9	8	0.6485	0.7226	0.7711	0.8053	0.8501	0.8781	0.9371	0.9680	0.9786	0.9839	0.9871
10	9	0.6228	0.7011	0.7527	0.7892	0.8373	0.8676	0.9314	0.9651	0.9766	0.9824	0.9859
11	10	0.5984	0.6803	0.7348	0.7735	0.8248	0.8572	0.9259	0.9622	0.9746	0.9809	0.9847
12	11	0.5751	0.6604	0.7175	0.7583	0.8126	0.8471	0.9204	0.9594	0.9427	0.9795	0.9835
13	12	0.5528	0.6411	0.7007	0.7435	0.8007	0.8371	0.9149	0.9565	0.9708	0.9780	0.9824
14	13	0.5316	0.6225	0.6844	0.7291	0.7890	0.8273	0.9096	0.9537	0.9689	0.9766	0.9812
15	14	0.5112	0.6046	0.6686	0.7150	0.7776	0.8177	0.9043	0.9509	0.9670	0.9752	0.9801
16	15	0.4918	0.5872	0.6532	0.7013	0.7663	0.8082	0.8990	0.9482	0.9651	0.9737	0.9789
17	16	0.4731	0.5705	0.6382	0.6878	0.7553	0.7989	0.8938	0.9454	0.9633	0.9723	0.9778
18	17	0.4553	0.5542	0.6237	0.6747	0.7445	0.7897	0.8887	0.9427	0.9614	0.9709	0.9767
19	18	0.4381	0.5385	0.6095	0.6619	0.7338	0.7807	0.8836	0.9400	0.9596	0.9695	0.9755
20	19	0.4217	0.5233	0.5957	0.6494	0.7234	0.7718	0.8785	0.9373	0.9577	0.9681	0.9744
21	20	0.4060	0.5086	0.5822	0.6372	0.7132	0.7630	0.8735	0.9346	0.9559	0.9668	0.9733
22	21	0.3908	0.4943	0.5691	0.6252	0.7031	0.7544	0.8686	0.9320	0.9541	0.9654	0.9722
23	22	0.3763	0.4805	0.5563	0.6134	0.6932	0.7459	0.8636	0.9293	0.9523	0.9640	0.9711
24	23	0.3624	0.4670	0.5439	0.6020	0.6834	0.7375	0.8588	0.9267	0.9505	0.9626	0.9700
25	24	0.3490	0.4540	0.5317	0.5907	0.6738	0.7292	0.8539	0.9241	0.9487	0.9613	0.9689
30	29	0.2894	0.3946	0.4752	0.5380	0.6281	0.6894	0.8303	0.9112	0.9399	0.9546	0.9635
35	34	0.2404	0.3433	0.4252	0.4903	0.5859	0.6520	0.8075	0.8986	0.9312	0.9479	0.9581
40	39	0.2000	0.2990	0.3807	0.4472	0.5468	0.6170	0.7855	0.8863	0.9227	0.9414	0.9529
45	44	0.1665	0.2607	0.3411	0.4080	0.5105	0.5840	0.7642	0.8742	0.9143	0.9350	0.9476
50	49	0.1388	0.2274	0.3058	0.3725	0.4768	0.5529	0.7436	0.8623	0.9060	0.9286	0.9425
60	59	0.0966	0.1733	0.2460	0.3108	0.4163	0.4960	0.7043	0.8392	0.8897	0.9161	0.9323
70	69	0.0674	0.1323	0.1982	0.2596	0.3637	0.4452	0.6673	0.8169	0.8738	0.9038	0.9223
80	79	0.0471	0.1011	0.1599	0.2171	0.3180	0.3999	0.6324	0.7952	0.8583	0.8918	0.9124
90	89	0.0330	0.0774	0.1291	0.1816	0.2782	0.3594	0.5995	0.7743	0.8432	0.8799	0.9027
100	99	0.0231	0.0593	0.1043	0.1521	0.2435	0.3230	0.5683	0.7539	0.8283	0.8683	0.8931
120	119	0.0114	0.0349	0.0682	0.1068	0.1868	0.2612	0.5111	0.7149	0.7995	0.8455	0.8744
140	139	0.0056	0.0206	0.0447	0.0751	0.0434	0.2115	0.4599	0.6781	0.7719	0.8235	0.8561
160	159	0.0028	0.0122	0.0294	0.0529	0.1102	0.1713	0.4139	0.6434	0.7453	0.8021	0.8383
180	179	0.0014	0.0072	0.0193	0.0373	0.0848	0.1389	0.3727	0.6105	0.7197	0.7813	0.8209
200	199	0.0007	0.0043	0.0127	0.0263	0.0653	0.1127	0.3357	0.5794	0.6950	0.7612	0.8039

TABLE C.3d   Reliability determined from equivalent missions and number of failures at 95 percent confidence

Number of failures		Number of equivalent missions									
		0.5	1.0	1.5	2.0	2.5	3.0	3.5	4.0	4.5	5.0
A	B										
1	0	0.0025	0.0500	0.1357	0.2236	0.3017	0.3684	0.4249	0.4729	0.5139	0.5493
2	1	0.0001	0.0087	0.0423	0.0933	0.1499	0.2057	0.2578	0.3054	0.3485	0.3872
3	2	0.0000	0.0018	0.0150	0.0429	0.0806	0.1226	0.1655	0.2072	0.2468	0.2839
4	3	0.0000	0.0004	0.0057	0.0207	0.0450	0.0754	0.1091	0.1439	0.1785	0.2121
5	4	0.0000	0.0001	0.0022	0.0103	0.0257	0.0473	0.0731	0.1014	0.1308	0.1603
		5.5	6.0	6.5	7.0	7.5	8.0	8.5	9.0	9.5	10.0
1	0	0.5801	0.6070	0.6307	0.6519	0.6707	0.6877	0.7030	0.7169	0.7296	0.7412
2	1	0.4221	0.4535	0.4820	0.5078	0.5312	0.5527	0.5723	0.5903	0.6069	0.6223
3	2	0.3183	0.3502	0.3796	0.4068	0.4319	0.4552	0.4768	0.4968	0.5154	0.5328
4	3	0.2442	0.2747	0.3034	0.3303	0.3557	0.3794	0.4016	0.4225	0.4421	0.4605
5	4	0.1893	0.2175	0.2446	0.2705	0.2951	0.3185	0.3407	0.3617	0.3815	0.4004
6	5	0.1479	0.1734	0.1984	0.2227	0.2462	0.2687	0.2903	0.3110	0.3307	0.3495
7	6	0.1161	0.1389	0.1617	0.1842	0.2062	0.2276	0.2483	0.2683	0.2875	0.3060
8	7	0.0916	0.1118	0.1323	0.1529	0.1732	0.1933	0.2129	0.2320	0.2506	0.2685
9	8	0.0725	0.0902	0.1085	0.1272	0.1459	0.1646	0.1830	0.2011	0.2188	0.2361
10	9	0.0575	0.0730	0.0893	0.1061	0.1232	0.1404	0.1576	0.1746	0.1914	0.2079
		11.0	120.	13.0	14.0	15.0	16.0	17.0	18.0	19.0	20.0
1	0	0.7616	0.7791	0.7942	0.8074	0.8190	0.8293	0.8384	0.8467	0.8541	0.8609
3	1	0.6497	0.6735	0.6943	0.7126	0.7289	0.7434	0.7565	0.7683	0.7790	0.7888
3	2	0.5642	0.5918	0.6161	0.6378	0.6572	0.6747	0.6905	0.7048	0.7179	0.7299
4	3	0.4942	0.5241	0.5508	0.5747	0.5964	0.6159	0.6338	0.6500	0.6649	0.6786
5	4	0.4351	0.4664	0.4945	0.5201	0.5432	0.5643	0.5837	0.6014	0.6177	0.6328
6	5	0.3845	0.4164	0.4454	0.4719	0.4962	0.5184	0.5388	0.5576	0.5750	0.5912
7	6	0.3408	0.3727	0.4021	0.4292	0.4541	0.4770	0.4983	0.5179	0.5362	0.5532
8	7	0.3026	0.3343	0.3637	0.3910	0.4162	0.4397	0.4614	0.4817	0.5006	0.5182
9	8	0.2692	0.3003	0.3294	0.3566	0.3820	0.4057	0.4278	0.4485	0.4678	0.4859
10	9	0.2399	0.2702	0.2988	0.3257	0.3510	0.3747	0.3970	0.4179	0.4375	0.4560
11	10	0.2140	0.2433	0.2712	0.2977	0.3228	0.3464	0.3687	0.3897	0.4095	0.4282
12	11	0.1910	0.2193	0.2465	0.2724	0.2971	0.3205	0.3427	0.3637	0.3835	0.4024
13	12	0.1708	0.1979	0.2241	0.2494	0.2736	0.2967	0.3186	0.3395	0.3594	0.3783
14	13	0.1527	0.1786	0.2039	0.2285	0.2521	0.2748	0.2965	0.3172	0.3370	0.3558
15	14	0.1367	0.1614	0.1857	0.2094	0.2324	0.2546	0.2760	0.2964	0.3160	0.3348
16	15	0.1225	0.1459	0.1692	0.1921	0.2144	0.2361	0.2570	0.2772	0.2965	0.3151
17	16	0.1098	0.1320	0.1542	0.1763	0.1979	0.2190	0.2394	0.2592	0.2783	0.2967
18	17	0.0985	0.1194	0.1407	0.1618	0.1827	0.2032	0.2231	0.2425	0.2613	0.2794
19	18	0.0883	0.1081	0.1283	0.1486	0.1687	0.1886	0.2080	0.2270	0.2454	0.2633
20	19	0.0793	0.0980	0.1171	0.1365	0.1559	0.1751	0.1940	0.2125	0.2305	0.2481

Number of failures		Number of equivalent missions										
		30.0	40.0	50.0	60.0	80.0	100.0	200.0	400.0	600.0	800.0	1,000.0
A	B											
1	0	0.9050	0.9278	0.9418	0.9513	0.9632	0.9705	0.9851	0.9925	0.9950	0.9963	0.9970
2	1	0.8537	0.8882	0.9095	0.9240	0.9424	0.9537	0.9766	0.9882	0.9921	0.9941	0.9953
3	2	0.8107	0.8544	0.8817	0.9004	0.9243	0.9390	0.9690	0.9844	0.9896	0.9922	0.9937
4	3	0.7722	0.8238	0.8564	0.8788	0.9076	0.9254	0.9620	0.9808	0.9872	0.9904	0.9923
5	4	0.7370	0.7955	0.8327	0.8585	0.8919	0.9125	0.9553	0.9774	0.9849	0.9886	0.9909
6	5	0.7044	0.7689	0.8104	0.8393	0.8769	0.9002	0.9488	0.9741	0.9826	0.9869	0.9895
7	6	0.6738	0.7437	0.7891	0.8209	0.8624	0.8883	0.9425	0.9708	0.9805	0.9853	0.9882
8	7	0.6452	0.7199	0.7688	0.8032	0.8484	0.8768	0.9364	0.9677	0.9783	0.9837	0.9869
9	8	0.6181	0.6971	0.7492	0.7862	0.8349	0.8656	0.9304	0.9646	0.9762	0.9821	0.9857
10	9	0.5924	0.6753	0.7304	0.7697	0.8218	0.8547	0.9245	0.9615	0.9742	0.9806	0.9844
11	10	0.5681	0.6544	0.7123	0.7537	0.8089	0.8440	0.9187	0.9585	0.9721	0.9790	0.9832
12	11	0.5450	0.6343	0.6948	0.7383	0.7964	0.8335	0.9130	0.9555	0.9701	0.9775	0.9820
13	12	0.5230	0.6150	0.6778	0.7232	0.7842	0.8233	0.9074	0.9526	0.9681	0.9760	0.9807
14	13	0.5021	0.5965	0.6614	0.7086	0.7723	0.8133	0.9018	0.9496	0.9661	0.9745	0.9795
15	14	0.4821	0.5786	0.6455	0.6944	0.7607	0.8034	0.8963	0.9468	0.9642	0.9730	0.9784
16	15	0.4631	0.5613	0.6301	0.6805	0.7492	0.7938	0.8909	0.9439	0.9622	0.9715	0.9772
17	16	0.4448	0.5447	0.6151	0.6670	0.7380	0.7843	0.8856	0.9411	0.9603	0.9701	0.9760
18	17	0.4274	0.5286	0.6005	0.6538	0.7271	0.7749	0.8803	0.9382	0.9584	0.9686	0.9748
19	18	0.4108	0.5131	0.5863	0.6409	0.7163	0.7657	0.8751	0.9354	0.9565	0.9672	0.9737
20	19	0.3948	0.4981	0.5726	0.6284	0.7058	0.7567	0.8699	0.9327	0.9546	0.9658	0.9725
21	20	0.3796	0.4836	0.5592	0.6161	0.6954	0.7478	0.8648	0.9299	0.9527	0.9643	0.9714
22	21	0.3649	0.4695	0.5462	0.6041	0.6852	0.7390	0.8597	0.9272	0.9508	0.9629	0.9702
23	22	0.3509	0.4559	0.5335	0.5924	0.6752	0.7304	0.8546	0.9245	0.9490	0.9615	0.9691
24	23	0.3375	0.4428	0.5212	0.5809	0.6654	0.7219	0.8497	0.9218	0.9471	0.9601	0.9679
25	24	0.3246	0.4301	0.5091	0.5698	0.6558	0.7135	0.8447	0.9191	0.9453	0.9587	0.9668
30	29	0.2677	0.3721	0.4535	0.5174	0.6100	0.6734	0.8206	0.9059	0.9362	0.9518	0.9612
35	34	0.2212	0.3225	0.4044	0.4703	0.5679	0.6359	0.7975	0.8930	0.9273	0.9450	0.9557
40	39	0.1831	0.2799	0.3610	0.4278	0.5290	0.6009	0.7752	0.8804	0.9186	0.9383	0.9503
45	44	0.1517	0.2431	0.3226	0.3895	0.4930	0.5679	0.7536	0.8681	0.9100	0.9317	0.9450
50	49	0.1259	0.2113	0.2884	0.3548	0.4597	0.5370	0.7328	0.8560	0.9016	0.9252	0.9397
60	59	0.0869	0.1601	0.2309	0.2948	0.4001	0.4805	0.6932	0.8326	0.8850	0.9125	0.9293
70	69	0.0602	0.1215	0.1852	0.2453	0.3486	0.4304	0.6560	0.8100	0.8689	0.9000	0.9191
80	79	0.0418	0.0924	0.1488	0.2044	0.3040	0.3858	0.6211	0.7881	0.8532	0.8878	0.9091
90	89	0.0291	0.0704	0.1197	0.1705	0.2653	0.3460	0.5882	0.7669	0.8379	0.8757	0.8993
100	99	0.0202	0.0537	0.0963	0.1423	0.2317	0.3104	0.5571	0.7464	0.8228	0.8639	0.8896
120	119	0.0099	0.0313	0.0626	0.0993	0.1769	0.2502	0.5002	0.7072	0.7938	0.8410	0.8706
140	139	0.0048	0.0183	0.0407	0.0695	0.1353	0.2019	0.4493	0.6703	0.7659	0.8187	0.8521
160	159	0.0024	0.0107	0.0266	0.0487	0.1036	0.1631	0.4038	0.6355	0.7392	0.7972	0.8341
180	179	0.0012	0.0063	0.0174	0.0342	0.0794	0.1318	0.3631	0.6026	0.7134	0.7763	0.8166
200	199	0.0006	0.0037	0.0114	0.0240	0.0610	0.1067	0.3266	0.5715	0.6886	0.7560	0.7995

TABLE C.3e   Reliability determined from equivalent missions and
number of failures at 99 percent confidence

Number of failures		Number of equivalent missions									
A	B	0.5	1.0	1.5	2.0	2.5	3.0	3.5	4.0	4.5	5.0
1	0	0.0001	0.0100	0.0464	0.1000	0.1585	0.3155	0.2683	0.3162	0.3594	0.3981
2	1	0.0000	0.0013	0.0120	0.0362	0.0703	0.1094	0.1501	0.1902	0.2287	0.2651
3	2	0.0000	0.0002	0.0037	0.0150	0.0347	0.0607	0.0906	0.1223	0.1544	0.1862
4	3	0.0000	0.0000	0.0012	0.0066	0.0180	0.0351	0.0567	0.0812	0.1073	0.1341
5	4	0.0000	0.0000	0.0004	0.0030	0.0096	0.0209	0.0363	0.0550	0.0759	0.0982
		5.5	6.0	6.5	7.0	7.5	8.0	8.5	9.0	9.5	10.0
1	0	0.4329	0.4642	0.4924	0.5180	0.5412	0.5624	0.5817	0.5995	0.6159	0.6310
2	1	0.2991	0.3307	0.3601	0.3874	0.4127	0.4361	0.4579	0.4783	0.4972	0.5149
3	2	0.2169	0.2464	0.2744	0.3009	0.3260	0.3497	0.3720	0.3930	0.4128	0.4315
4	3	0.1610	0.1875	0.2132	0.2381	0.2620	0.2849	0.3067	0.3276	0.3474	0.3662
5	4	0.1212	0.1446	0.1677	0.1906	0.2128	0.2344	0.2553	0.2754	0.2948	0.3133
6	5	0.0922	0.1125	0.1331	0.1537	0.1742	0.1943	0.2139	0.2331	0.2516	0.2696
7	6	0.0707	0.0882	0.1063	0.1347	0.1433	0.1618	0.1801	0.1981	0.2157	0.2329
8	7	0.0545	0.0695	0.0853	0.1017	0.1184	0.1353	0.1522	0.1690	0.1856	0.2019
9	8	0.0423	0.0550	0.0687	0.0832	0.0982	0.1136	0.1291	0.1446	0.1601	0.1755
10	9	0.0329	0.0437	0.0556	0.0683	0.0817	0.0956	0.1097	0.1241	0.1385	0.1528
		11.0	12.0	13.0	14.0	15.0	16.0	17.0	18.0	19.0	20.0
1	0	0.6579	0.6813	0.7017	0.7197	0.7357	0.7499	0.7627	0.7743	0.7848	0.7943
2	1	0.5469	0.5751	0.6001	0.6224	0.6424	0.6604	0.6767	0.6916	0.7051	0.7175
3	2	0.4657	0.4963	0.5238	0.5486	0.5710	0.5913	0.6099	0.6269	0.6425	0.6568
4	3	0.4012	0.4330	0.4618	0.4880	0.5119	0.5338	0.5538	0.5723	0.5894	0.6052
5	4	0.3482	0.3802	0.4096	0.4365	0.4613	0.4842	0.5053	0.5248	0.5429	0.5598
6	5	0.3037	0.3354	0.3648	0.3921	0.4173	0.4407	0.4625	0.4828	0.5016	0.5192
7	6	0.2659	0.2969	0.3260	0.3532	0.3786	0.4023	0.4244	0.4451	0.4645	0.4826
8	7	0.2335	0.2636	0.2921	0.3189	0.3442	0.3679	0.3902	0.4111	0.4308	0.4493
9	8	0.2056	0.2345	0.2622	0.2885	0.3134	0.3370	0.3593	0.3803	0.4001	0.4189
10	9	0.1813	0.2090	0.2358	0.2614	0.2859	0.3091	0.3312	0.3522	0.3721	0.3910
11	10	0.1602	0.1866	0.2123	0.2372	0.2611	0.2839	0.3058	0.3266	0.3464	0.3652
12	11	0.1418	0.1668	0.1915	0.2155	0.2387	0.2610	0.2825	0.3030	0.3227	0.3415
13	12	0.1256	0.1493	0.1728	0.1959	0.2184	0.2402	0.2612	0.2814	0.3009	0.3195
14	13	0.1114	0.1338	0.1562	0.1783	0.2000	0.2212	0.2417	0.2616	0.2807	0.2991
15	14	0.0989	0.1200	0.1412	0.1624	0.1833	0.2038	0.2238	0.2432	0.2620	0.2802
16	15	0.0879	0.1077	0.1278	0.1480	0.1682	0.1880	0.2074	0.2263	0.2447	0.2626
17	16	0.0782	0.0967	0.1158	0.1350	0.1543	0.1734	0.1923	0.2107	0.2287	0.2462
18	17	0.0696	0.0869	0.1049	0.1233	0.1417	0.1601	0.1783	0.1963	0.2138	0.2310
19	18	0.0620	0.0782	0.0951	0.1126	0.1302	0.1479	0.1655	0.1829	0.2000	0.2167
20	19	0.0553	0.0704	0.0863	0.1028	0.1197	0.1366	0.1536	0.1705	0.1871	0.2035

Number of failures		Number of equivalent missions										
		30.0	40.0	50.0	60.0	80.0	100.0	200.0	400.0	600.0	800.0	1,000.0
A	B											
1	0	0.8577	0.8913	0.9120	0.9261	0.9441	0.9550	0.9772	0.9886	0.9924	0.9943	0.9954
2	1	0.8015	0.8471	0.8757	0.8953	0.9204	0.9358	0.9674	0.9835	0.9890	0.9917	0.9934
3	2	0.7556	0.8105	0.8453	0.8693	0.9003	0.9194	0.9588	0.9792	0.9861	0.9895	0.9916
4	3	0.7155	0.7779	0.8180	0.8458	0.8820	0.9044	0.9510	0.9752	0.9834	0.9875	0.9900
5	4	0.6792	0.7482	0.7929	0.8241	0.8650	0.8904	0.9436	0.9714	0.9808	0.9856	0.9885
6	5	0.6460	0.7206	0.7694	0.8037	0.8489	0.8771	0.9366	0.9678	0.9784	0.9837	0.9870
7	6	0.6153	0.6947	0.7472	0.7844	0.8335	0.8644	0.9297	0.9642	0.9760	0.9820	0.9855
8	7	0.5866	0.6703	0.7261	0.7659	0.8187	0.8521	0.9231	0.9608	0.9737	0.9802	0.9841
9	8	0.5599	0.6472	0.7061	0.7482	0.8045	0.8403	0.9167	0.9574	0.9714	0.9785	0.9827
10	9	0.5347	0.6253	0.6868	0.7312	0.7907	0.8288	0.9104	0.9541	0.9692	0.9768	0.9814
11	10	0.5110	0.6043	0.6684	0.7148	0.7774	0.8175	0.9042	0.9509	0.9670	0.9751	0.9801
12	11	0.4885	0.5844	0.6506	0.6990	0.7644	0.8066	0.8981	0.9477	0.9648	0.9735	0.9787
13	12	0.4673	0.5652	0.6335	0.6836	0.7518	0.7960	0.8922	0.9445	0.9627	0.9719	0.9774
14	13	0.4473	0.5469	0.6171	0.6688	0.7395	0.7855	0.8863	0.9414	0.9606	0.9703	0.9762
15	14	0.4282	0.5293	0.6011	0.6544	0.7275	0.7753	0.8805	0.9384	0.9585	0.9687	0.9749
16	15	0.4101	0.5124	0.5858	0.6404	0.7158	0.7653	0.8748	0.9353	0.9564	0.9671	0.9736
17	16	0.3928	0.4962	0.5709	0.6268	0.7044	0.7556	0.8692	0.9323	0.9544	0.9656	0.9724
18	17	0.3764	0.4806	0.5564	0.6136	0.6932	0.7460	0.8637	0.9293	0.9523	0.9640	0.9711
19	18	0.3608	0.4656	0.5425	0.6007	0.6823	0.7365	0.8582	0.9264	0.9503	0.9625	0.9699
20	19	0.3459	0.4511	0.5289	0.5882	0.6716	0.7273	0.8528	0.9235	0.9483	0.9610	0.9687
21	20	0.3317	0.4371	0.5158	0.5760	0.6611	0.7182	0.8475	0.9206	0.9463	0.9595	0.9674
22	21	0.3182	0.4236	0.5030	0.5641	0.6509	0.7092	0.8422	0.9177	0.9444	0.9580	0.9662
23	22	0.3052	0.4107	0.4907	0.5525	0.6408	0.7005	0.8369	0.9148	0.9424	0.9565	0.9650
24	23	0.2929	0.3981	0.4786	0.5412	0.6310	0.6918	0.8318	0.9120	0.9404	0.9550	0.9638
25	24	0.2810	0.3860	0.4669	0.5301	0.6213	0.6833	0.8266	0.9092	0.9385	0.9535	0.9626
30	29	0.2292	0.3313	0.4132	0.4788	0.5756	0.6428	0.8018	0.8954	0.9290	0.9463	0.9568
35	34	0.1875	0.2850	0.3663	0.4331	0.5338	0.6052	0.7780	0.8820	0.9197	0.9392	0.9510
40	39	0.1538	0.2456	0.3252	0.3922	0.4956	0.5703	0.7552	0.8690	0.9106	0.9322	0.9454
45	44	0.1264	0.2119	0.2890	0.3555	0.4604	0.5376	0.7332	0.8563	0.9017	0.9254	0.9398
50	49	0.1040	0.1831	0.2572	0.3225	0.4279	0.5071	0.7121	0.8439	0.8930	0.9186	0.9344
60	59	0.0707	0.1371	0.2040	0.2659	0.3703	0.4517	0.6721	0.8198	0.8759	0.9054	0.9236
70	69	0.0483	0.1030	0.1623	0.2197	0.3209	0.4028	0.6347	0.7967	0.8594	0.8926	0.9131
80	79	0.0331	0.0776	0.1294	0.1819	0.2786	0.3597	0.5997	0.7744	0.8433	0.8800	0.9028
90	89	0.0227	0.0586	0.1033	0.1508	0.2420	0.3214	0.5669	0.7529	0.8276	0.8677	0.8927
100	99	0.0156	0.0442	0.0825	0.1251	0.2103	0.2873	0.5360	0.7321	0.8123	0.8556	0.8827
120	119	0.0075	0.0254	0.0529	0.0864	0.1593	0.2301	0.4796	0.6926	0.7828	0.8322	0.8633
140	139	0.0036	0.0146	0.0341	0.0598	0.1209	0.1845	0.4279	0.6554	0.7545	0.8096	0.8445
160	159	0.0017	0.0085	0.0220	0.0415	0.0920	0.1483	0.3850	0.6205	0.7275	0.7877	0.8262
180	179	0.0008	0.0049	0.0142	0.0289	0.0701	0.1192	0.3453	0.5876	0.7016	0.7666	0.8084
200	199	0.0004	0.0029	0.0092	0.0201	0.0534	0.0960	0.3098	0.5566	0.6766	0.7461	0.7911

The curves in Table C.4 are reproduced, by permission, from Reliability Assessment Guides for Apollo Suppliers (SID 64-1447A) by B. L. Amstadter and T. A. Siciliano, published by North American-Rockwell Corporation. The formula used for the development appears in the article Use of Variables in Acceptance Sampling for Percent Defective by W. A. Wallis in "Techniques of Statistical Analysis," Columbia University Statistical Research Group, McGraw-Hill Book Company, New York, 1947, and in "Statistical Theory with Engineering Applications" by A. Hald, published by John Wiley & Sons, Inc., New York, 1952.

TABLE C.4a Reliabilities determined from safety margins computed from 3 sample units

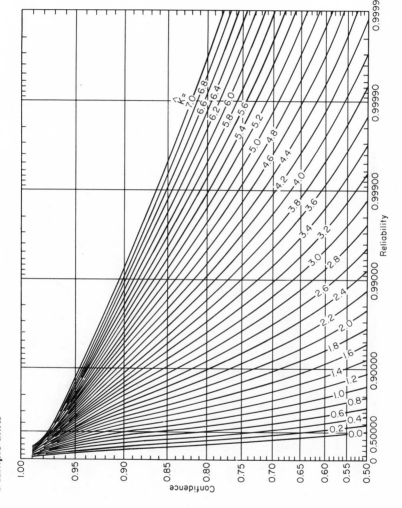

TABLE C.4b Reliabilities determined from safety margins computed from 4 sample units

TABLE C.4c Reliabilities determined from safety margins computed from 5 sample units

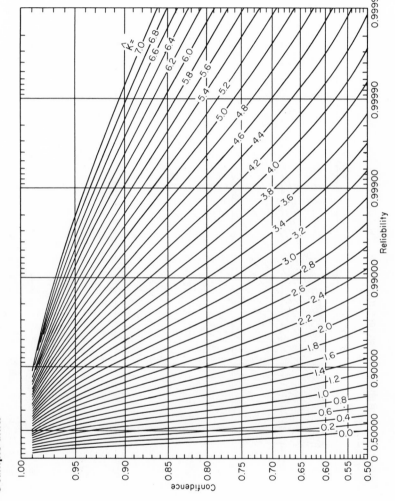

TABLE C.4d Reliabilities determined from safety margins computed from 6 sample units

TABLE C.4e Reliabilities determined from safety margins computed from 7 sample units

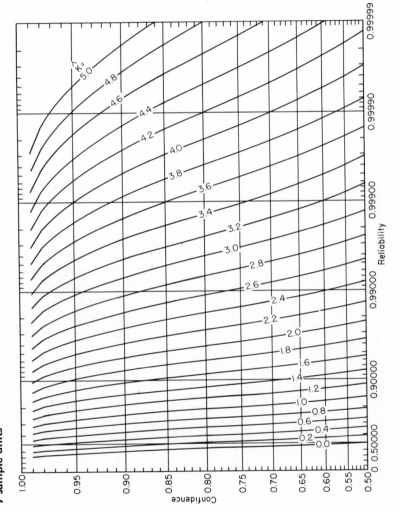

TABLE C.4f Reliabilities determined from safety margins computed from 8 sample units

TABLE C.4g Reliabilities determined from safety margins computed from 9 sample units

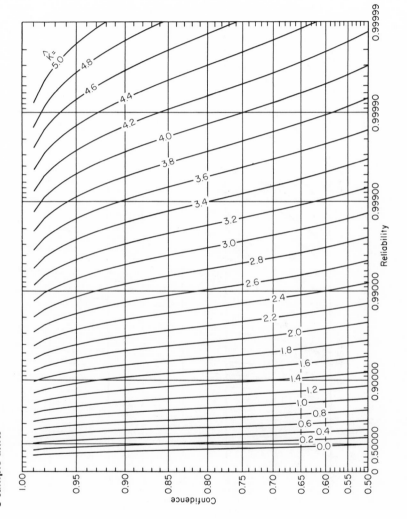

TABLE C.4h Reliabilities determined from safety margins computed from 10 sample units

TABLE C.4i Reliabilities determined from safety margins computed from 12 sample units

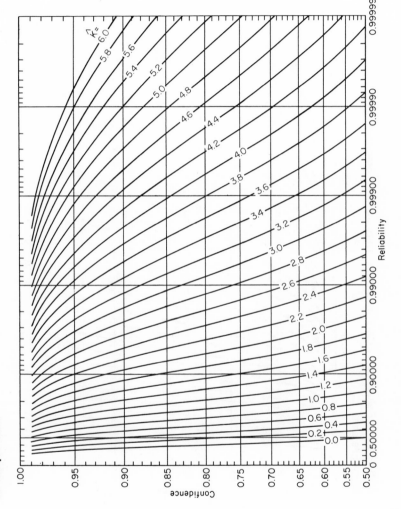

TABLE C.4j Reliabilities determined from safety margins computed from 15 sample units

TABLE C.4k   Reliabilities determined from safety margins computed from 20 sample units

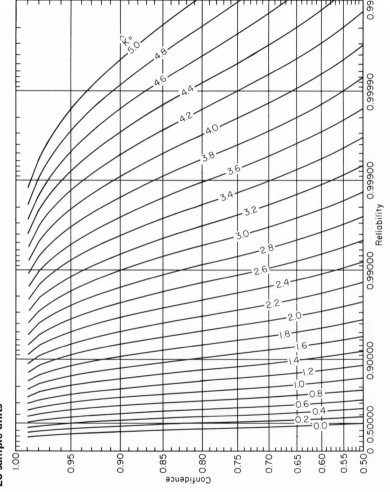

Confidence

Reliability

385

TABLE C.41 Reliabilities determined from safety margins computed from 25 sample units

386

TABLE C.4m Reliabilities determined from safety margins computed from 30 sample units

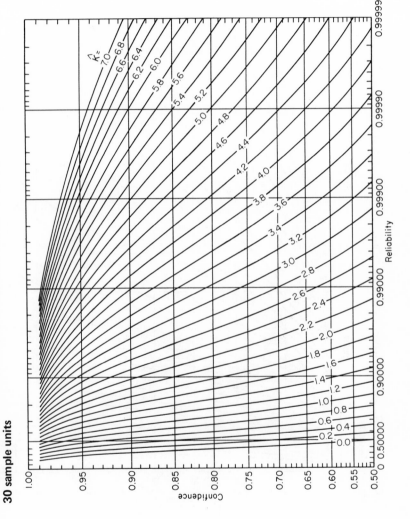

Confidence

Reliability

TABLE C.4n Reliabilities determined from safety margins computed from 40 sample units

TABLE C.4o Reliabilities determined from safety margins computed from 50 sample units

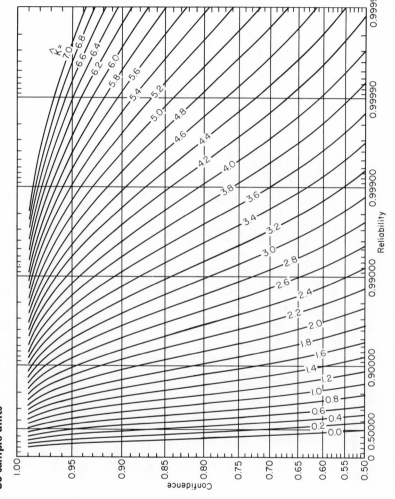

TABLE C.4p Reliabilities determined from safety margins computed from 60 sample units

390

TABLE C.4q Reliabilities determined from safety margins computed from 80 sample units

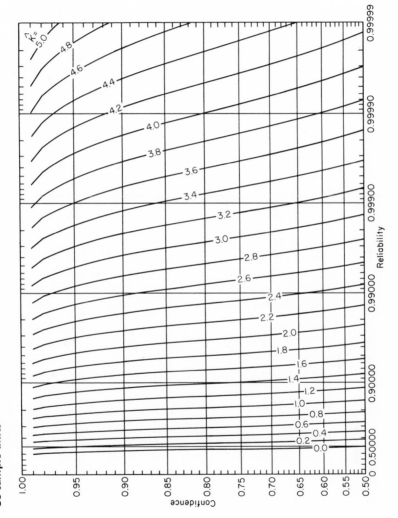

TABLE C.4r Reliabilities determined from safety margins computed from 100 sample units

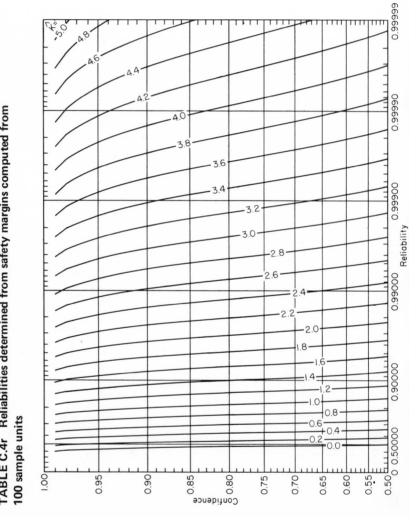

TABLE C.4s   Reliabilities determined from safety margins computed from 250 sample units

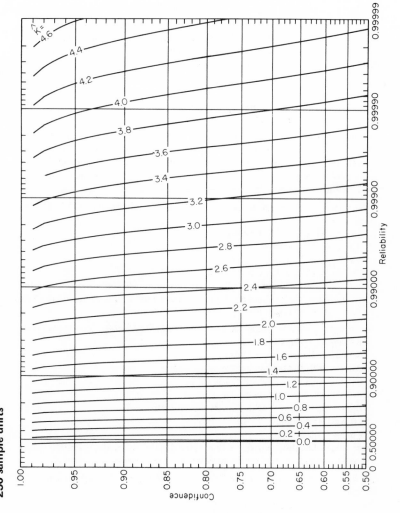

TABLE C.5   Failure-rate-data sources

"Failure Rate Data Handbook" (FARADA)—Several volumes, revisions, and
supplements   Current
   U.S. Naval Fleet Missile Systems
   Analysis and Evaluation Group
   Corona, Calif.
Provides actual failure rates experienced in tri-service and NASA programs for
all types of components; also includes, in separate volume, failure percentages
by mode. Principal limitation is extremely wide range of failure rates leading to
difficulty in selection.

Interagency Data Exchange Program (IDEP)   Current
   Air Force Space Systems Division
   Los Angeles Air Force Station
   Los Angeles, Calif. 90045
Reports contained in this program are provided on microfilm and cover a
multitude of engineering reports on components. Failure-rate data are covered
principally in Sec. 347.25. Reports on environmental and radiation effects are
listed under Secs. 347.60 and 347.65, respectively. (Available only to
participating contractors.)

Data Collection for Nonelectronic Reliability Handbook—Five volumes
(NEDCO II)   1968
Technical Report RADC-TR-68-114
   Rome Air Development Center
   Air Force Systems Command
   Griffiss Air Force Base, N.Y. 13440
Provides actual experienced failure rates for non-electronic parts (including
electromechanical), application data, operating hours, and populations. Also
provides percentage of failures by mode.

Microelectronic Failure Data; Microelectronic Failure Rates—Separate volumes   1968
   Reliability Analysis Center
   Rome Air Development Center
   Air Force Systems Command
   Griffiss Air Force Base, N.Y. 13440
      Also available through:
   Illinois Institute of Technology Research Institute
   10 West 35th Street
   Chicago, Ill. 60616
These two publications provide failure rates, failures by mode, operating hours,
and environments for various microelectronic components.

Dormant Operating and Storage Effects on Electronic Equipment and Part   1967
Reliability
Technical Report RADC-TR-67-307
   Rome Air Development Center
   Air Force Systems Command
   Griffiss Air Force Base, N.Y. 13440
Provides dormant and storage failure rates of electronic parts. (Not available to
the general public.)

Reliability and Maintainability Data Source Guide   1967
   System Effectiveness Branch

Electronics Division
U.S. Naval Applied Science Laboratory
Brooklyn, N.Y.
Identifies government reliability and maintainability data sources. (Does not provide actual failure-rate data.)

"Electronics Reliability: Calculation and Design"   1966
Text written by G. W. Dummer and N. B. Griffin
Pergamon Press, Ltd.
London
Provides application and experienced failure rates, including usage and environments, for electronic parts; provides operating hours data. Includes some data on derated operating failure rates.

Parts Reliability and Applications Notebook – Volumes I and II   1966
Technical Report RADC-TR-65-528
Reliability Analysis Center
Rome Air Development Center
Research and Technology Division
Air Force Systems Command
Griffiss Air Force Base, N.Y. 13440
Also available through:
Defense Documentation Center for Scientific and Technical Information
Cameron Station
Alexandria, Va.
Provides application failure rates and related data for electronic parts used in Minuteman I. (Not available to the general public.)

Investigation of Reliability of Mechanical Systems   1965
Lockheed-Georgia Company
Marietta, Ga.
for the Bureau of Naval Weapons
Provides unscheduled replacement and maintenance rates and their standard deviations, system quantities, and related data for mechanical components and assemblies used on the C-130B aircraft.

MIL-HDBK-217A: Reliability Stress and Failure Rate Data for Electronic   1965
Equipment
Revision of MIL-HDBK-217
Bureau of Naval Weapons
Department of the Navy
Washington, D.C. 20360
Provides application failure rates and environmental multipliers for electronic parts. Use of this handbook is sometimes specified contractually. Also provides information in other areas of reliability such as mathematical modeling.

Bureau of Ships Reliability Design Handbook   1963
NAVSHIPS 94501
Superintendent of Documents
U.S. Government Printing Office
Washington, D.C. 20302
Provides application failure rates for electronic parts. (Some of the data may be outdated and inapplicable to present equipment.)

MIL-STD-756A: Military Standard, Reliability Prediction   1963
Defense Supply Agency
Washington, D.C.

TABLE C.5   Failure-rate-data sources (Continued)

Provides environmental multipliers to be used in conjunction with failure-rate data in MIL-HDBK-217.

MIL-HDBK-217: Reliability Stress and Failure Rate Data for Electronic   1962
Equipment
    Bureau of Naval Weapons
    Department of the Navy
    Washington, D.C. 20360
Provides application failure rates for electronic parts. Use of this handbook is sometimes specified contractually. (Some of the data may be outdated and inapplicable to present equipment.) Environmental multipliers are given in MIL-STD-756A.

Minuteman Wing I Environmental Control System Reliability Analysis Report   1962
    HQ Ballistic Systems Division
    Air Force Systems Command
    United States Air Force
    Norton Air Force Base, Calif.
      Also available through:
    Defense Documentation Center for Scientific and Technical Information
    Cameron Station
    Alexandria, Va.
Provides failure-rate data for electronic and mechanical components for Minuteman environmental control system.

Operational-Reliability Estimates and Part-failure Rates for Naval Avionics   1962
Equipments
Publ. 202-1-331
    Defense Documentation Center
    Defense Supply Agency
    Cameron Station
    Alexandria, Va.
      Also available through:
    ARINC Research Corporation
    1700 K Street, N.W.
    Washington, D.C.
Provides failure-rate data and operating hours information on electrical and electromechanical components. Also includes some failure percentages by mode and comparisons of failure rates from different sources. (Some of the data may be outdated and inapplicable to present equipment.)

Reliability Engineering Data Series–Failure Rates   1962
    Avco Corporation
    Research and Advanced Development Division
    201 Lowell Street
    Wilmington, Mass. 01887
      Also provided in the:
    Proceedings of the 8th National Symposium on Reliability and Quality
      Control, pp. 259-267.
    Institute of Electrical and Electronics Engineers, Inc.
    345 East 47th Street
    New York, N.Y. 10017

Provides generic failure rates, environmental multipliers, and application factors for electronic and mechanical parts. (Some of the data may be outdated and inapplicable to present equipment.)

Handbook of Reliability Analysis Data for Systems and Component Design   1961
Engineers
TRA-873-74
    U.S. Department of Commerce
    Business and Defense Services Administration
    Office of Technical Services
    Washington, D.C.
Provides failure percentages by mode for typical electronic parts. (Some of the data may be outdated and inapplicable to present equipment.)

Proceedings of the Annual Symposia on Reliability   year as noted
(Prior to 1966, the Symposium included Quality Control)
    Institute of Electrical and Electronics Engineers, Inc.
    345 East 47th Street
    New York, N.Y. 10017

Reliability of Beam-Lead Sealed-junction Devices (1969) pp. 191-201
    Provides median life data for some transistors and integrated circuits as a function of junction temperature.
Reliability of Epoxy Transistors (1969) pp. 202-210
    Provides failure-rate data for some epoxy transistors at various operating and environmental conditions.
Reliability Data from In-flight Spacecraft (1968) pp. 271-279
    Provides some failure rates and causes for electronic parts and components for orbital environment.
Reliability Prediction Techniques (1967) pp. 17-29
    Provides some failure-rate data and relative quantities of electronic parts in various applications.
High-power High-frequency Reliability Techniques (1966) pp. 180-197
    Provides mean time to removal data for high-power, high-frequency tubes.
Operational Reliability of Components in Selected Systems (1966) pp. 198-211
    Provides replacement rates of some electronic components in selected applications.
Parts Reliability Problems in Aerospace Systems (1966) pp. 212-220
    Provides failure-rate data for some electronic parts in spacecraft applications.
Reliability of Integrated Circuits—Analysis of a Survey (1966) pp. 450-463
    Provides some failure-rate data for monolithic integrated circuits.
Quantitative Reliability Prediction (1966) pp. 670-677
    Provides failure-rate data for electronic components in various applications.
Reliability and Failure Distributions of Inertial Sensors (1965) pp. 144-153
    Provides some failure distribution data for gyros and related components.
Correlation of Ground, Air, and Space Failure Rates (1965) pp. 273-278
    Provides environmental multipliers for failure rates of electronic and mechanical components for laboratory, ground, air, and missile applications.
Parts Reliability versus Redundance Trade-offs (1965) pp. 285-292
    Compares reliabilities of some electronic parts made to commercial, QPL, and high-reliability specifications.

INDEX